Grundlehren der mathematischen Wissenschaften 334

A Series of Comprehensive Studies in Mathematics

Series editors

M. Berger B. Eckmann P. de la Harpe
F. Hirzebruch N. Hitchin L. Hörmander
M.-A. Knus A. Kupiainen G. Lebeau
M. Ratner D. Serre Ya. G. Sinai
N.J.A. Sloane B. Totaro
A. Vershik M. Waldschmidt

Editor-in-Chief

A. Chenciner J. Coates S.R.S. Varadhan

Edoardo Sernesi

Deformations of Algebraic Schemes

Springer

Edoardo Sernesi
Universitá "Roma Tre"
Department of Mathematics
Largo San Leonardo Murialdo 1
00146 Roma, Italy
e-mail: *sernesi@mat.uniroma3.it*

Library of Congress Control Number: 2006924565

Mathematics Subject Classification (2000): 14D15, 14B12

ISSN 0072-7830
ISBN-10 3-540-30608-0 Springer Berlin Heidelberg New York
ISBN-13 978-3-540-30608-5 Springer Berlin Heidelberg New York

Springer is a part of Springer Science+Business Media
springer.com

Printed in The Netherlands

Typesetting: by the author and SPI Publisher Services using a Springer LaTeX macro package
Cover design: *design & production* GmbH, Heidelberg

Printed on acid-free paper SPIN: 11371335 41/SPI 5 4 3 2 1 0

Preface

In one sense, deformation theory is as old as algebraic geometry itself: this is because all algebro-geometric objects can be "deformed" by suitably varying the coefficients of their defining equations, and this has of course always been known by the classical geometers. Nevertheless, a correct understanding of what "deforming" means leads into the technically most difficult parts of our discipline. It is fair to say that such technical obstacles have had a vast impact on the crisis of the classical language and on the development of the modern one, based on the theory of schemes and on cohomological methods.

The modern point of view originates from the seminal work of Kodaira and Spencer on small deformations of complex analytic manifolds and from its formalization and translation into the language of schemes given by Grothendieck. I will not recount the history of the subject here since good surveys already exist (e.g. [27], [138], [145], [168]). Today, while this area is rapidly developing, a self-contained text covering the basic results of what we can call "classical deformation theory" seems to be missing. Moreover, a number of technicalities and "well-known" facts are scattered in a vast literature as folklore, sometimes with proofs available only in the complex analytic category. This book is an attempt to fill such a gap, at least partially. More precisely, it aims at giving an account with complete proofs of the results and techniques which are needed to understand the local deformation theory of algebraic schemes over an algebraically closed field, thus providing the tools needed, for example, in the local study of Hilbert schemes and moduli problems. The existing monographs, like [14], [93], [105], [109], [124], [163], [175], [176], [184], all aim at goals different from the above.

For these reasons my approach has been to work exclusively in the category of locally noetherian schemes over a fixed algebraically closed field **k**, to avoid switching back and forth between the algebraic and the analytic category. I tried to make the text self-contained as much as possible, but without forgetting that all the technical ideas and prerequisites can be found in [3] and [2]: therefore the reader is advised to keep a copy of them to hand while reading this text. In any case a good familiarity with [84] and with a standard text in commutative algebra like [48] or [127] will be generally sufficient; the classical [167] and [190] will be also useful. A good acquaintance with homological algebra is assumed throughout.

One of the difficulties of writing about this subject is that it needs a great number of technical results, which make it hard to maintain a proper balance between generality and understandability. In order to overcome this problem I tried to keep the technicalities to a minimum, and I introduced the main deformation problems in an elementary fashion in Chapter 1; they are then reconsidered as functors of Artin rings in Chapter 2, where the main results of the theory are proved. The first two chapters therefore give a self-contained treatment of formal deformation theory via the "classical" approach; cotangent complexes and functors are not introduced, nor the method of differential graded Lie algebras. Another chapter treats in more detail the most important deformation functors, with the single exception of vector bundles; this was motivated by reasons of space and because good monographs on the subject are already available (e.g. [90], [118], [59]). Although they are not the central issue of the book, I considered it necessary to include a chapter on Hilbert schemes and Quot schemes, since it would be impossible to give meaningful examples and applications without them, and because of the lack of an appropriate reference. Deformation theory is closely tied with classical algebraic geometry because some of the issues which had remained controversial and unclear in the old language have found a natural explanation using the methods discussed here. I have included a section on plane curves which gives a good illustration of this point.

Unfortunately, important topics and results have been omitted because of lack of space, energy and competence. In particular, I did not include the construction of any global moduli spaces/stacks, which would have taken me too far from the main theme.

The book is organized in the following way. Chapter 1 starts with a concise treatment of algebra extensions which are fundamental in deformation theory. It then discusses locally trivial infinitesimal deformations of algebraic schemes. Chapter 2 deals with "functors of Artin rings", the abstract tool for the study of formal deformation theory. The main result of this theory is Schlessinger's theorem. A section on obstruction theory, an elementary but crucial technical point, is included. We discuss the relation between formal and algebraic deformations and the algebraization problem. This part is not entirely self-contained since Artin's algebraization theorem is not proved in general and the approximation theorem is only stated. The last section explains the role of automorphisms and the related notion of "isotriviality". Chapter 3 is an introduction to the most important deformation problems. By applying Schlessinger's theorem to them, we derive the existence of formal (semi)universal deformations. Many examples are discussed in detail so that all the basic principles of deformation theory become visible. This chapter can be used as a reference for several standard facts of deformation theory, and it can be also helpful in supplementing the study of the more abstract Chapter 2. Chapter 4 is devoted to the construction and general properties of Hilbert schemes, Quot schemes and their variants, the "flag Hilbert schemes". It ends with a section on plane curves, where the main properties of Severi varieties are discussed. My approach to the proof of existence of nodal curves with any number of nodes uses multiple point schemes and is apparently new.

In the Appendices I have collected several topics which are well known and standard but I felt it would be convenient for the reader to have them available here.

Acknowledgements. Firstly, I would like to express my deepest gratitude to D. Lieberman, who introduced me to the study of deformations of complex manifolds a long time ago. More recently, R. Hartshorne, after a careful reading of a previous draft of this book, made a number of comments and suggestions which significantly contributed in improving it. I warmly thank him for his generous help. For many extremely useful remarks on another draft of the book I am also indebted to M. Brion. I gratefully acknowledge comments and suggestions from other colleagues and students, in particular: L. Badescu, I. Bauer, A. Bruno, M. Gonzalez, A. Lopez, M. Manetti, A. Molina Rojas, D. Tossici, A. Verra, A. Vistoli.

I would also like to thank L. Caporaso, F. Catanese, C. Ciliberto, L. Ein, D. Laksov and H. Lange for encouragement and support which have been of great help.

Contents

Terminology and notation

All rings will be commutative with 1. A ring homomorphism $A \to B$ is called *essentially of finite type* (e.f.t.) if B is a localization of an A-algebra of finite type. We will also say that B is e.f.t. over A.

We will always denote by $\mathbf{k}$ a fixed algebraically closed field. All schemes will be assumed to be defined over $\mathbf{k}$, locally noetherian and separated, and all algebraic sheaves will be quasi-coherent unless otherwise specified. If X and Y are schemes we will write $X \times Y$ instead of $X \times_{\mathbf{k}} Y$. If S is a scheme and $s \in S$ we denote by $\mathbf{k}(s) = \mathcal{O}_{S,s}/m_{S,s}$ the residue field of S at s.

As is customary, various categories will be denoted by indicating their objects within parentheses when it will be clear what the morphisms in the category are. For instance (sets), (A-modules), etc. The class of objects of a category $\mathcal{C}$ will be denoted by ob($\mathcal{C}$). The dual of a category $\mathcal{C}$ will be denoted by $\mathcal{C}^\circ$. Given categories $\mathcal{C}$ and $\mathcal{D}$, a contravariant functor F from $\mathcal{C}$ to $\mathcal{D}$ will be always denoted as a covariant functor $F : \mathcal{C}^\circ \to \mathcal{D}$.

We will consider the following categories of $\mathbf{k}$-algebras:

$\mathcal{A}$ = the category of local artinian $\mathbf{k}$-algebras with residue field $\mathbf{k}$

$\hat{\mathcal{A}}$ = the category of complete local noetherian $\mathbf{k}$-algebras with residue field $\mathbf{k}$

$\mathcal{A}^*$ = the category of local noetherian $\mathbf{k}$-algebras with residue field $\mathbf{k}$

($\mathbf{k}$-algebras) = the category of noetherian $\mathbf{k}$-algebras

Morphisms are unitary $\mathbf{k}$-homomorphisms, which are local in $\mathcal{A}$, $\hat{\mathcal{A}}$ and $\mathcal{A}^*$. For a given Λ in ob($\mathcal{A}^*$) we will consider the following:

$\mathcal{A}_\Lambda$ = the category of local artinian Λ-algebras with residue field $\mathbf{k}$

$\mathcal{A}^*_\Lambda$ = the category of local noetherian Λ-algebras with residue field $\mathbf{k}$

They are subcategories of $\mathcal{A}$ and $\mathcal{A}^*$ respectively. If Λ is in ob($\hat{\mathcal{A}}$) then we will let

$$\hat{\mathcal{A}}_\Lambda = \text{the category of complete local noetherian } \Lambda\text{-algebras with residue field } \mathbf{k}$$

which is a subcategory of $\hat{\mathcal{A}}$. Moreover, we will set:

$$(\text{schemes}) = \text{the category of schemes}$$

(i.e. of locally noetherian separated **k**-schemes) and

$$(\text{algschemes}) = \text{the category of algebraic schemes}$$

For a given scheme Z we set

$$\begin{aligned}(\text{schemes}/Z) &= \text{the category of } Z\text{-schemes}\\ (\text{algschemes}/Z) &= \text{the category of algebraic } Z\text{-schemes}\end{aligned}$$

$h^i(X,\mathcal{F})$ denotes $\dim[H^i(X,\mathcal{F})]$ where $\mathcal{F}$ is a coherent sheaf on the complete scheme X. When no confusion is possible we will sometimes write $H^i(\mathcal{F})$ and $h^i(\mathcal{F})$ instead of $H^i(X,\mathcal{F})$ and $h^i(X,\mathcal{F})$ respectively.

$\coprod_i X_i$ denotes the disjoint union of the schemes X_i.

If E is a vector space or a locally free sheaf we will always denote its dual by $E^\vee$. If V is a **k**-vector space we will denote by $\mathbb{P}(V)$ the projective space $\mathrm{Proj}(\mathrm{Sym}(V^\vee))$ (where Sym(−) is the symmetric algebra of −): thus the closed points of $\mathbb{P}(V)$ are the one-dimensional subspaces of V. Similarly, if E is a locally free sheaf on an algebraic scheme S, *the projective bundle associated to* E will be defined as

$$\mathbb{P}(E) = \mathrm{Proj}(\mathrm{Sym}(E^\vee))$$

Note that this definition is dual to the one given in [84], p. 162.

For all definitions not explicitly given we will refer to [84].

Introduction

La méthode générale consiste toujours à faire des constructions formelles*, ce qui consiste essentiellement à faire de la* géométrie algébrique *sur un* anneau artinien*, et à en tirer des conclusions de nature "algébrique" en utilisant les trois théorèmes fondamentaux (Grothendieck [71], p. 11).*

Deformation theory is a formalization of the Kodaira–Nirenberg–Spencer–Kuranishi (KNSK) approach to the study of small deformations of complex manifolds. Its main ideas are clearly outlined in the series of Bourbaki seminar expositions by Grothendieck which go under the name of "Fondements de la Géométrie Algébrique" [2]; in particular they are explained in detail in [72] (see especially page 17), while the technical foundations are laid in [71]. The quotation at the top of this page gives a concise description of the method employed.

The first step of this formalization consists in studying infinitesimal deformations, and this is accomplished via the notion of "functor of Artin rings"; the study of such functors leads to the construction of "formal deformations". This method enhances the analogies between the analytic and the algebraic cases, and at the same time hides some delicate phenomena typical of the algebraic geometrical world. These phenomena become visible when one tries to pass from formal to algebraic deformations. The techniques of deformation theory have a variety of applications which make them an extremely useful tool, especially in understanding the local structure of schemes defined by geometrical conditions or by functorial constructions.

In this introduction we shall explain in outline the logical structure of deformation theory; for this purpose we will start by outlining the KNSK theory of small deformations of compact complex manifolds.

Given a compact complex manifold X, a *family of deformations* of X is a commutative diagram of holomorphic maps between complex manifolds

$$\begin{array}{cccc} & X & \subset & \mathcal{X} \\ \xi : & \downarrow & & \downarrow \pi \\ & \star & \xrightarrow{t_0} & B \end{array}$$

with π proper and smooth (i.e. with everywhere surjective differential), B connected and where $\star$ denotes the singleton space. We denote by $\mathcal{X}_t$ the fibre $\pi^{-1}(t)$, $t \in B$. It is a standard fact that, locally on B, $\mathcal{X}$ is differentiably a product so that π can be viewed locally as a family of complex structures on the differentiable manifold X_{diff}.

The family ξ is *trivial* at t_o if there is a neighbourhood $U \subset B$ of t_o such that we have $\pi^{-1}(U) \cong X \times U$ analytically.

Kodaira and Spencer started by defining, for every tangent vector $\frac{\partial}{\partial t} \in T_{t_o}B$, *the derivative of the family* π *along* $\frac{\partial}{\partial t}$ as an element

$$\frac{\partial \mathcal{X}_t}{\partial t} \in H^1(X, T_X)$$

thus giving a linear map

$$\kappa : T_{t_o}B \to H^1(X, T_X)$$

called the *Kodaira–Spencer map* of the family π. They showed that if π is trivial at t_o then $\kappa(\frac{\partial}{\partial t}) = 0$ for all $\frac{\partial}{\partial t} \in T_{t_o}B$. Then they investigated the problem of classifying all small deformations of X, by constructing a "complete family" of deformations of X. A family ξ as above is called *complete* if for every other family of deformations of X:

$$\begin{array}{cccc} & X & \subset & \mathcal{Y} \\ \eta : & \downarrow & & \downarrow p \\ & \star & \xrightarrow{m_o} & M \end{array}$$

there is an open neighbourhood $V \subset M$ and a commutative diagram

$$\begin{array}{ccccc} & & X & & \\ & \swarrow & & \searrow & \\ p^{-1}(V) & & \to & & \mathcal{X} \\ \downarrow & & & & \downarrow \\ V & & \to & & B \end{array}$$

inducing an isomorphism $p^{-1}(V) \cong V \times_B \mathcal{X}$. The family is called *universal* if it is complete and moreover, the morphism $V \to B$ is unique locally around m_o for each family η as above. Kodaira and Spencer proved that if κ is surjective then the family ξ is complete. The following existence result was then proved:

Theorem 0.0.1 (Kodaira–Nirenberg–Spencer [106]). *If* $H^2(X, T_X) = 0$ *then there exists a complete family of deformations of* X *whose Kodaira–Spencer map is an isomorphism. If, moreover,* $H^0(X, T_X) = 0$ *then such complete family is universal.*

Later Kuranishi [111] generalized this result by showing that a complete family of deformations of X such that κ is an isomorphism exists without assumptions on $H^2(X, T_X)$ provided the base B is allowed to be an analytic space.

We want to rephrase everything algebraically as far as possible. Let's fix an algebraically closed field $\mathbf{k}$ and consider an algebraic $\mathbf{k}$-scheme X. A *local deformation*, or a *local family of deformations*, of X is a cartesian diagram

$$\begin{array}{cccc} & X & \to & \mathcal{X} \\ \xi : & \downarrow & & \downarrow \pi \\ & \mathrm{Spec}(\mathbf{k}) & \subset & S \end{array}$$

where π is a flat morphism, $S = \mathrm{Spec}(A)$ where A is a local $\mathbf{k}$-algebra with residue field $\mathbf{k}$, and X is identified with the fibre over the closed point. If X is nonsingular and/or projective we will require π to be smooth and/or projective. We say that ξ is *a deformation over* Spec(A) or over A. If in particular A is an artinian local $\mathbf{k}$-algebra then we speak of an *infinitesimal deformation.*

The notion of local family has the fundamental property of being *funtorial.* Given two deformations of X:

$$\xi: \begin{array}{ccc} X & \to & \mathcal{X} \\ \downarrow & & \downarrow \pi \\ \mathrm{Spec}(\mathbf{k}) & \subset & \mathrm{Spec}(A) \end{array} \qquad \text{and} \qquad \eta: \begin{array}{ccc} X & \to & \mathcal{Y} \\ \downarrow & & \downarrow \rho \\ \mathrm{Spec}(\mathbf{k}) & \subset & \mathrm{Spec}(A) \end{array}$$

parametrized by the same Spec(A), an isomorphism $\xi \cong \eta$ is defined to be a morphism $f : \mathcal{X} \to \mathcal{Y}$ of schemes over Spec(A) inducing the identity on the closed fibre, i.e. such that the following diagram

$$\begin{array}{ccccc} & & X & & \\ & \swarrow & & \searrow & \\ \mathcal{X} & & \xrightarrow{f} & & \mathcal{Y} \\ & \searrow & & \swarrow & \\ & & \mathrm{Spec}(A) & & \end{array}$$

is commutative. Consider the category

$$\mathcal{A}^* = (\text{noetherian local } \mathbf{k}\text{-algebras with residue field } \mathbf{k})$$

and its full subcategory

$$\mathcal{A} = (\text{artinian local } \mathbf{k}\text{-algebras with residue field } \mathbf{k})$$

One defines a covariant functor

$$\mathrm{Def}_X : \mathcal{A}^* \to (\text{sets})$$

by

$$\mathrm{Def}_X(A) = \{\text{local deformations of } X \text{ over Spec(A)}\}/(\text{isomorphism})$$

This is *the functor of local deformations* of X; its restriction to $\mathcal{A}$ is *the functor of infinitesimal deformations* of X. One may now ask whether Def_X is representable, namely if there is a noetherian local $\mathbf{k}$-algebra $\mathcal{O}$ and a local deformation

$$\upsilon: \begin{array}{ccc} X & \to & \mathcal{X}^\circ \\ \downarrow & & \downarrow p \\ \mathrm{Spec}(\mathbf{k}) & \subset & \mathrm{Spec}(\mathcal{O}) \end{array}$$

which is universal, i.e. such that any other local deformation ξ is obtained by pulling back υ under a unique Spec$(A) \to$ Spec$(\mathcal{O})$.

The approach of Grothendieck to this problem was to formalize the method of Kodaira and Spencer, which consists in a formal construction followed by a proof of convergence. In the search for the universal deformation υ the formal construction corresponds to the construction of the sequence of its restrictions to the truncations $\mathrm{Spec}(\mathcal{O}/m_{\mathcal{O}}^{n+1})$:

$$u_n : \quad \begin{array}{ccc} X & \to & \mathcal{X}_n^{\circ} \\ \downarrow & & \downarrow \\ \mathrm{Spec}(\mathbf{k}) & \to & \mathrm{Spec}(\mathcal{O}/m_{\mathcal{O}}^{n+1}) \end{array} \qquad n \geq 0$$

These are infinitesimal deformations of X because the rings $\mathcal{O}/m_{\mathcal{O}}^{n+1}$ are in $\mathcal{A}$. The sequence $\hat{u} = \{u_n\}$ can be considered as a formal approximation of υ. It is a special case of a formal deformation: more precisely, a *formal deformation* of X is given by a complete local $\mathbf{k}$-algebra R with residue field $\mathbf{k}$ and by a sequence of infinitesimal deformations

$$\xi_n : \quad \begin{array}{ccc} X & \to & \mathcal{X}_n \\ \downarrow & & \downarrow \\ \mathrm{Spec}(\mathbf{k}) & \to & \mathrm{Spec}(R/m_R^{n+1}) \end{array} \qquad n \geq 0$$

such that $\xi_n \mapsto \xi_{n-1}$ under the truncation $R/m_R^{n+1} \to R/m_R^n$. In our case $R = \hat{\mathcal{O}}$. The goal of the formal step in deformation theory is the construction of $\hat{u}$ for a given X, i.e. of a formal deformation having a suitable universal property which is inherited from the corresponding property of υ, and which we do not need to specify now.

Observe that in trying to perform the formal step we will at best succeed in describing $\hat{\mathcal{O}}$ and not $\mathcal{O}$. Since a formal deformation consists of infinitesimal deformations, for the construction of $\hat{u}$ we will only need to work with the covariant functor

$$\mathrm{Def}_X : \mathcal{A} \to (\text{sets})$$

A covariant functor $F : \mathcal{A} \to (\text{sets})$ is called a *functor of Artin rings*. To every complete local $\mathbf{k}$-algebra R we can associate a functor of Artin rings h_R by

$$h_R(A) = \mathrm{Hom}_{\mathcal{A}}(R, A)$$

A functor of this form is called *prorepresentable*. By categorical general nonsense one shows that a formal deformation $\hat{\xi}$ defines a morphism of functors (a natural transformation) $h_R \to \mathrm{Def}_X$ and that this morphism is an isomorphism precisely when $\hat{\xi}$ is universal. Therefore we see that the search for $\hat{u}$ is a problem of prorepresentability of Def_X. More generally, to every local deformation problem there corresponds a functor of Artin rings F analogous to Def_X; the task of constructing a formal universal deformation for the given problem consists in showing that F is prorepresentable, producing the ring R prorepresenting F and the formal universal deformation defining the isomorphism $h_R \to F$. This is the scheme of approach to the formal part of every local deformation problem as it was outlined by Grothendieck. What one needs is to find criteria for the prorepresentability of a functor of Artin rings; we will also need to consider properties weaker than prorepresentability *(semiuniversality)* satisfied by more general classes of functors coming from interesting deformation theoretic problems. Necessary and sufficient conditions of prorepresentability and of semiuniversality are given by Schlessinger's theorem.

After having solved the problem of existence of a formal universal (or semiuniversal) deformation by means of necessary and sufficient conditions for its existence one still has to decide whether $\mathcal{O}$ and v exist and to find them. To pass from $\hat{\mathcal{O}}$ to $\mathcal{O}$ is the analogue of the convergence step in the Kodaira–Spencer theory, and it is a very difficult problem, the *algebraization problem*. Under reasonably general assumptions one shows that there exists a deformation v over an *algebraic local ring* (i.e. the henselization of a local **k**-algebra essentially of finite type) which does not quite represent the functor Def_X but at least has a universal (or semiuniversal) associated formal deformation. The further property of representing Def_X is not in general satisfied by $(\mathcal{O}, v)$, being related to the existence of nontrivial automorphisms of X. This part of the theory is largely due to the work of M. Artin, and based on the notions of *effectivity* of a formal deformation and of *local finite presentation* of a functor, already introduced by Grothendieck. The main technical tool is Artin's approximation theorem.

1

Infinitesimal deformations

The purpose of this chapter is to introduce the reader to deformation theory in an elementary and direct fashion. We will be especially interested in *first-order deformations* and *obstructions* and in giving them appropriate interpretation mostly by elementary Čech cohomology computations. We will start by introducing some algebraic tools needed. For other notions used we refer the reader to the appendices.

1.1 Extensions

1.1.1 Generalities

Let $A \to R$ be a ring homomorphism. An *A-extension* of R (or of R by I) is an exact sequence

$$(R', \varphi): \quad 0 \to I \to R' \xrightarrow{\varphi} R \to 0$$

where R' is an A-algebra and φ is a homomorphism of A-algebras whose kernel I is an ideal of R' satisfying $I^2 = (0)$. This condition implies that I has a structure of R-module. (R', φ) is also called *an extension of A-algebras.*

If (R', φ) and (R'', ψ) are A-extensions of R by I, an A-homomorphism $\xi : R' \to R''$ is called an *isomorphism of extensions* if the following diagram commutes:

$$\begin{array}{ccccccc} 0 \to & I & \to & R' & \to & R & \to 0 \\ & \| & & \downarrow \xi & & \| & \\ 0 \to & I & \to & R'' & \to & R & \to 0 \end{array}$$

Such a ξ is necessarily an isomorphism of A-algebras. More generally, given A-extensions (R', φ) and (R'', ψ) of R, not necessarily having the same kernel, a homomorphism of A-algebras $r : R' \to R''$ such that $\psi r = \varphi$ is called a *homomorphism of extensions.*

The following lemma is immediate.

Lemma 1.1.1. *Let (R', φ) be an extension as above. Given an A-algebra B and two A-homomorphisms $f_1, f_2 : B \to R'$ such that $\varphi f_1 = \varphi f_2$ the induced map*

$f_2 - f_1 : B \to I$ is an A-derivation. In particular, given two homomorphisms of extensions

$$r_1, r_2 : (R', \varphi) \to (R'', \psi)$$

the induced map $r_2 - r_1 : R' \to \ker(\psi)$ is an A-derivation.

The A-extension (R', φ) is called *trivial* if it has a *section*, that is, if there exists a homomorphism of A-algebras $\sigma : R \to R'$ such that $\varphi\sigma = 1_R$. We also say that (R', φ) *splits*, and we call σ a *splitting*.

Given an R-module I, a trivial A-extension of R by I can be constructed by considering the A-algebra $R\tilde{\oplus}I$ whose underlying A-module is $R \oplus I$ and with multiplication defined by:

$$(r, i)(s, j) = (rs, rj + si)$$

The first projection

$$p : R\tilde{\oplus}I \to R$$

defines an A-extension of R by I which is trivial: a section q is given by $q(r) = (r, 0)$.

The sections of p can be identified with the A-derivations $d : R \to I$. Indeed, if we have a section $\sigma : R \to R\tilde{\oplus}I$ with $\sigma(r) = (r, d(r))$ then for all $r, r' \in R$:

$$\sigma(rr') = (rr', d(rr')) = \sigma(r)\sigma(r') = (r, d(r))(r', d(r')) = (rr', rd(r') + r'd(r))$$

and if $a \in A$ then:

$$\sigma(ar) = (ar, d(ar)) = a\sigma(r) = a(r, d(r)) = (ar, ad(r))$$

hence $d : R \to I$ is an A-derivation. Conversely, every A-derivation $d : R \to I$ defines a section $\sigma_d : R \to R\tilde{\oplus}I$ by $\sigma_d(r) = (r, d(r))$.

Every trivial A-extension (R', φ) of R by I is isomorphic to $(R\tilde{\oplus}I, p)$. If $\sigma : R \to R'$ is a section an isomorphism $\xi : R\tilde{\oplus}I \to R'$ is given by:

$$\xi((r, i)) = \sigma(r) + i$$

and its inverse is

$$\xi^{-1}(r') = (\varphi(r'), r' - \sigma\varphi(r'))$$

An A-extension (P, f) of R will be called *versal* if for every other A-extension (R', φ) of R there is a homomorphism of extensions $r : (P, f) \to (R', \varphi)$. If $R = P/I$ where P is a polynomial algebra over A then

$$0 \to I/I^2 \to P/I^2 \to R \to 0$$

is a versal A-extension of R. Therefore, since such a P always exists, we see that every A-algebra R has a versal extension.

Examples 1.1.2. (i) Every A-extension of A is trivial because by definition it has a section. Therefore it is of the form $A\tilde{\oplus}V$ for an A-module V. In particular, if t is an indeterminate the A-extension $A[t]/(t^2)$ of A is trivial, and is denoted by $A[\epsilon]$ (where $\epsilon = t \bmod (t^2)$ satisfies $\epsilon^2 = 0$). The corresponding exact sequence is:

$$0 \to (\epsilon) \to A[\epsilon] \to A \to 0$$

$A[\epsilon]$ is called the *algebra of dual numbers* over A.

(ii) Assume that K is a field. If R is a local K-algebra with residue field K a K-extension of R by K is called a *small extension* of R. Let

$$(R', f): \quad 0 \to (t) \to R' \xrightarrow{f} R \to 0$$

be a small K-extension; in other words $t \in m_{R'}$ is annihilated by $m_{R'}$ so that (t) is a K-vector space of dimension one.

(R', f) is trivial if and only if the surjective linear map induced by f:

$$f_1 : \frac{m_{R'}}{m_{R'}^2} \to \frac{m_R}{m_R^2}$$

is not bijective.
Indeed for the trivial K-extension

$$0 \to (t) \to R\tilde{\oplus}(t) \to R \to 0$$

we have $t \in m_{R\tilde{\oplus}(t)} \setminus m^2_{R\tilde{\oplus}(t)}$, hence the map f_1 is not injective because $f_1(\bar{t}) = 0$. Conversely, if f_1 is not injective then $f_1(\bar{t}) = \bar{0}$; choose a vector subspace $U \subset m_{R'}/m^2_{R'}$ such that $m_{R'}/m^2_{R'} = U \oplus (\bar{t})$ and let $V \subset R'$ be the subring generated by U. Then V is a subring mapped isomorphically onto R by f. The inverse of $f_{|V}$ is a section of f, therefore (R', f) is trivial.

For example, it follows from this criterion that the extension of K-algebras

$$0 \to \frac{(t^n)}{(t^{n+1})} \to \frac{K[t]}{(t^{n+1})} \to \frac{K[t]}{(t^n)} \to 0$$

$n \geq 2$, is nontrivial.

(iii) Let K be a field. The K-algebra

$$K[\epsilon, \epsilon'] := K[t, t']/(t, t')^2$$

is a K-extension of $K[\epsilon]$ by K in two different ways. The first

$$0 \to (\epsilon') \to K[\epsilon, \epsilon'] \xrightarrow{p_\epsilon} K[\epsilon] \to 0$$

is a trivial extension, isomorphic to $p^*((K[\epsilon'], p'))$:

$$\begin{array}{ccccccccc} 0 \to & (\epsilon') & \to & K[\epsilon]\times_K K[\epsilon'] & \to & K[\epsilon] & \to 0 \\ & \| & & \downarrow & & \downarrow p & \\ 0 \to & (\epsilon') & \to & K[\epsilon'] & \xrightarrow{p'} & K & \to 0 \end{array}$$

The isomorphism is given by

$$\begin{array}{ccc} K[\epsilon,\epsilon'] & \longrightarrow & K[\epsilon]\times_K K[\epsilon'] \\ a+b\epsilon+b'\epsilon' & \longmapsto & (a+b\epsilon, a+b'\epsilon') \end{array}$$

The second way is by "sum":

$$\begin{array}{cccccc} 0 \to & (\epsilon-\epsilon') & \to K[\epsilon,\epsilon'] & \xrightarrow{+} & K[\epsilon] & \to 0 \\ & & a+b\epsilon+b'\epsilon' & \mapsto & a+(b+b')\epsilon & \end{array}$$

We leave it as an exercise to show that $(K[\epsilon,\epsilon'],+)$ is isomorphic to $(K[\epsilon,\epsilon'], p_\epsilon)$.

1.1.2 The module $\mathrm{Ex}_A(R, I)$

Let $A \to R$ be a ring homomorphism. In this subsection we will show how to give an R-module structure to the set of isomorphism classes of extensions of an A-algebra R by a module I, closely following the analogous theory of extensions in an abelian category as explained, for example, in Chapter III of [123].

Let (R', φ) be an A-extension of R by I and $f : S \to R$ a homomorphism of A-algebras. We can define an A-extension $f^*(R', \varphi)$ of S by I, called the *pullback* of (R', φ) by f, in the following way:

$$\begin{array}{lccccccc} f^*(R',\varphi): & 0\to & I & \to & R'\times_R S & \to & S & \to 0 \\ & & \| & & \downarrow & & \downarrow f & \\ (R',\varphi): & 0\to & I & \to & R' & \to & R & \to 0 \end{array}$$

where $R' \times_R S$ denotes the fibred product defined in the usual way.

Let $\lambda : I \to J$ be a homomorphism of R-modules. The *pushout* of (R', φ) by λ is the A-extension $\lambda_*(R', \varphi)$ of R by J defined by the following commutative diagram:

$$\begin{array}{ccccccc} 0\to & I & \xrightarrow{\alpha} & R' & \xrightarrow{\varphi} & R & \to 0 \\ & \downarrow\lambda & & \downarrow & & \| & \\ 0\to & J & \to & R'\coprod_I J & \to & R & \to 0 \end{array}$$

where

$$R'\coprod_I J = \frac{R'\tilde{\oplus} J}{\{(-\alpha(i), \lambda(i)), i \in I\}}$$

Definition 1.1.3. *For every A-algebra R and for every R-module I we define $\mathrm{Ex}_A(R, I)$ to be the set of isomorphism classes of A-extensions of R by I. If (R', φ) is such an extension we will denote by $[R', \varphi] \in \mathrm{Ex}_A(R, I)$ its class.*

Using the operations of pullback and pushout it is possible to define an R-module structure on $\mathrm{Ex}_A(R, I)$.

If $r \in R$ and $[R', \varphi] \in \mathrm{Ex}_A(R, I)$ we define

$$r[R', \varphi] = [r_*(R', \varphi)]$$

where $r : I \to I$ is the multiplication by r.

Given $[R', \varphi], [R'', \psi] \in \mathrm{Ex}_A(R, I)$, to define their sum we use the following diagram:

$$\begin{array}{ccccccccc}
0 & & & & 0 & & 0 & & \\
 & \searrow & & & \downarrow & & \downarrow & & \\
 & & I \oplus I & & I & = & I & & \\
 & & & \searrow & \downarrow & & \downarrow & & \\
0 & \to & I & \to & R' \times_R R'' & \to & R' & \to & 0 \\
 & & \| & & \downarrow & \searrow & \downarrow & & \\
0 & \to & I & \to & R'' & \to & R & \to & 0 \\
 & & & & \downarrow & & \downarrow & \searrow & \\
 & & & & 0 & & 0 & & 0
\end{array}$$

which defines an A-extension:

$$(R' \times_R R'', \zeta) : \quad 0 \to I \oplus I \to R' \times_R R'' \xrightarrow{\zeta} R \to 0$$

We define

$$[R', \varphi] + [R'', \psi] := [\delta_*(R' \times_R R'', \zeta)]$$

where $\delta : I \oplus I \to I$ is the "sum homomorphism": $\delta(i \oplus j) = i + j$.

Proposition 1.1.4. *Let $A \to R$ be a ring homomorphism and I an R-module. With the operations defined above $\mathrm{Ex}_A(R, I)$ is an R-module whose zero element is $[R\tilde{\oplus}I, p]$. This construction defines a covariant functor:*

$$\begin{array}{ccc}
(R\text{-modules}) & \longrightarrow & (R\text{-modules}) \\
I & \longmapsto & \mathrm{Ex}_A(R, I) \\
(f : I \to J) & \longmapsto & (f_* : \mathrm{Ex}_A(R, I) \to \mathrm{Ex}_A(R, J))
\end{array}$$

Proof. Straightforward. □

It is likewise straightforward to check that if $f : R \to S$ is a homomorphism of A-algebras and I is an S-module, then the operation of pullback induces an application:

$$f^* : \mathrm{Ex}_A(S, I) \to \mathrm{Ex}_A(R, I)$$

which is a homomorphism of R-modules.

We have the following useful result.

Proposition 1.1.5. *Let A be a ring, $f : S \to R$ a homomorphism of A-algebras and let I be an R-module. Then there is an exact sequence of R-modules:*

$$0 \to \mathrm{Der}_S(R, I) \to \mathrm{Der}_A(R, I) \to \mathrm{Der}_A(S, I) \otimes_S R \xrightarrow{\rho}$$
$$\to \mathrm{Ex}_S(R, I) \xrightarrow{\upsilon} \mathrm{Ex}_A(R, I) \xrightarrow{f^*} \mathrm{Ex}_A(S, I) \otimes_S R$$

Proof. υ is the obvious application sending an S-extension to itself considered as an A-extension. An A-extension

$$0 \to I \to R' \xrightarrow{\varphi} R \to 0$$

is also an S-extension if and only if there exists $f' : S \to R$ such that the triangle

$$\begin{array}{ccc} R' & \to & R \\ & \nwarrow & \uparrow \\ & & S \end{array}$$

commutes, and this is equivalent to saying that $f^*(R', \varphi)$ is trivial. This proves the exactness in $\mathrm{Ex}_A(R, I)$.

The homomorphism ρ is defined by letting $\rho(d) = (R\tilde{\oplus}I, p)$ where the structure of S-algebra on $R\tilde{\oplus}I$ is given by the homomorphism

$$s \mapsto (f(s), d(s))$$

Clearly, $\upsilon\rho = 0$. On the other hand, for

$$\begin{array}{cccccccc} (R', \varphi): & 0 \to & I & \to & R' & \xrightarrow{\varphi} & R & \to 0 \\ & & & & \uparrow & & & \\ & & & & S & & & \end{array}$$

to define an element of $\ker(\upsilon)$ there must exist an isomorphism of A-algebras $R' \to R\tilde{\oplus}I$ inducing the identity on I and on R. Hence the composition $S \to R' \to R\tilde{\oplus}I$ is of the form

$$s \mapsto (f(s), d(s))$$

for some $d \in \mathrm{Der}_A(S, I)$: therefore the sequence is exact at $\mathrm{Ex}_S(R, I)$. To prove the exactness at $\mathrm{Der}_A(S, I)$ note that $\rho(d) = 0$ if and only if $p : R\tilde{\oplus}I \to R$ has a section as a homomorphism of S-algebras, if and only if there exists an A-derivation $R \to I$ whose restriction to S is d: this proves the assertion. The exactness at $\mathrm{Der}_S(R, I)$ and $\mathrm{Der}_A(R, I)$ is straightforward. □

Definition 1.1.6. *The R-module* $\mathrm{Ex}_A(R, R)$ *is called the* first cotangent module *of R over A and it is denoted by $T^1_{R/A}$. In the case $A = \mathbf{k}$ we will write T^1_R instead of $T^1_{R/\mathbf{k}}$.*

Proposition 1.1.7. *Let $A \to B$ be an e.f.t. ring homomorphism and let $B = P/J$ where P is a smooth A-algebra. Then for every B-module M we have an exact sequence:*

$$\mathrm{Der}_A(P, M) \to \mathrm{Hom}_B(J/J^2, M) \to \mathrm{Ex}_A(B, M) \to 0 \qquad (1.1)$$

If $A \to B$ is a smooth homomorphism then $\mathrm{Ex}_A(B, M) = 0$ *for every B-module M.*

Proof. We have a natural surjective homomorphism

$$\begin{array}{ccc} \mathrm{Hom}_B(J/J^2, M) & \to & \mathrm{Ex}_A(B, M) \\ \lambda & \mapsto & \lambda_*(\eta) \end{array}$$

where

$$\eta : 0 \to J/J^2 \to P/J^2 \to B \to 0$$

The surjectivity follows from the fact that η is versal. The extension $\lambda_*(\eta)$ is trivial if and only if we have a commutative diagram

$$\begin{array}{ccccccc} 0 \to & J/J^2 & \to & P/J^2 & \to & B & \to 0 \\ & \downarrow \lambda & & \downarrow & & \| & \\ 0 \to & M & \to & B\tilde{\oplus}M & \to & B & \to 0 \end{array}$$

if and only if λ extends to an A-derivation $\bar{D} : P/J^2 \to M$, equivalently to an A-derivation $D : P \to M$. The last assertion is immediate (see Theorem C.9). □

Corollary 1.1.8. *If $A \to B$ is an e.f.t. ring homomorphism and M is a finitely generated B-module then* $\mathrm{Ex}_A(B, M)$ *is a finitely generated B-module. In particular $T^1_{B/A}$ is a finitely generated B-module and we have an exact sequence:*

$$\begin{aligned} 0 \to \mathrm{Hom}_B(\Omega_{B/A}, M) \to \mathrm{Hom}_B(\Omega_{P/A} \otimes_P B, M) \to \\ \to \mathrm{Hom}_B(I/I^2, M) \to \mathrm{Ex}_A(B, M) \to 0 \end{aligned} \tag{1.2}$$

if $B = P/J$ for a smooth A-algebra P and an ideal $J \subset P$.

Proof. It is a direct consequence of the exact sequence (1.1). □

1.1.3 Extensions of schemes

Let $X \to S$ be a morphism of schemes. An *extension* of X/S is a closed immersion $X \subset X'$, where X' is an S-scheme, defined by a sheaf of ideals $\mathcal{I} \subset \mathcal{O}_{X'}$ such that $\mathcal{I}^2 = 0$. It follows that $\mathcal{I}$ is, in a natural way, a sheaf of $\mathcal{O}_X$-modules, which coincides with the conormal sheaf of $X \subset X'$. To give an extension $X \subset X'$ of X/S is equivalent to giving an exact sequence on X:

$$\mathcal{E} : \ 0 \to \mathcal{I} \to \mathcal{O}_{X'} \xrightarrow{\varphi} \mathcal{O}_X \to 0$$

where $\mathcal{I}$ is an $\mathcal{O}_X$-module, φ is a homomorphism of $\mathcal{O}_S$-algebras and $\mathcal{I}^2 = 0$ in $\mathcal{O}_{X'}$; we call $\mathcal{E}$ *an extension of X/S by $\mathcal{I}$* or *with kernel $\mathcal{I}$*. Two such extensions $\mathcal{O}_{X'}$ and $\mathcal{O}_{X''}$ are called *isomorphic* if there is an $\mathcal{O}_S$-homomorphism $\alpha : \mathcal{O}_{X'} \to \mathcal{O}_{X''}$ inducing the identity on both $\mathcal{I}$ and $\mathcal{O}_X$. It follows that α must necessarily be an S-isomorphism.

We denote by $\mathrm{Ex}(X/S, \mathcal{I})$ the set of isomorphism classes of extensions of X/S with kernel $\mathcal{I}$. In the case where $\mathrm{Spec}(B) \to \mathrm{Spec}(A)$ is a morphism of affine schemes and $\mathcal{I} = \tilde{M}$ we have an obvious identification:

$$\mathrm{Ex}_A(B, M) = \mathrm{Ex}(X/S, \mathcal{I})$$

If $S = \mathrm{Spec}(A)$ is affine we will sometimes write $\mathrm{Ex}_A(X, \mathcal{I})$ instead of $\mathrm{Ex}(X/\mathrm{Spec}(A), \mathcal{I})$. Exactly as in the affine case one proves that $\mathrm{Ex}(X/S, \mathcal{I})$ is a $\Gamma(X, \mathcal{O}_X)$-module with identity element the class of the *trivial extension*:

$$0 \to \mathcal{I} \to \mathcal{O}_X\tilde{\oplus}\mathcal{I} \to \mathcal{O}_X \to 0$$

where $\mathcal{O}_X\tilde{\oplus}\mathcal{I}$ is defined as in the affine case (see Section 1.1). The correspondence

$$\mathcal{I} \mapsto \mathrm{Ex}(X/S, \mathcal{I})$$

defines a covariant functor from $\mathcal{O}_X$-modules to $\Gamma(X, \mathcal{O}_X)$-modules.

In deformation theory the case $\mathcal{I} = \mathcal{O}_X$ is the most important one, being related to first-order deformations. If, more generally, $\mathcal{I}$ is a locally free sheaf we get the notions of *ribbon, carpet* etc. (see [17]).

Using the fact that the exact sequence (1.2) of page 15 localizes, it is immediate to check that the cotangent module localizes. More specifically, it is straightforward to show that given a morphism of finite type of schemes $f : X \to S$ one can define a quasi-coherent sheaf $T^1_{X/S}$ on X with the following properties. If $U = \mathrm{Spec}(A)$ is an affine open subset of S and $V = \mathrm{Spec}(B)$ is an affine open subset of $f^{-1}(U)$, then

$$\Gamma(V, T^1_{X/S}) = T^1_{B/A}$$

It follows from the properties of the cotangent modules that $T^1_{X/S}$ is coherent. $T^1_{X/S}$ is called the *first cotangent sheaf* of X/S. We will write T^1_X if $S = \mathrm{Spec}(\mathbf{k})$. For future reference it will be convenient to state the following:

Proposition 1.1.9. *(i) If X is an algebraic scheme then T^1_X is supported on the singular locus of X. More generally, if $X \to S$ is a morphism of finite type of algebraic schemes then $T^1_{X/S}$ is supported on the locus where X is not smooth over S.*

(ii) If we have a closed embedding $X \subset Y$ with Y nonsingular, then we have an exact sequence of coherent sheaves on X:

$$0 \to T_X \to T_{Y|X} \to N_{X/Y} \to T^1_X \to 0 \tag{1.3}$$

so that, letting $N'_{X/Y} = \ker[N_{X/Y} \to T^1_X]$, we have the short exact sequence

$$0 \to T_X \to T_{Y|X} \to N'_{X/Y} \to 0 \tag{1.4}$$

$N'_{X/Y}$ *is called the* equisingular normal sheaf of X in Y.

(iii) For every scheme S and morphism of S-schemes $f : X \to Y$ we have an exact sequence of sheaves

$$0 \to \mathrm{T}_{X/Y} \to T_{X/S} \to Hom(f^*\Omega^1_{Y/S}, \mathcal{O}_X) \to T^1_{X/Y} \to T^1_{X/S} \to f^*T^1_{Y/S} \tag{1.5}$$

(iv) When $S = \mathrm{Spec}(\mathbf{k})$ and f is a closed embedding of algebraic schemes, with Y nonsingular, we have $\mathrm{T}_{X/Y} = 0$ and $N_{X/Y} = T^1_{X/Y}$. Moreover, (1.3) is a special case of (1.5) in this case.

Proof. (i) Use Proposition 1.1.7.

(ii) (1.3) globalizes the exact sequence (1.2).

(iii) (1.5) globalizes the exact sequence of Proposition 1.1.5.

(iv) Follows from (1.2) and 1.1.5. □

Note that the first half of the exact sequence (1.5) is the dual of the cotangent sequence of f. For more about (1.5) see also (3.43), page 162. The following is a basic result:

Theorem 1.1.10. *Let $X \to S$ be a morphism of finite type of algebraic schemes and $\mathcal{I}$ a coherent locally free sheaf on X. Assume that X is reduced and S-smooth on a dense open subset. Then there is a canonical identification*

$$\mathrm{Ex}(X/S, \mathcal{I}) = \mathrm{Ext}^1_{\mathcal{O}_X}(\Omega^1_{X/S}, \mathcal{I})$$

which to the isomorphism class of an extension of X/S:

$$\mathcal{E} : 0 \to \mathcal{I} \to \mathcal{O}_{X'} \to \mathcal{O}_X \to 0$$

associates the isomorphism class of the relative conormal sequence of $X \subset X'$:

$$c_{\mathcal{E}} : 0 \to \mathcal{I} \stackrel{\delta}{\to} (\Omega^1_{X'/S})_{|X} \to \Omega^1_{X/S} \to 0$$

(which is exact also on the left).

Proof. Suppose given an extension $\mathcal{E}$. Since $\mathcal{I}$ is locally free and X is reduced in order to show that $c_{\mathcal{E}}$ is exact on the left it suffices to prove that $\ker(\delta)$ is torsion, equivalently that $c_{\mathcal{E}}$ is exact at every general closed point x of any irreducible component of X. Since X is smooth over S at x it follows from 1.1.7 that there is an affine open neighbourhood U of x such that $\mathcal{E}_{|U}$ is trivial. From Theorem B.3 we deduce that the relative conormal sequence of $\mathcal{E}_{|U}$ is split exact. Since it coincides with the restriction of $c_{X'}$ to U we see that $\delta_{|U}$ is injective; this shows that $\ker(\delta)$ is torsion and $c_{\mathcal{E}}$ is exact. Since isomorphic extensions have isomorphic relative cotangent sequences we have a well-defined map

$$c_{-} : \mathrm{Ex}(X/S, \mathcal{I}) \to \mathrm{Ext}^1_{\mathcal{O}_X}(\Omega^1_{X/S}, \mathcal{I})$$

Now let

$$\eta : 0 \to \mathcal{I} \to \mathcal{A} \stackrel{p}{\to} \Omega^1_{X/S} \to 0$$

define an element of $\mathrm{Ext}^1_{\mathcal{O}_X}(\Omega^1_{X/S}, \mathcal{I})$. Letting $d : \mathcal{O}_X \to \Omega^1_{X/S}$ be the canonical derivation, consider the sheaf of $\mathcal{O}_S$ algebras $\mathcal{O} = \mathcal{A} \times_{\Omega^1_{X/S}} \mathcal{O}_X$: over an open subset $U \subset X$ we have $\Gamma(U, \mathcal{O}) = \{(a, f) : p(a) = d(f)\}$ and the multiplication rule is

$$(a, f)(a', f') = (fa' + f'a, ff')$$

Then we have an exact commutative diagram

$$\begin{array}{ccccccc} 0 \to & \mathcal{I} & \to & \mathcal{O} & \to & \mathcal{O}_X & \to 0 \\ & \| & & \downarrow \bar{d} & & \downarrow d & \\ 0 \to & \mathcal{I} & \to & \mathcal{A} & \to & \Omega^1_{X/S} & \to 0 \end{array}$$

where one immediately checks that the projection $\bar{d}$ is an $\mathcal{O}_S$-derivation and therefore it must factor as

$$\mathcal{O} \to \Omega^1_{\mathcal{O}/\mathcal{O}_S} \otimes_{\mathcal{O}} \mathcal{O}_X \to \mathcal{A}$$

and we have an exact commutative diagram

$$\begin{array}{ccccccc} 0 \to & \mathcal{I} & \to & \mathcal{O} & \to & \mathcal{O}_X & \to 0 \\ & \| & & \downarrow & & \downarrow d & \\ & \mathcal{I} & \to & \Omega^1_{\mathcal{O}/\mathcal{O}_S} \otimes_{\mathcal{O}} \mathcal{O}_X & \to & \Omega^1_{X/S} & \to 0 \\ & \| & & \downarrow & & \| & \\ 0 \to & \mathcal{I} & \to & \mathcal{A} & \to & \Omega^1_{X/S} & \to 0 \end{array} \tag{1.6}$$

which implies $\Omega^1_{\mathcal{O}/\mathcal{O}_S} \otimes_{\mathcal{O}} \mathcal{O}_X \cong \mathcal{A}$. Therefore, letting e_η be the extension given by the first row of (1.6), we see that $c_{e_\eta} = \eta$. Similarly, one shows that $e_{c_{\mathcal{E}}} = \mathcal{E}$ for any $[\mathcal{E}] \in \mathrm{Ex}(X/S, \mathcal{I})$. Therefore c_- and e_- are inverse of each other and the conclusion follows. □

Corollary 1.1.11. *Let $X \to S$ be a morphism of finite type of algebraic schemes, smooth on a dense open subset of X. Assume X reduced. Then there is a canonical isomorphism of coherent sheaves on X:*

$$T^1_{X/S} \cong \mathcal{E}xt^1_{\mathcal{O}_X}(\Omega^1_{X/S}, \mathcal{O}_X)$$

In particular, if X is a reduced algebraic scheme then

$$T^1_X \cong \mathcal{E}xt^1_{\mathcal{O}_X}(\Omega^1_X, \mathcal{O}_X)$$

and if, moreover, $X = \mathrm{Spec}(B_0)$ then

$$T^1_{B_0} \cong \mathrm{Ext}^1_{\mathbf{k}}(\Omega_{B_0/\mathbf{k}}, B_0)$$

Proof. An immediate consequence of the above theorem. □

A closer analysis of the proof of Theorem 1.1.10 shows that without assuming X reduced we only have an inclusion

$$\mathcal{E}xt^1_{\mathcal{O}_X}(\Omega^1_X, \mathcal{O}_X) \subset T^1_X$$

NOTES

1. An alternative approach to the topics treated in this section can be obtained by means of the so-called "truncated cotangent complex", first introduced in [76]. A more general version of the cotangent complex was introduced in [120] and later incorporated in general theories of André [4], Quillen [146] and Tate [181]. Here we will just recall the main facts about the truncated cotangent complex, without entering into any details, with the only purpose of showing the relation to the notions introduced in this section. For details we refer to [94].

Let A be a ring and R an A-algebra. To every A-extension

$$\eta : 0 \to I \to R' \xrightarrow{\varphi} R \to 0$$

we associate a complex $c_\bullet(\eta)$ of R-modules (also denoted $c_\bullet(\varphi)$) defined as follows:

$$\begin{array}{rcll} c_0(\eta) & = & \Omega_{R'/A} \otimes_{R'} R & \\ c_1(\eta) & = & I & \\ c_n(\eta) & = & (0) & n \neq 0, 1 \end{array}$$

$d_1 : c_1(\eta) \to c_0(\eta)$ is the map $x \mapsto d(x) \otimes 1$. In other words $c_\bullet(\eta)$ consists of the first map in the conormal sequence of φ. If $r : (R', \varphi) \to (R'', \psi)$ is a homomorphism of A-extensions then r induces a homomorphism of complexes

$$c_\bullet(r) : c_\bullet(\varphi) \to c_\bullet(\psi)$$

in an obvious way. The following is easy to establish:

- Let $r_1, r_2 : (R', \varphi) \to (R'', \psi)$ be two homomorphisms of A-extensions of R. Then $c_\bullet(r_1)$ and $c_\bullet(r_2)$ are homotopic. As an immediate consequence we have:
- if (E, p) and (F, q) are two versal A-extensions of R then the complexes $c_\bullet(p)$ and $c_\bullet(q)$ are homotopically equivalent.

Definition 1.1.12. *Let A be a ring, R an A-algebra and (E, p) a versal A-extension of R. The homotopy class of the complex $c_\bullet(p)$ is called the* (truncated) cotangent complex *of R over A and denoted by $\check{T}(R/A)$.*

If $R = P/I$ for a polynomial A-algebra P the A-extension

$$0 \to I/I^2 \to P/I^2 \to R \to 0 \tag{1.7}$$

is versal and therefore the complex

$$I/I^2 \xrightarrow{\delta} \Omega_{P/A} \otimes_P R \tag{1.8}$$

where δ is the map appearing in the conormal sequence of $A \to P \to R$, represents $\check{T}(R/A)$. If R is e.f.t. and P is replaced by a smooth A-algebra then (1.7) is again a versal A-extension and (1.8) again represents $\check{T}(R/A)$. From the fact that every A-algebra R can be obtained as the quotient of a polynomial A-algebra P it follows that the cotangent complex $\check{T}(R/A)$ exists for every A-algebra R.

The cotangent complex can be used to define "upper and lower cotangent functors", as follows.

Definition 1.1.13. *Let $A \to B$ be a ring homomorphism, M a B-module, and let $c_\bullet = \{C_1 \xrightarrow{d_1} C_0\}$ represent $\check{T}(B/A)$. Then for $i = 0, 1$ the* lower cotangent module *of B over A relative to M is:*

$$\check{T}_i(B/A, M) = H_i(c_\bullet \otimes_B M)$$

and the upper cotangent module *of B over A relative to M is:*

$$\check{T}^i(B/A, M) = H^i(\mathrm{Hom}(c_\bullet, M))$$

Because of the definition of cotangent complex it follows that the cotangent modules are independent on the choice of the complex $c_\bullet$ representing $\check{T}(B/A)$, but only depend on

A, B, M. Moreover, one immediately checks that the definition is functorial in M and therefore we have covariant functors:

$$\check{T}_i(B/A, -) : (B\text{-modules}) \to (B\text{-modules}) \qquad i = 0, 1$$

and

$$\check{T}^i(B/A, -) : (B\text{-modules}) \to (B\text{-modules}) \qquad i = 0, 1$$

One immediately sees that for $i = 0$ the cotangent functors are:

$$\check{T}_0(B/A, M) = \Omega_{B/A} \otimes_B M$$

and

$$\check{T}^0(B/A, M) = \mathrm{Der}_A(B, M)$$

From the extension (1.7) we obtain the exact sequences:

$$0 \to \check{T}_1(B/A, M) \to I/I^2 \otimes_B M \to \Omega_{P/A} \otimes_P M \to \Omega_{B/A} \otimes_B M \to 0$$

and

$$0 \to \mathrm{Hom}_B(\Omega_{B/A}, M) \to \mathrm{Hom}_B(\Omega_{P/A} \otimes_P B, M) \to$$
$$\mathrm{Hom}_B(I/I^2, M) \to \check{T}^1(B/A, M) \to 0$$

If B is e.f.t. then in (1.7) P can be chosen to be a smooth A-algebra; in this case it follows that $\check{T}_i(B/A, M)$ and $\check{T}^i(B/A, M)$ are finitely generated B-modules if M is finitely generated. Moreover, recalling Corollary 1.1.8, we see that we have an identification

$$\check{T}^1(B/A, M) = \mathrm{Ex}_A(B, M)$$

2. The topics of this section originate from [76]. See also [1], Ch. 0_{IV}, § 18. The proof of Theorem 1.1.10 has been taken from [17]; see also [66].

1.2 Locally trivial deformations

1.2.1 Generalities on deformations

Let X be an algebraic scheme. A cartesian diagram of morphisms of schemes

$$\eta: \quad \begin{array}{ccc} X & \to & \mathcal{X} \\ \downarrow & & \downarrow \pi \\ \mathrm{Spec}(\mathbf{k}) & \xrightarrow{s} & S \end{array}$$

where π is flat and surjective, and S is connected, is called a *family of deformations*, or simply a *deformation*, of X parametrized by S, or over S; we call S and $\mathcal{X}$ respectively the *parameter scheme* and the *total scheme* of the deformation. If S is algebraic, for each $\mathbf{k}$-rational point $t \in S$ the scheme-theoretic fibre $\mathcal{X}(t)$ is also called a *deformation* of X. When $S = \mathrm{Spec}(A)$ with A in $\mathrm{ob}(\mathcal{A}^*)$ and $s \in S$ is the closed point we have a *local family of deformations* (shortly a *local deformation*) of X over A. The deformation η will be also denoted by (S, η) or (A, η) when $S = \mathrm{Spec}(A)$.

The local deformation (A, η) is *infinitesimal* (resp. *first-order*) if $A \in \text{ob}(\mathcal{A})$ (resp. $A = \mathbf{k}[\epsilon]$). Given another deformation

$$\xi : \begin{array}{ccc} X & \to & \mathcal{Y} \\ \downarrow & & \downarrow \\ \text{Spec}(\mathbf{k}) & \to & S \end{array}$$

of X over S, an isomorphism of η with ξ is an S-isomorphism $\phi : \mathcal{X} \to \mathcal{Y}$ inducing the identity on X, i.e. such that the following diagram is commutative:

$$\begin{array}{ccccc} & & X & & \\ & \swarrow & & \searrow & \\ \mathcal{X} & & \xrightarrow{\phi} & & \mathcal{Y} \\ & \searrow & & \swarrow & \\ & & S & & \end{array}$$

By a *pointed scheme* we will mean a pair (S, s) where S is a scheme and $s \in S$. If K is a field we call (S, s) a *K-pointed scheme* if $K \cong \mathbf{k}(s)$.

Observe that for every X and for every $\mathbf{k}$-pointed scheme (S, s) there exists at least one family of deformation of X over S, namely the *product family*:

$$\begin{array}{ccc} X & \to & X \times S \\ \downarrow & & \downarrow \\ \text{Spec}(\mathbf{k}) & \xrightarrow{s} & S \end{array}$$

A deformation of X over S is called *trivial* if it is isomorphic to the product family. It will be also called a *trivial family with fibre X*. All fibres over $\mathbf{k}$-rational points of a trivial deformation of X parametrized by an algebraic scheme are isomorphic to X. The converse is not true: there are deformations which are not trivial but have isomorphic fibres over all the $\mathbf{k}$-rational points (see Example 1.2.2(ii) below). The scheme X is called *rigid* if every infinitesimal deformation of X over A is trivial for every A in $\text{ob}(\mathcal{A})$.

Given a deformation η of X over S as above and a morphism $(S', s') \to (S, s)$ of $\mathbf{k}$-pointed schemes there is induced a commutative diagram by base change

$$\begin{array}{ccc} X & \to & \mathcal{X} \times_S S' \\ \downarrow & & \downarrow \\ \text{Spec}(\mathbf{k}) & \to & S' \end{array}$$

which is clearly a deformation of X over S'. This operation is functorial, in the sense that it commutes with composition of morphisms and the identity morphism does not change η. Moreover, it carries isomorphic deformations to isomorphic ones.

An infinitesimal deformation η of X is called *locally trivial* if every point $x \in X$ has an open neighbourhood $U_x \subset X$ such that

$$\begin{array}{ccc} U_x & \to & \mathcal{X}_{|U_x} \\ \downarrow & & \downarrow \\ \text{Spec}(\mathbf{k}) & \to & S \end{array}$$

is a trivial deformation of U_x.

Remark 1.2.1. Let

$$\begin{array}{cccc} & X & \xrightarrow{j} & \mathcal{X} \\ \eta : & \downarrow & & \downarrow \pi \\ & \mathrm{Spec}(\mathbf{k}) & \xrightarrow{s} & S \end{array}$$

be a family of deformations of an algebraic scheme X parametrized by an algebraic scheme S and let $Z \subset X$ be a proper closed subset. Then

$$\begin{array}{ccc} X\backslash Z & \xrightarrow{j} & \mathcal{X}\backslash j(Z) \\ \downarrow & & \downarrow \pi' \\ \mathrm{Spec}(\mathbf{k}) & \xrightarrow{s} & S \end{array}$$

is a family of deformations of $X\backslash Z$ having the same fibres as π over $t \in S$ for $t \neq s$: thus such fibres are deformations both of X and of $X\backslash Z$. This shows that the definition of family of deformations given above is somewhat ambiguous unless we assume that π is projective or that the deformation is infinitesimal. In what follows we will restrict to the consideration of deformations of projective schemes and/or of infinitesimal deformations when discussing the general theory, so that such ambiguity will be removed; only occasionally will we consider non-infinitesimal deformations of affine schemes.

Examples 1.2.2. (i) The quadric $Q \subset \mathbf{A}^3$ of equation $xy - t = 0$ defines, via the projection

$$\begin{array}{ccc} \mathbf{A}^3 & \to & \mathbf{A}^1 \\ (x, y, t) & \mapsto & t \end{array}$$

a flat family $Q \to \mathbf{A}^1$ whose fibres are affine conics. This family is not trivial since the fibre $Q(0)$ is singular, hence not isomorphic to the fibres $Q(t)$, $t \neq 0$, which are nonsingular.

(ii) Consider, for a given integer $m \geq 0$, the rational ruled surface

$$F_m = I\!P(\mathcal{O}_{I\!P^1}(m) \oplus \mathcal{O}_{I\!P^1})$$

The structural morphism $\pi : F_m \to I\!P^1$ defines a flat family whose fibres are all isomorphic to $I\!P^1$; but if $m > 0$ then π is not a trivial family because $F_m \not\cong F_0 = I\!P^1 \times I\!P^1$ (see Example B.11(iii)).

(iii) Let $0 \leq n < m$ be two distinct nonnegative integers having the same parity and let $k = \frac{1}{2}(m - n)$. Consider two copies of $\mathbf{A}^2 \times I\!P^1$ given as $\mathrm{Proj}(\mathbf{k}[t, z, \xi_0, \xi_1]) =: W$ and $\mathrm{Proj}(\mathbf{k}[t, z', \xi'_0, \xi'_1]) =: W'$ (here the rings are graded with respect to the variables ξ_i and ξ'_i). Letting $\xi = \xi_1/\xi_0$ and $\xi' = \xi'_1/\xi'_0$ consider the open subsets

$$\mathrm{Spec}(\mathbf{k}[t, z, \xi]) \subset W, \quad \mathrm{Spec}(\mathbf{k}[t, z', \xi']) \subset W'$$

and glue them together along the open subsets

$$\mathrm{Spec}(\mathbf{k}[t, z, z^{-1}, \xi]) \subset \mathrm{Spec}(\mathbf{k}[t, z, \xi])$$

and

$$\operatorname{Spec}(\mathbf{k}[t, z', z'^{-1}, \xi']) \subset \operatorname{Spec}(\mathbf{k}[t, z', \xi'])$$

according to the following rules:

$$z' = z^{-1}, \quad \xi' = z^m \xi + t z^k \tag{1.9}$$

This induces a gluing of W and W' along

$$\operatorname{Proj}(\mathbf{k}[t, z, z^{-1}, \xi_0, \xi_1]) \quad \text{and} \quad \operatorname{Proj}(\mathbf{k}[t, z', z'^{-1}, \xi'_0, \xi'_1])$$

Call the resulting scheme $\mathcal{W}$ and $f : \mathcal{W} \to \mathbf{A}^1 = \operatorname{Spec}(\mathbf{k}[t])$ the morphism induced by the projections. Then f is a flat morphism because it is locally a projection; moreover,

$$\mathcal{W}(0) \cong F_m$$

Let $\mathcal{W}^\circ = f^{-1}(\mathbf{A}^1 \backslash \{0\})$ and $f^\circ : \mathcal{W}^\circ \to \mathbf{A}^1 \backslash \{0\}$ the restriction of f.

In $\mathbf{k}[t, t^{-1}, z, \xi]$ define

$$\zeta = \frac{z^k \xi - t}{t\xi}$$

and in $\mathbf{k}[t, t^{-1}, z', \xi']$

$$\zeta' = \frac{\xi'}{t z'^{m-k} \xi' + t^2}$$

It is straightforward to verify that the gluing (1.9) induces the relation

$$\zeta' = z^n \zeta$$

This means that we have an isomorphism

$$\mathcal{W}^\circ \cong F_n \times (\mathbf{A}^1 \backslash \{0\})$$

compatible with the projections to $\mathbf{A}^1 \backslash \{0\}$. Therefore the family f° is trivial, in particular all its fibres are isomorphic to F_n, but the family f is not trivial because $\mathcal{W}(0) \cong F_m$.

(iv) Let $f : X \to Y$ be a surjective morphism of algebraic schemes, with X integral and Y an irreducible and nonsingular curve. Then f is flat. This is a special case of Prop. III.9.7 of [84]. Therefore f defines a family of deformations of any of its closed fibres.

1.2.2 Infinitesimal deformations of nonsingular affine schemes

We will start by considering infinitesimal deformations of affine schemes. We need the following:

Lemma 1.2.3. *Let Z_0 be a closed subscheme of a scheme Z, defined by a sheaf of nilpotent ideals $N \subset \mathcal{O}_Z$. If Z_0 is affine then Z is affine as well.*

Proof. Let $r \geq 2$ be the smallest integer such that $N^r = (0)$. Since we have a chain of inclusions

$$Z \supset V(N^{r-1}) \supset V(N^{r-2}) \supset \cdots \supset V(N) = Z_0$$

it suffices to prove the assertion in the case $r = 2$. In this case N is a coherent $\mathcal{O}_{Z_0}$-module, and therefore

$$H^1(Z, N) = H^1(Z_0, N) = 0$$

Let R_0 be the **k**-algebra such that $Z_0 = \text{Spec}(R_0)$. We have the exact sequence:

$$0 \to H^0(Z, N) \to H^0(Z, \mathcal{O}_Z) \to R_0 \to 0$$

Put $R = H^0(Z, \mathcal{O}_Z)$ and let $Z' = \text{Spec}(R)$. We have a commutative diagram:

$$\begin{array}{ccc} Z & \xrightarrow{\theta} & Z' \\ & \nwarrow \nearrow & \\ & Z_0 & \end{array}$$

The sheaf homomorphism $\theta^{-1}\mathcal{O}_{Z'} \to \mathcal{O}_Z$ is clearly injective and θ is a homeomorphism. It will therefore suffice to prove that $\theta^{-1}\mathcal{O}_{Z'} \to \mathcal{O}_Z$ is surjective.

Let $z \in Z$ and $f \in \Gamma(U, \mathcal{O}_Z)$ for some affine open neighbourhood U of z. Let $f_0 = f_{|U \cap Z_0}$. It is possible to find $\varphi_0, \psi_0 \in R_0$ such that $f_0 = \frac{\varphi_0}{\psi_0}$, $\psi_0(z) \neq 0$ and $\psi_0 = 0$ on $Z_0 \backslash U$, because Z_0 is affine. Let $\psi \in R$ be such that $\psi_{|Z_0} = \psi_0$ (it exists by the surjectivity of $R \to R_0$). Then $\psi(z) \neq 0$ and $\psi = 0$ on $Z \backslash U$. There exists $n \gg 0$ such that $\psi^n f =: g \in R$ (it suffices to cover Z with affines). Then $f = \frac{g}{\psi^n} \in \theta^{-1}\mathcal{O}_{Z'}$. □

Let B_0 be a **k**-algebra, and let $X_0 = \text{Spec}(B_0)$. Consider an infinitesimal deformation of X_0 parametrized by $\text{Spec}(A)$, where A is in $\text{ob}(\mathcal{A})$. By definition this is a cartesian diagram

$$\begin{array}{ccc} X_0 & \to & \mathcal{X} \\ \downarrow & & \downarrow \\ \text{Spec}(\mathbf{k}) & \to & \text{Spec}(A) \end{array}$$

where $\mathcal{X}$ is a scheme flat over $\text{Spec}(A)$. By Lemma 1.2.3 $\mathcal{X}$ is necessarily affine. Therefore, equivalently, we can talk about an *infinitesimal deformation of B_0 over A* as a cartesian diagram of **k**-algebras:

$$\begin{array}{ccc} B & \to & B_0 \\ \uparrow & & \uparrow \\ A & \to & \mathbf{k} \end{array} \tag{1.10}$$

with $A \to B$ flat. Note that to give this diagram is the same as to give $A \to B$ flat and a **k**-isomorphism $B \otimes_A \mathbf{k} \to B_0$. We will sometimes abbreviate by calling $A \to B$ the deformation.

Given another deformation $A \to B'$ of B_0 over A, an isomorphism of deformations of $A \to B$ to $A \to B'$ is a homomorphism $\varphi : B \to B'$ of A-algebras inducing a commutative diagram:

$$\begin{array}{ccccc} & & B_0 & & \\ & \nearrow & & \nwarrow & \\ B & & \stackrel{\varphi}{\longrightarrow} & & B' \\ & \nwarrow & & \nearrow & \\ & & A & & \end{array}$$

It follows from Lemma A.4 that such a φ is an isomorphism.

An infinitesimal deformation of B_0 over A is trivial if it is isomorphic to the product deformation

$$\begin{array}{ccc} B_0 \otimes_{\mathbf{k}} A & \to & B_0 \\ \uparrow & & \uparrow \\ A & \to & \mathbf{k} \end{array}$$

The $\mathbf{k}$-algebra B_0 is called *rigid* if $\mathrm{Spec}(B_0)$ is rigid.

Theorem 1.2.4. *Every smooth* $\mathbf{k}$*-algebra is rigid. In particular, every affine nonsingular algebraic variety is rigid.*

Proof. Suppose $\mathbf{k} \to B_0$ is smooth, and suppose given a first-order deformation of B_0:

$$\eta_0 : \begin{array}{ccc} B & \to & B_0 \\ \uparrow f & & \uparrow \\ \mathbf{k}[\epsilon] & \to & \mathbf{k} \end{array}$$

Consider the commutative diagram:

$$\begin{array}{ccc} B & \to & B_0 \\ \uparrow f & & \uparrow \\ \mathbf{k}[\epsilon] & \to & B_0[\epsilon] \end{array}$$

where $B_0[\epsilon] = B_0 \otimes \mathbf{k}[\epsilon]$. Since f is smooth (because flat with smooth fibre, see [84], ch. III, Th. 10.2) and the right vertical morphism is a $\mathbf{k}[\epsilon]$-extension, by Theorem C.9 there exists a $\mathbf{k}[\epsilon]$-homomorphism $\phi : B \to B_0[\epsilon]$ making the diagram

$$\begin{array}{ccc} B & \to & B_0 \\ \uparrow f & \searrow & \uparrow \\ \mathbf{k}[\epsilon] & \to & B_0[\epsilon] \end{array}$$

commutative. Therefore ϕ is an isomorphism of deformations and η_0 is trivial.

Consider more generally a deformation of B_0

$$\eta : \begin{array}{ccc} B & \to & B_0 \\ \uparrow f & & \uparrow \\ A & \to & \mathbf{k} \end{array}$$

parametrized by A in $\mathrm{ob}(\mathcal{A})$. To show that η is trivial we proceed by induction on $d = \dim_{\mathbf{k}}(A)$. The case $d = 2$ has been already proved; assume $d \geq 3$ and let

$$0 \to (t) \to A \to A' \to 0$$

be a small extension. Consider the commutative diagram:

$$\begin{array}{ccc} B & \rightarrow & B\otimes_A A' \cong B_0\otimes_{\mathbf{k}} A' \\ \uparrow f & & \uparrow \\ A & \rightarrow & B_0\otimes_{\mathbf{k}} A \end{array}$$

f is smooth, the upper right isomorphism is by the inductive hypothesis, and the right vertical homomorphism is an A-extension. By the smoothness of f and by Theorem C.9 we deduce the existence of an A-homomorphism $B \rightarrow B_0 \otimes_{\mathbf{k}} A$ which is an isomorphism of deformations. □

Example 1.2.5. Let $\lambda \in \mathbf{k}$ and $B_0 = \mathbf{k}[X, Y]/(Y^2 - X(X-1)(X-\lambda))$. If $\lambda \neq 0, 1$ then B_0 is a smooth $\mathbf{k}$-algebra, being the coordinate ring of a nonsingular plane cubic curve. By Theorem 1.2.4, B_0 is rigid. On the other hand, the elementary theory of elliptic curves (see [84]) shows that the following flat family of affine curves

$$\begin{array}{c} \mathrm{Spec}\mathbf{k}[X, Y]/(Y^2 - X(X-1)(X-(\lambda+t))) \\ \downarrow \\ \mathrm{Spec}(\mathbf{k}[t]) \end{array}$$

is not trivial around the origin $t = 0$ so that it defines a nontrivial (non-infinitesimal) deformation of B_0. This example shows that by studying infinitesimal deformations of affine schemes we are losing some information. In this specific case we will see that this information is recovered by considering the infinitesimal deformations of the projective closure of $\mathrm{Spec}(B_0)$ (see Corollary 2.6.6, page 94).

1.2.3 Extending automorphisms of deformations

In deformation theory it is very important to have a good control of automorphisms of deformations and of their extendability properties. We will now begin to introduce such matters and to recall some terminology. In Section 2.6 we will consider these problems again in general, and we will relate them with the property of "prorepresentability". Let's start with a basic lemma.

Lemma 1.2.6. *Let B_0 be a $\mathbf{k}$-algebra, and*

$$e : 0 \rightarrow (t) \rightarrow \tilde{A} \rightarrow A \rightarrow 0$$

a small extension in $\mathcal{A}$. Then there is a canonical isomorphism of groups:

$$\left\{ \begin{array}{c} \text{automorphisms of the trivial deformation } B_0 \otimes_{\mathbf{k}} \tilde{A} \\ \text{inducing the identity on } B_0 \otimes_{\mathbf{k}} A \end{array} \right\} \rightarrow \mathrm{Der}_{\mathbf{k}}(B_0, B_0)$$

In particular the group on the left is abelian.

Proof. Every automorphism $\theta : B_0 \otimes_{\mathbf{k}} \tilde{A} \rightarrow B_0 \otimes_{\mathbf{k}} \tilde{A}$ belonging to the first group must be $\tilde{A}$-linear and induce the identity mod t. Therefore:

$$\theta(x) = x + t dx$$

where $d : B_0 \otimes_{\mathbf{k}} \tilde{A} \to B_0$ is a $\tilde{A}$-derivation (Lemma 1.1.1). But

$$\mathrm{Der}_{\tilde{A}}(B_0 \otimes_{\mathbf{k}} \tilde{A}, B_0) = \mathrm{Hom}_{B_0 \otimes_{\mathbf{k}} \tilde{A}}(\Omega_{B_0 \otimes_{\mathbf{k}} \tilde{A}/\tilde{A}}, B_0)$$

$$= \mathrm{Hom}_{B_0}(\Omega_{B_0/\mathbf{k}}, B_0) = \mathrm{Der}_{\mathbf{k}}(B_0, B_0)$$

By sending $\theta \mapsto d$ we define the correspondence of the statement. Since θ is determined by d the correspondence is one to one. Clearly the identity corresponds to the zero derivation. If we compose two automorphisms:

$$B_0 \otimes_{\mathbf{k}} \tilde{A} \xrightarrow{\theta} B_0 \otimes_{\mathbf{k}} \tilde{A} \xrightarrow{\sigma} B_0 \otimes_{\mathbf{k}} \tilde{A}$$

where $\theta(x) = x + tdx$, $\sigma(x) = x + t\delta x$, we obtain:

$$\sigma(\theta(x)) = \theta(x) + t\delta(\theta(x)) = x + tdx + t(\delta x + t\delta(dx)) = x + t(dx + \delta x)$$

therefore the correspondence is a group isomorphism. □

Recall the following well-known definition.

Definition 1.2.7. *Let G be a group acting on a set T and let*

$$\pi : G \times T \to T$$

be the map defining the action. T is called a homogeneous space *under (the action of) G if π is transitive, i.e. if*

$$\pi(G \times \{t\}) = T$$

for some (equivalently for any) $t \in T$ (i.e. if there is only one orbit). The action is called free *if for every point $t \in T$ the stabilizer $G_t = \{g \in G : gt = t\}$ is trivial, i.e. $gt = t$ implies $g = 1_G$ for all $t \in T$. If the action is both transitive and free then T is called a* principal homogeneous space *(or a* torsor*) under (the action of) G.*

To an action $\pi : G \times T \to T$ we can associate the map:

$$\begin{array}{cccc} p: & G \times T & \to & T \times T \\ & (g, t) & \mapsto & (gt, t) \end{array}$$

The condition that the action is transitive (resp. free) is equivalent to p being surjective (resp. injective); therefore T is a torsor under G if and only if p is bijective. Note that $\pi = \mathrm{pr}_1 p$ is determined by p.

More generally, suppose that we have a map of sets $f : T \to T'$. Then the map p factors through $T \times_{T'} T \subset T \times T$ if and only if the action π is compatible with f, i.e. if $f(t) = f(gt)$ for all $t \in T$, $g \in G$. As before, the map

$$p : G \times T \to T \times_{T'} T$$

is surjective (resp. injective) if and only if the action of G is transitive (resp. free) on all the non-empty fibres of f. In particular, p is bijective if and only if all the non-empty fibres of f are torsors under G. In this case one also says, according to [3],

p. 114, that T over T' is a *formal principal homogeneous space* (or a *pseudo-torsor*) under G.

Now we come back to deformations and we prove a generalization of Lemma 1.2.6.

Lemma 1.2.8. *Let B_0 be a* $\mathbf{k}$*-algebra,*

$$e : 0 \to (t) \to \tilde{A} \to A \to 0$$

a small extension in $\mathcal{A}$, $\tilde{A} \to \tilde{B}$ a deformation of B_0 and $A \to B = \tilde{B} \otimes_{\tilde{A}} A$ the induced deformation of B_0 over A. Let $\sigma : B \to B$ be an automorphism of the deformation. Then:

(i) If

$$\mathrm{Aut}_\sigma(\tilde{\mathrm{B}}) := \left\{\text{automorphisms } \tau : \tilde{B} \to \tilde{B} \text{ such that } \tau \otimes_{\tilde{A}} A = \sigma\right\} \neq \emptyset$$

then there is a free and transitive action

$$\mathrm{Der}_{\mathbf{k}}(B_0, B_0) \times \mathrm{Aut}_\sigma(\tilde{\mathrm{B}}) \to \mathrm{Aut}_\sigma(\tilde{\mathrm{B}})$$

defined by

$$(d, \tau) \mapsto \tau + td$$

(ii) If B_0 is a smooth $\mathbf{k}$*-algebra then* $\mathrm{Aut}_\sigma(\tilde{\mathrm{B}}) \neq \emptyset$ *for any σ.*

Proof. (i) Recall that we have a chain of natural identifications

$$\mathrm{Der}_{\tilde{A}}(\tilde{B}, B_0) = \mathrm{Hom}_{\tilde{B}}(\Omega_{\tilde{B}/\tilde{A}}, B_0) = \mathrm{Hom}_{B_0}(\Omega_{\tilde{B}/\tilde{A}} \otimes_{\tilde{A}} \mathbf{k}, B_0)$$

$$= \mathrm{Hom}_{B_0}(\Omega_{B_0/\mathbf{k}}, B_0) = \mathrm{Der}_{\mathbf{k}}(B_0, B_0)$$

Therefore the action in the statement is well defined once we consider a $d \in \mathrm{Der}_{\mathbf{k}}(B_0, B_0)$ as an $\tilde{A}$-derivation of $\tilde{B}$ into B_0. Given any two elements $\tau, \eta \in \mathrm{Aut}_\sigma(\tilde{\mathrm{B}})$ we have by definition:

$$q\tau = \sigma q = q\eta$$

where $q : \tilde{B} \to B$ is the projection; hence by Lemma 1.1.1,

$$\eta - \tau : \tilde{B} \to tB_0 = \ker(q)$$

is an $\tilde{A}$-derivation which is 0 if and only if $\eta = \tau$. This implies that the action is free and transitive.

(ii) Since B_0 is smooth the deformation $\tilde{A} \to \tilde{B}$ is trivial (Theorem 1.2.4), so that we have an $\tilde{A}$-isomorphism $\tilde{B} \cong B_0 \otimes_{\mathbf{k}} \tilde{A}$; moreover, $B_0 \otimes_{\mathbf{k}} \tilde{A}$ is a smooth $\tilde{A}$-algebra (Proposition C.2(iii)) hence $\tilde{B}$ is $\tilde{A}$-smooth. Let $\sigma : B \to B$ be any automorphism of the deformation and consider the diagram of $\tilde{A}$-algebras:

$$\begin{array}{ccc} \tilde{B} \xrightarrow{q} B \xrightarrow{\sigma} & B \\ & \uparrow q \\ & \tilde{B} \end{array}$$

Since $\ker(q) = tB_0$ is a square-zero ideal, and since $\tilde{B}$ is $\tilde{A}$-smooth, we deduce that there is $\tilde{\sigma} : \tilde{B} \to \tilde{B}$ such that $q\tilde{\sigma} = \sigma q$. It is immediate to check that $\tilde{\sigma}$ is an isomorphism and therefore $\tilde{\sigma} \in \mathrm{Aut}_\sigma(\tilde{\mathrm{B}})$. □

In (ii) the condition that B_0 is smooth cannot be removed. A simple example is given in 2.6.8(i). The extendability of automorphisms of deformations of not necessarily affine schemes will be studied in § 2.6.

1.2.4 First-order locally trivial deformations

We will now apply 1.2.6 to first-order deformations of any algebraic variety.

Proposition 1.2.9. *Let X be an algebraic variety. There is a 1–1 correspondence:*

$$\kappa : \left\{ \begin{array}{c} \text{isomorphism classes of first order} \\ \text{locally trivial deformations of } X \end{array} \right\} \to H^1(X, T_X)$$

called the Kodaira–Spencer correspondence, *where* $T_X = Hom(\Omega^1_X, \mathcal{O}_X) = Der_{\mathbf{k}}(\mathcal{O}_X, \mathcal{O}_X)$, *such that* $\kappa(\xi) = 0$ *if and only if* ξ *is the trivial deformation class. In particular if X is nonsingular then* κ *is a 1–1 correspondence*

$$\kappa : \left\{ \begin{array}{c} \text{isomorphism classes of} \\ \text{first-order deformations of } X \end{array} \right\} \to H^1(X, T_X)$$

Proof. Given a first-order locally trivial deformation

$$\begin{array}{ccc} X & \to & \mathcal{X} \\ \downarrow & & \downarrow \\ \mathrm{Spec}(\mathbf{k}) & \to & \mathrm{Spec}(\mathbf{k}[\epsilon]) \end{array}$$

choose an affine open cover $\mathcal{U} = \{U_i\}_{i \in I}$ of X such that $\mathcal{X}_{|U_i}$ is trivial for all i. For each index i we therefore have an isomorphism of deformations:

$$\theta_i : U_i \times \mathrm{Spec}(\mathbf{k}[\epsilon]) \to \mathcal{X}_{|U_i}$$

by 1.2.4. Then for each $i, j \in I$

$$\theta_{ij} := \theta_i^{-1}\theta_j : U_{ij} \times \mathrm{Spec}(\mathbf{k}[\epsilon]) \to U_{ij} \times \mathrm{Spec}(\mathbf{k}[\epsilon])$$

is an automorphism of the trivial deformation $U_{ij} \times \mathrm{Spec}(\mathbf{k}[\epsilon])$. By Lemma 1.2.6, θ_{ij} corresponds to a $d_{ij} \in \Gamma(U_{ij}, T_X)$. Since on each U_{ijk} we have

$$\theta_{ij}\theta_{jk}\theta_{ik}^{-1} = 1_{U_{ijk} \times \mathrm{Spec}(\mathbf{k}[\epsilon])} \tag{1.11}$$

it follows that

$$d_{ij} + d_{jk} - d_{ik} = 0$$

i.e. $\{d_{ij}\}$ is a Čech 1-cocycle and therefore defines an element of $H^1(X, T_X)$. It is easy to check that this element does not depend on the choice of the open cover $\mathcal{U}$. If we have another deformation

$$\begin{array}{ccc} X & \rightarrow & \mathcal{X}' \\ \downarrow & & \downarrow \\ \mathrm{Spec}(\mathbf{k}) & \rightarrow & \mathrm{Spec}(\mathbf{k}[\epsilon]) \end{array}$$

and $\Phi : \mathcal{X} \rightarrow \mathcal{X}'$ is an isomorphism of deformations then for each $i \in I$ there is induced an automorphism:

$$\alpha_i : U_i \times \mathrm{Spec}(\mathbf{k}[\epsilon]) \xrightarrow{\theta_i} \mathcal{X}_{|U_i} \xrightarrow{\Phi_{|U_i}} \mathcal{X}'_{|U_i} \xrightarrow{\theta_i'^{-1}} U_i \times \mathrm{Spec}(\mathbf{k}[\epsilon])$$

and therefore a corresponding $a_i \in \Gamma(U_i, T_X)$. We have $\theta_i'\alpha_i = \Phi_{|U_i}\theta_i$ and therefore

$$(\theta_i'\alpha_i)^{-1}(\theta_j'\alpha_j) = \theta_i^{-1}\Phi_{|U_{ij}}^{-1}\Phi_{|U_{ij}}\theta_j = \theta_i^{-1}\theta_j$$

thus

$$\alpha_i^{-1}\theta_{ij}'\alpha_j = \theta_{ij}$$

Equivalently:

$$d_{ij}' + a_j - a_i = d_{ij}$$

namely, $\{d_{ij}\}$ and $\{d_{ij}'\}$ are cohomologous, and therefore define the same element of $H^1(X, T_X)$.

Conversely, given $\theta \in H^1(X, T_X)$ we can represent it by a Čech 1-cocycle $\{d_{ij}\} \in \mathcal{Z}^1(\mathcal{U}, T_X)$ with respect to some affine open cover $\mathcal{U}$. To each d_{ij} we can associate an automorphism θ_{ij} of the trivial deformation $U_{ij} \times \mathrm{Spec}(\mathbf{k}[\epsilon])$ by Lemma 1.2.6. They satisfy the identities (1.11). We can therefore use these automorphisms to patch the schemes $U_i \times \mathrm{Spec}(\mathbf{k}[\epsilon])$ by the well-known procedure (see [84], p. 69). We obtain a $\mathrm{Spec}(\mathbf{k}[\epsilon])$-scheme $\mathcal{X}$ which, by construction, defines a locally trivial first-order deformation of X. The equivalence between $\kappa(\xi) = 0$ and the triviality of ξ is easily proved. The last assertion follows from the first one because all deformations of a nonsingular variety are locally trivial by Theorem 1.2.4. □

Definition 1.2.10. *For every locally trivial first-order deformation ξ of a variety X the cohomology class $\kappa(\xi) \in H^1(X, T_X)$ is called the* Kodaira–Spencer class *of ξ.*

Let

$$\xi : \begin{array}{ccc} X & \rightarrow & \mathcal{X} \\ \downarrow & & \downarrow f \\ \mathrm{Spec}(\mathbf{k}) & \xrightarrow{s} & S \end{array} \tag{1.12}$$

be a family of deformations of a nonsingular variety X. By pulling back this family by morphisms $\mathrm{Spec}(\mathbf{k}[\epsilon]) \rightarrow S$ with image s and applying the Kodaira–Spencer correspondence (Proposition 1.2.9) we define a linear map

$$\kappa_\xi : T_{S,s} \to H^1(X, T_X)$$

also denoted by $\kappa_{f,s}$ or $\kappa_{\mathcal{X}/S,s}$, which is called the *Kodaira–Spencer map* of the family ξ.

Examples 1.2.11. (i) Let $m \geq 1$ and let $\pi : F_m \to I\!P^1$ be the structural morphism of the rational ruled surface F_m (see B.11(iii)). Then π is not a trivial family but has a trivial restriction around each closed point $s \in I\!P^1$, thus $\kappa_{\pi,s} = 0$.

(ii) Consider an unramified covering $\pi : X \to S$ of degree $n \geq 2$ where X and S are projective nonsingular and irreducible algebraic curves. All fibres of π over the closed points consist of n distinct points, hence they are all isomorphic. Moreover, each such fibre is rigid and unobstructed as an abstract variety. In particular the Kodaira–Spencer map is zero at each closed point $s \in S$. On the other hand, $\pi^{-1}(U)$ is irreducible for each open subset $U \subset S$ and therefore the restriction $\pi_U : \pi^{-1}(U) \to U$ is a nontrivial family; this follows also from the fact that π does not have rational sections.

This example exhibits a phenomenon which is not detected by infinitesimal considerations and in some sense opposite to the one described in Example 1.2.5: we can have a flat projective family of deformations, all of whose geometric fibres are isomorphic, but which is nevertheless nontrivial over every Zariski open subset of the base. Note that this is different from what happens with the projections $F_m \to I\!P^1$, $m \geq 1$ of Example (i), which are nontrivial but have trivial restriction to a Zariski open neighbourhood of every point of $I\!P^1$. See Subsection 2.6.2 for more about this.

(iii) Let $0 \leq n < m$ be integers having the same parity, and let $k = \frac{1}{2}(m - n)$. Consider the smooth proper morphism $f : \mathcal{W} \to \mathbf{A}^1$ introduced in Example 1.2.2(iii), whose fibres are $\mathcal{W}(0) \cong F_m$, and $\mathcal{W}(t) \cong F_n$ for $t \neq 0$. Recall that the family f is given as the gluing of two copies of $\mathbf{A}^2 \times I\!P^1$:

$$W = \mathrm{Proj}(\mathbf{k}[t, z, \xi_0, \xi_1]), \qquad W' = \mathrm{Proj}(\mathbf{k}[t, z', \xi_0', \xi_1'])$$

along $\mathrm{Proj}(\mathbf{k}[t, z, z^{-1}, \xi_0, \xi_1])$ and $\mathrm{Proj}(\mathbf{k}[t, z', z'^{-1}, \xi_0', \xi_1'])$ according to the rules:

$$z' = z^{-1}, \quad \xi' = z^m \xi + t z^k$$

where $\xi = \xi_1/\xi_0$ and $\xi' = \xi_1'/\xi_0'$ are nonhomogeneous coordinates on the corresponding copies of $I\!P^1$.

Let's compute the local Kodaira–Spencer map $\kappa_{f,0}$ of f at 0. The image $\kappa_{f,0}(\frac{d}{dt})$ is the element of $H^1(F_m, T_{F_m})$ corresponding to the first-order deformation of F_m obtained by gluing

$$W_0 := \mathrm{Proj}(\mathbf{k}[\epsilon, z, \xi_0, \xi_1]) \qquad W_0' := \mathrm{Proj}(\mathbf{k}[\epsilon, z', \xi_0', \xi_1'])$$

along $\mathrm{Proj}(\mathbf{k}[\epsilon, z, z^{-1}, \xi_0, \xi_1])$ and $\mathrm{Proj}(\mathbf{k}[\epsilon, z', z'^{-1}, \xi_0', \xi_1'])$ according to the rules

$$z' = z^{-1}, \quad \xi' = z^m \xi + \epsilon z^k$$

By definition we have that $\kappa_{f,0}(\frac{d}{dt})$ is the element of $H^1(\mathcal{U}, T_{F_m})$, where $\mathcal{U} = \{W_0, W_0'\}$, defined by the 1-cocycle corresponding to the vector field on $W_0 \cap W_0'$

$$\left\{z^k \frac{\partial}{\partial \xi}\right\}$$

According to Example B.11(iii) this element is nonzero; therefore $\kappa_{f,0}$ is injective.

Similarly, we can consider a smooth proper family $F : \mathcal{Y} \to \mathbf{A}^{m-1}$ defined as follows. $\mathcal{Y}$ is the gluing of

$$Y := \mathrm{Proj}(\mathbf{k}[t_1, \dots, t_{m-1}, z, \xi_0, \xi_1])$$

and

$$Y' := \mathrm{Proj}(\mathbf{k}[t_1, \dots, t_{m-1}, z', \xi_0', \xi_1'])$$

along $\mathrm{Proj}(\mathbf{k}[t_1, \dots, t_{m-1}, z, z^{-1}, \xi_0, \xi_1])$ and $\mathrm{Proj}(\mathbf{k}[t_1, \dots, t_{m-1}, z', z'^{-1}, \xi_0', \xi_1'])$ according to the rules:

$$z' = z^{-1}, \quad \xi' = z^m \xi + \sum_{j=1}^{m-1} t_j z^j$$

The morphism F is defined by the projections onto $\mathrm{Spec}(\mathbf{k}[t_1, \dots, t_{m-1}])$; the fibre of F over $\underline{0}$ is $\mathcal{Y}(\underline{0}) \cong F_m$. The computation we just did immediately implies that the local Kodaira–Spencer map

$$\kappa_{F,\underline{0}} : T_{\underline{0}}\mathbf{A}^{m-1} \to H^1(F_m, T_{F_m})$$

is an isomorphism.

1.2.5 Higher-order deformations – obstructions

Let X be a nonsingular algebraic variety. Consider a small extension

$$e : 0 \to (t) \to \tilde{A} \to A \to 0$$

in $\mathcal{A}$ and let

$$\begin{array}{cccc} & X & \to & \mathcal{X} \\ \xi : & \downarrow & & \downarrow \\ & \mathrm{Spec}(\mathbf{k}) & \to & \mathrm{Spec}(A) \end{array}$$

be an infinitesimal deformation of X. A *lifting of* ξ *to* $\tilde{A}$ consists in a deformation

$$\begin{array}{cccc} & X & \to & \tilde{\mathcal{X}} \\ \tilde{\xi} : & \downarrow & & \downarrow \\ & \mathrm{Spec}(\mathbf{k}) & \to & \mathrm{Spec}(\tilde{A}) \end{array}$$

and an isomorphism of deformations

$$\begin{array}{ccccc} & & X & & \\ & \swarrow & & \searrow & \\ \mathcal{X} & & \xrightarrow{\phi} & & \tilde{\mathcal{X}} \times_{\mathrm{Spec}(\tilde{A})} \mathrm{Spec}(A) \\ & \searrow & & \swarrow & \\ & & \mathrm{Spec}(A) & & \end{array}$$

If we want to study arbitrary infinitesimal deformations, and not only first-order ones, it is important to know whether, given ξ and e, a lifting of ξ to $\tilde{A}$ exists, and how many there are. Such information can then be used to build an inductive procedure for the description of infinitesimal deformations. The following proposition addresses this question.

Proposition 1.2.12. *Given A in* ob$(\mathcal{A})$ *and an infinitesimal deformation ξ of X over A:*

(i) To every small extension e of A there is associated an element $o_\xi(e) \in H^2(X, T_X)$ called the obstruction to lifting ξ to $\tilde{A}$, *which is 0 if and only if a lifting of ξ to $\tilde{A}$ exists.*
(ii) If $o_\xi(e) = 0$ then there is a natural transitive action of $H^1(X, T_X)$ on the set of isomorphism classes of liftings of ξ to $\tilde{A}$.
(iii) The correspondence $e \mapsto o_\xi(e)$ defines a **k***-linear map*

$$o_\xi : \mathrm{Ex}_{\mathbf{k}}(A, \mathbf{k}) \to H^2(X, T_X)$$

Proof. Let $\mathcal{U} = \{U_i\}_{i \in I}$ be an affine open cover of X. We have isomorphisms

$$\theta_i : U_i \times \mathrm{Spec}(A) \to \mathcal{X}_{|U_i}$$

and consequently, $\theta_{ij} := \theta_i^{-1}\theta_j$ is an automorphism of the trivial deformation $U_{ij} \times \mathrm{Spec}(A)$. Moreover,

$$\theta_{ij}\theta_{jk} = \theta_{ik} \tag{1.13}$$

on $U_{ijk} \times \mathrm{Spec}(A)$. To give a lifting $\tilde{\xi}$ of ξ to $\tilde{A}$ it is necessary and sufficient to give a collection of automorphisms $\{\tilde{\theta}_{ij}\}$ of the trivial deformations $U_{ij} \times \mathrm{Spec}(\tilde{A})$ such that

(a) $\tilde{\theta}_{ij}\tilde{\theta}_{jk} = \tilde{\theta}_{ik}$
(b) $\tilde{\theta}_{ij}$ restricts to θ_{ij} on $U_{ij} \times \mathrm{Spec}(A)$

In fact from such data we will be able to define $\tilde{\mathcal{X}}$ by patching the local pieces $U_i \times \mathrm{Spec}(\tilde{A})$ along the open subsets $U_{ij} \times \mathrm{Spec}(\tilde{A})$ in the usual way. To establish the existence of the collection $\{\tilde{\theta}_{ij}\}$ let's choose arbitrarily automorphisms $\{\tilde{\theta}_{ij}\}$ satisfying condition (b); they exist by Lemma 1.2.8(ii). Let

$$\tilde{\theta}_{ijk} = \tilde{\theta}_{ij}\tilde{\theta}_{jk}\tilde{\theta}_{ik}^{-1}$$

This is an automorphism of the trivial deformation $U_{ijk} \times \mathrm{Spec}(\tilde{A})$. Since by (1.13) it restricts to the identity on $U_{ijk} \times \mathrm{Spec}(A)$, by Lemma 1.2.6 we can identify each $\tilde{\theta}_{ijk}$

with a $\tilde{d}_{ijk} \in \Gamma(U_{ijk}, T_X)$ and it is immediate to check that $\{\tilde{d}_{ijk}\} \in \mathcal{Z}^2(\mathcal{U}, T_X)$. If we choose different automorphisms $\{\Phi_{ij}\}$ of the trivial deformations $U_{ij} \times \mathrm{Spec}(\tilde{A})$ satisfying the analogue of condition (b) then

$$\Phi_{ij} = \tilde{\theta}_{ij} + t d_{ij} \tag{1.14}$$

for some $d_{ij} \in \Gamma(U_{ij}, T_X)$, by Lemma 1.2.8(i). For each i, j, k the automorphism

$$\Phi_{ij}\Phi_{jk}\Phi_{ik}^{-1}$$

corresponds to the derivation

$$\delta_{ijk} = \tilde{d}_{ijk} + (d_{ij} + d_{jk} - d_{ik})$$

and therefore we see that the 2-cocycles $\{\tilde{d}_{ijk}\}$ and $\{\delta_{ijk}\}$ are cohomologous. Their cohomology class

$$o_\xi(e) \in H^2(X, T_X)$$

depends only on ξ and e and is 0 if and only if we can find a collection of automorphisms $\{\Phi_{ij}\}$ such that $\delta_{ijk} = 0$ for all $i, j, k \in I$. In such a case $\{\Phi_{ij}\}$ defines a lifting $\tilde{\xi}$ of ξ. This proves (i).

Assume that $o_\xi(e) = 0$, i.e. that the lifting $\tilde{\xi}$ of ξ exists. Then we can choose the collection $\{\tilde{\theta}_{ij}\}$ of automorphisms satisfying conditions (a) and (b) as above, in particular $\tilde{d}_{ijk} = 0$, all i, j, k. Any other choice of a lifting $\bar{\tilde{\xi}}$ of ξ to $\tilde{A}$ corresponds to a choice of automorphisms $\{\Phi_{ij}\}$ satisfying (1.14) and the analogue of condition (b). Therefore, for all i, j, k, we have

$$0 = \delta_{ijk} = d_{ij} + d_{jk} - d_{ik}$$

so that $\{d_{ij}\} \in \mathcal{Z}^1(\mathcal{U}, T_X)$ defines an element $\bar{d} \in H^1(X, T_X)$. As before, one checks that this element depends only on the isomorphism class of $\bar{\tilde{\xi}}$; it follows in a straightforward way that the correspondence $(\tilde{\xi}, \bar{d}) \mapsto \bar{\tilde{\xi}}$ defines a transitive action of $H^1(X, T_X)$ on the set of isomorphism classes of liftings of ξ to $\tilde{A}$. This proves (ii).

(iii) is left to the reader. □

Definition 1.2.13. *The deformation ξ is called* unobstructed *if o_ξ is the zero map; otherwise ξ is called* obstructed. *X is* unobstructed *if every infinitesimal deformation of X is unobstructed; otherwise X is* obstructed.

Corollary 1.2.14. *A nonsingular variety X is unobstructed if*

$$H^2(X, T_X) = (0)$$

The proof is obvious.

Corollary 1.2.15. *A nonsingular variety X is rigid if and only if*

$$H^1(X, T_X) = (0)$$

Proof. The hypothesis implies, by Proposition 1.2.9, that all first-order deformations of X are trivial; moreover, by Proposition 1.2.12(ii), it implies that every infinitesimal deformation of X over any A in ob($\mathcal{A}$) has at most one lifting to any small extension of A. These two facts together easily give the conclusion. □

Examples 1.2.16. (i) If X is a projective nonsingular curve of genus g then from the Riemann–Roch theorem it follows that

$$h^1(X, T_X) = \begin{cases} 0 & \text{if } g = 0 \\ 1 & \text{if } g = 1 \\ 3g - 3 & \text{if } g \geq 2 \end{cases}$$

and $h^2(X, T_X) = 0$. In particular, projective nonsingular curves are unobstructed.

(ii) If X is a projective, irreducible and nonsingular surface X then

$$H^2(X, T_X) \cong H^0(X, \Omega^1_X \otimes K_X)^\vee$$

by Serre duality, and this rarely vanishes. For example, a nonsingular surface of degree ≥ 5 in $\mathbb{P}^3$ satisfies $H^2(X, T_X) \neq (0)$, but it is nevertheless unobstructed (see Example 3.2.11(i)); therefore the sufficient condition of Corollary 1.2.14 is not necessary. In general a surface such that $H^2(X, T_X) \neq (0)$ can be obstructed, but explicit examples are not elementary (see [96], [24], [87]). We will describe a class of such examples in Theorem 3.4.26, page 185. In § 2.4 we will show how to construct examples of obstructed 3-folds (see remarks following Proposition 3.4.25, page 185). The first examples of obstructed compact complex manifolds where given in Kodaira–Spencer [107], § 16: they are of the form $T \times \mathbb{P}^1$, where T is a two-dimensional complex torus.

(iii) The projective space $\mathbb{P}^n$ is rigid for every $n \geq 1$. In fact it follows immediately from the Euler sequence:

$$0 \to \mathcal{O}_{\mathbb{P}^n} \to \mathcal{O}_{\mathbb{P}^n}(1)^{n+1} \to T_{\mathbb{P}^n} \to 0$$

that $H^1(\mathbb{P}^n, T_{\mathbb{P}^n}) = 0$. Similarly, one shows that finite products

$$\mathbb{P}^{n_1} \times \cdots \times \mathbb{P}^{n_k}$$

of projective spaces are rigid.

(iv) The ruled surfaces F_m are unobstructed because

$$h^2(F_m, T_{F_m}) = 0$$

(see (B.13), page 292).

2

Formal deformation theory

In this chapter we develop the theory of "functors of Artin rings". The main result of this theory is a theorem of Schlessinger giving necessary and sufficient conditions for a functor of Artin rings to have a semiuniversal or a universal formal element. We then apply the functorial machinery to the construction of formal semiuniversal, or universal, deformations, which is the final goal of formal deformation theory, and we explain the relation between formal and algebraic deformations. In order to make the arguments easier to follow, these applications are given only in the case of the deformation functors Def_X and Def'_X of an algebraic scheme. Only in Chapter 3 will we apply the general theory to other deformation functors.

2.1 Obstructions

In this section we investigate the notion of formal smoothness in the category $\mathcal{A}^*$ using the language of extensions. The results we prove are crucial for the understanding of obstructions in deformation theory. Our treatment is an expansion of [157]; for a more systematic treatment we refer to [54].

Let $\Lambda \in \mathrm{ob}(\mathcal{A}^*)$ and $\mu : \Lambda \to R$ be in $\mathrm{ob}(\mathcal{A}^*_\Lambda)$. The *relative obstruction space* of R/Λ is

$$o(R/\Lambda) := \mathrm{Ex}_\Lambda(R, \mathbf{k})$$

If $\Lambda = \mathbf{k}$ then $o(R/\mathbf{k})$ is called the (absolute) *obstruction space* of R and simply denoted by $o(R)$. We say that R is *unobstructed* (resp. *obstructed*) *over* Λ if $o(R/\Lambda) = (0)$ (resp. if $o(R/\Lambda) \neq (0)$); R is said to be *unobstructed* (resp. *obstructed*) if $o(R) = (0)$ (resp. if $o(R) \neq (0)$). Given a homomorphism $f : R \to S$ in $\mathcal{A}^*_\Lambda$ we denote by

$$o(f/\Lambda) : o(S/\Lambda) \to o(R/\Lambda)$$

the linear map induced by pullback:

$$o(f/\Lambda)([\eta]) = [f^*\eta] \in \mathrm{Ex}_\Lambda(R, \mathbf{k})$$

for all $[\eta] \in \mathrm{Ex}_\Lambda(S, \mathbf{k})$. Since this definition is functorial we have a contravariant functor:

$$o(-/\Lambda) : \mathcal{A}^*_\Lambda \to (\text{vector spaces}/\mathbf{k})$$

When $\Lambda = \mathbf{k}$ we write $o(f)$ instead of $o(f/\mathbf{k})$. If μ is such that $o(\mu)$ is injective one sometimes says that R is *less obstructed* than Λ. By applying Proposition 1.1.5 we obtain an exact sequence for each $f : R \to S$ in $\mathcal{A}^*_\Lambda$:

$$0 \to t_{S/R} \to t_{S/\Lambda} \to t_{R/\Lambda} \to o(S/R) \to o(S/\Lambda) \xrightarrow{o(f/\Lambda)} o(R/\Lambda) \tag{2.1}$$

In the case $\Lambda = \mathbf{k}$ we obtain the exact sequence:

$$0 \to t_{S/R} \to t_S \to t_R \to o(S/R) \to o(S) \xrightarrow{o(f)} o(R) \tag{2.2}$$

which relates the absolute and the relative obstruction spaces.

The next result gives a description of $o(R/\Lambda)$ and an interpretation of formal smoothness of a Λ-algebra (R, m) in $\mathcal{A}^*_\Lambda$.

Proposition 2.1.1. *Assume that Λ is in* $\mathrm{ob}(\hat{\mathcal{A}})$.

(i) Let (R, m) be in $\mathrm{ob}(\mathcal{A}^*_\Lambda)$ and let $\chi : R \to \hat{R}$ be the natural homomorphism of R into its m-adic completion $\hat{R}$. Then the induced map:

$$o(\chi/\Lambda) : o(\hat{R}/\Lambda) \to o(R/\Lambda)$$

is an isomorphism.

(ii) For every (R, m) in $\mathrm{ob}(\mathcal{A}^*_\Lambda)$ let $d = \dim_{\mathbf{k}}(t_{R/\Lambda})$ and let

$$\hat{R} = \Lambda[[X_1, \ldots, X_d]]/J$$

with $J \subset (\underline{X})^2$, be a presentation of the m-adic completion $\hat{R}$. Then there is a natural isomorphism:

$$o(R/\Lambda) \cong \left(J/(\underline{X})J\right)^\vee$$

In particular, R is unobstructed over Λ if and only if it is a formally smooth Λ-algebra.

Proof. (i) Let

$$\eta : 0 \to \mathbf{k} \to S \to \hat{R} \to 0$$

be a small extension; denote by m' the maximal ideal of S.

Claim: S is complete.

Let $\{f_n\} \subset S$ be a Cauchy sequence; then the image sequence $\{\bar{f_n}\}$ in $\hat{R}$ is Cauchy, hence it converges to a limit which we may assume to be zero, after possibly subtracting a constant sequence from $\{f_n\}$. We have $\bar{f}_n \in \hat{m}^{e(n)}$, with $\lim_n[e(n)] = \infty$. For every n we may find $g_n \in m'^{e(n)}$ lying above $\bar{f}_n$. The sequence $\{g_n\}$ in S is Cauchy and converges to zero, and $\{f_n - g_n\}$ is a Cauchy sequence in $\mathbf{k}$. Since $\mathbf{k}$ is complete as an S-module, because it is annihilated by the maximal ideal, $\{f_n - g_n\}$ converges to a limit $f \in \mathbf{k}$. This is also the limit of $\{f_n\}$ because

$$f_n - f = (f_n - g_n - f) + g_n$$

Therefore S is complete.

If $\chi^*(\eta/\Lambda)$ is trivial the section induces a homomorphism $g : R \to S$ which factors through $\hat{R}$ because S is complete. Hence η is trivial. This proves that $o(\chi/\Lambda)$ is injective.

Given a small Λ-extension of R:

$$(S, \varphi) : \quad 0 \to \mathbf{k} \to S \to R \to 0$$

the map $\hat{\varphi} : \hat{S} \to \hat{R}$ is surjective and $\ker(\hat{\varphi}) = \hat{\mathbf{k}} = \mathbf{k}$. Therefore $[\hat{S}, \hat{\varphi}] \in \mathrm{Ex}_\Lambda(\hat{R}, \mathbf{k})$ and $o(\chi/\Lambda)([\hat{S}, \hat{\varphi}]) = [S, \varphi]$: this means that $o(\chi/\Lambda)$ is also surjective.

(ii) R is a formally smooth Λ-algebra if and only if $\hat{R}$ is a power series ring over Λ, i.e. if and only if $J = (0)$. Therefore the last assertion follows from the fact that $J/(\underline{X})J = (0)$ if and only if $J = (0)$, by Nakayama's lemma.

In order to prove the first assertion we may assume that R is in $\mathrm{ob}(\hat{\mathcal{A}}_\Lambda)$, since $o(R/\Lambda) = o(\hat{R}/\Lambda)$ by the first part of the proposition. Hence $R = \Lambda[[\underline{X}]]/J$ with $J \subset (\underline{X})^2$. The extension of R:

$$\Phi : \quad 0 \to J/(\underline{X})J \to \Lambda[[\underline{X}]]/(\underline{X})J \to R \to 0$$

induces by pushouts a homomorphism:

$$\begin{array}{rccll} \alpha : & (J/(\underline{X})J)^\vee & \to & \mathrm{Ex}_\Lambda(R, \mathbf{k}) & = o(R) \\ & d & \longmapsto & [d_*\Phi] & \end{array}$$

Letting M be the maximal ideal of $\Lambda[[\underline{X}]]/(\underline{X})J$ we have $J/(\underline{X})J \subset M^2$. If $d \in (J/(\underline{X})J)^\vee$ is such that $[d_*\Phi] = 0$ then we have:

$$\begin{array}{lccccccc} \Phi : & 0 \to & J/(\underline{X})J & \to & \Lambda[[\underline{X}]]/(\underline{X})J & \to & R & \to 0 \\ & & \downarrow d & & \downarrow h & & \| & \\ d_*\Phi : & 0 \to & \mathbf{k} & \to & A & \to & R & \to 0 \end{array}$$

with $d_*\Phi$ trivial. From Example 1.1.2(ii) it follows that the generator ϵ of $\mathbf{k}$ in A is contained in $m_A \backslash m_A^2$. Since $h(J/(\underline{X})J) \subset m_A^2$ we deduce that $d = 0$. It follows that α is injective.

Conversely, given a Λ-extension (A, φ) of R by $\mathbf{k}$ it is possible to find a lifting:

$$\begin{array}{lccc} & \Lambda[[\underline{X}]] & & \\ & \downarrow \tilde{\varphi} & \searrow & \\ \varphi : & A & \to & R \end{array}$$

because A is complete (see C.3(ii)). From the fact that $\ker(\varphi) = \mathbf{k}$ it follows that $\ker(\tilde{\varphi}) \supset (\underline{X})J$ and therefore we have a commutative diagram:

$$\begin{array}{lccccccc} \Phi: & 0 \to & J/(\underline{X})J & \to & \Lambda[[\underline{X}]]/(\underline{X})J & \to & R & \to 0 \\ & & \downarrow d & & \downarrow \tilde{\varphi} & & \| & \\ (A,\varphi): & 0 \to & \mathbf{k} & \to & A & \to & R & \to 0 \end{array}$$

in which d is the map induced by $\tilde{\varphi}$. It follows that $(A,\varphi) = d_*\Phi$; hence α is surjective. □

Corollary 2.1.2. *For every R in $\mathrm{ob}(\mathcal{A}^*)$ the following are true:*

$$(i) \qquad \dim_{\mathbf{k}}[o(R)] < \infty$$

$$(ii) \qquad \dim_{\mathbf{k}}(t_R) \geq \dim(R) \geq \dim_{\mathbf{k}}(t_R) - \dim_{\mathbf{k}}[o(R)]$$

(where $\dim(R)$ *means* Krull dimension *of R). In (ii) the first equality holds if and only if R is formally smooth; the second equality holds if and only if $\hat{R} = \mathbf{k}[[X_1, \ldots, X_d]]/J$, with $J \subset (\underline{X})^2$ and J generated by a regular sequence.*

Proof. We may assume that R is in $\mathrm{ob}(\hat{\mathcal{A}})$; hence $R = \hat{R} = \mathbf{k}[[X_1, \ldots, X_d]]/J$, with $J \subset (\underline{X})^2$ and $o(R) \cong (J/(\underline{X})J)^\vee$. Then (i) and (ii) follow from the fact that $\dim_{\mathbf{k}}[J/(\underline{X})J]$ is the number of elements of a minimal set of generators of J. □

Remarks 2.1.3. The only formally smooth $\mathbf{k}$-algebra in $\mathcal{A}$ is $\mathbf{k}$ itself. By 2.1.1(ii) this means that $o(\mathbf{k}) = (0)$ and that $o(A) \neq (0)$ for every $A \neq \mathbf{k}$ in $\mathcal{A}$. The following are some special cases.

- If $A = \mathbf{k}\tilde{\oplus}V$, a trivial extension of $\mathbf{k}$ by a vector space V of dimension d, then $A \cong \mathbf{k}[X_1, \ldots, X_d]/(\underline{X})^2$, and $o(A) = [(\underline{X})^2/(\underline{X})^3]^\vee$.
- If $A = \mathbf{k}[\underline{X}]/(\underline{X})^k$ then $o(A) = [(\underline{X})^k/(\underline{X})^{k+1}]^\vee$. In particular, if $A = \mathbf{k}[t]/(t^n)$, $n \geq 2$, then $o(A) = [(t^n)/(t^{n+1})]^\vee$ is one-dimensional; from the proof of 2.1.1(ii) it follows immediately that $o(A)$ is generated by the class of the extension:
$$0 \to (t^n)/(t^{n+1}) \to \mathbf{k}[t]/(t^{n+1}) \to \mathbf{k}[t]/(t^n) \to 0$$

We will need the following:

Lemma 2.1.4. *(i) Let $\mu : \Lambda \to R$ be a homomorphism in $\mathcal{A}^*$. Given a small extension $\eta : B \to A$ in $\mathcal{A}$ and a homomorphism $\varphi : R \to A$, the condition $\varphi^*(\eta) \in \ker(o(\mu))$ is equivalent to the existence of a commutative diagram:*

$$\begin{array}{ccc} A & \overset{\varphi}{\longleftarrow} & R \\ \uparrow \eta & & \uparrow \mu \\ B & \underset{\tilde{\varphi}}{\longleftarrow} & \Lambda \end{array} \tag{2.3}$$

Moreover, $\varphi^*(\eta) = 0$ if and only if there exists $\varphi' : R \to B$ such that the resulting diagram

$$\begin{array}{ccc} A & \stackrel{\varphi}{\longleftarrow} & R \\ \uparrow \eta & \swarrow \varphi' & \uparrow \mu \\ B & \underset{\tilde{\varphi}}{\longleftarrow} & \Lambda \end{array} \quad (2.4)$$

is commutative.

(ii) *For every Λ in* $\mathrm{ob}(\hat{\mathcal{A}})$ *and $\mu : \Lambda \to R$ in* $\mathrm{ob}(\mathcal{A}^*_\Lambda)$ *there exists A in* $\mathrm{ob}(\mathcal{A}_\Lambda)$ *and a homomorphism $p : R \to A$ such that $o(p/\Lambda) : o(A/\Lambda) \to o(R/\Lambda)$ is surjective.*

Proof. (i) is left to the reader.

(ii) We will show something more precise, namely that if $p_n : R \to R/m^{n+1}$ is the natural map, then

$$o(p_n/\Lambda) : o((R/m^{n+1})/\Lambda) \to o(R/\Lambda)$$

is surjective for all $n \gg 0$.

Since $o(p_n/\Lambda)$ factors through $o(\hat{R}/\Lambda)$, which is isomorphic to $o(R/\Lambda)$, we may assume that R is in $\hat{\mathcal{A}}_\Lambda$. Let's write:

$$R = \Lambda[[\underline{X}]]/J$$

where $J = (g_1, \ldots, g_s) \subset (\underline{X})^2$. Let $n \gg 0$ be such that $g_j \notin (\underline{X})^{n+2}$ for all $j = 1, \ldots, s$. Then we have:

$$\frac{R}{m^{n+1}} = \frac{\Lambda[[\underline{X}]]}{\left(J, (\underline{X})^{n+1}\right)}$$

and therefore

$$o((R/m^{n+1})/\Lambda) = \left[\frac{(J, (\underline{X})^{n+1})}{((\underline{X})J, (\underline{X})^{n+2})}\right]^\vee$$

The map $o(p_n/\Lambda)$ is the transpose of

$$\iota_n : \frac{J}{(\underline{X})J} \to \left[\frac{(J, (\underline{X})^{n+1})}{((\underline{X})J, (\underline{X})^{n+2})}\right]$$

induced by the inclusion $J \subset (J, (\underline{X})^{n+1})$. From the hypothesis on n it follows that if $\gamma \in J \cap ((\underline{X})J, (\underline{X})^{n+2})$ then $\gamma \in (\underline{X})J$; this means that ι_n is injective, i.e. that $o(p_n/\Lambda)$ is surjective. □

The following theorem gives a characterization of formally smooth homomorphisms in $\mathcal{A}^*$.

Theorem 2.1.5. *Let $\mu : \Lambda \to R$ be a homomorphism in $\mathcal{A}^*$. The following conditions are equivalent:*

(i) *For every commutative diagram (2.3) with η a small extension in $\mathcal{A}^*$ there exists $\varphi' : R \to B$ such that diagram (2.4) is commutative.*

(ii) μ *is formally smooth.*
(iii) $d\mu : t_R \to t_\Lambda$ *is surjective and* $o(\mu)$ *is injective.*
(iv) $o(R/\Lambda) = (0)$.

Proof. $(i) \Rightarrow (ii)$ is trivial.

$(ii) \Rightarrow (iii)$ Let $v \in t_\Lambda$ be given as a **k**-algebra homomorphism $\Lambda \to \mathbf{k}[\epsilon]$. The formal smoothness of μ implies the existence of a homomorphism $w : R \to \mathbf{k}[\epsilon]$ which makes the following diagram commutative:

$$\begin{array}{ccc} \mathbf{k} & \leftarrow & R \\ \uparrow & \swarrow & \uparrow \\ \mathbf{k}[\epsilon] & \leftarrow & \Lambda \end{array}$$

and this means that $d\mu(w) = v$. Therefore $d\mu$ is surjective.

Consider a commutative diagram of **k**-algebra homomorphisms (2.3) with η a small extension in $\mathcal{A}$. Then $o(\varphi)([\eta]) \in \ker(o(\mu))$. By the formal smoothness of μ there exists $\varphi' : R \to B$ making (2.4) commutative: this implies that $o(\varphi)([\eta]) = 0$. Since, by Lemma 2.1.4, φ and η can be chosen so that $o(\varphi)([\eta])$ is an arbitrary element of $\ker(o(\mu))$, we deduce that $o(\mu)$ is injective.

$(iii) \Leftrightarrow (iv)$ follows from the exact sequence (2.2).

$(iii) \Rightarrow (i)$. Consider a diagram (2.3) with η a small extension in $\mathcal{A}^*$. Then

$$o(\varphi)([\eta]) \in \ker(o(\mu))$$

By assumption $o(\varphi)([\eta]) = 0$, and therefore there exists $\bar{\varphi} : R \to B$ such that $\eta\bar{\varphi} = \varphi$. It follows that $\tilde{\varphi} - \bar{\varphi}\mu : \Lambda \to \ker(\eta) = \mathbf{k}$ is a **k**-derivation. By assumption there exists a **k**-derivation $v : R \to \mathbf{k}$ such that $\tilde{\varphi} - \bar{\varphi}\mu = d\mu(v) = v\mu$.

Then $\varphi' := \bar{\varphi} + v : R \to B$ is a **k**-homomorphism which obviously satisfies $\eta\varphi' = \varphi$. Moreover,

$$\varphi'\mu = (\bar{\varphi} + v)\mu = \bar{\varphi}\mu + v\mu = \tilde{\varphi}$$

and therefore φ' makes (2.4) commutative. □

In the special case $\Lambda = \mathbf{k}$ we obtain that a **k**-algebra R in ob$(\mathcal{A}^*)$ is unobstructed if and only if R is formally smooth, a result already proven in 2.1.1. More generally, the theorem says that μ is formally smooth if and only if $d\mu$ is surjective and R is less obstructed than Λ. The following corollary is immediate.

Corollary 2.1.6. *(i) Let* $\mu : \Lambda \to R$ *be a homomorphism in* $\mathcal{A}^*$ *such that* $d\mu$ *is surjective and* R *is formally smooth. Then* Λ *and* μ *are formally smooth.*
(ii) A homomorphism $\mu : \Lambda \to R$ *in* $\mathcal{A}^*$ *is formally etale if and only if* $d\mu$ *is an isomorphism and* $o(\mu)$ *is injective. This happens in particular if* $d\mu$ *is an isomorphism and* R *is formally smooth.*
(iii) If $R \in$ ob$(\mathcal{A}^*)$ *then the natural homomorphism* $R \to \hat{R}$ *is formally etale.*

In practice it is seldom possible to compute the obstruction map $o(\mu)$ explicitly for a given $\mu : \Lambda \to R$ in $\mathcal{A}^*$. But for the purpose of studying the formal smoothness of μ all that counts is to have information about $\ker[o(\mu)]$. This can be achieved somewhat indirectly by means of the following simple result, which turns out to be very effective in practice.

Proposition 2.1.7. *Let $\mu : \Lambda \to R$ be a morphism in $\mathcal{A}^*$. Assume that there exists a* **k***-vector space $\upsilon(R/\Lambda)$ such that for every homomorphism $\varphi : R \to A$ in $\mathcal{A}^*_\Lambda$ with A in* ob$(\mathcal{A}_\Lambda)$ *there is a* **k***-linear map*

$$\varphi_\upsilon : \mathrm{Ex}_\Lambda(A, \mathbf{k}) \to \upsilon(R/\Lambda)$$

satisfying

$$\ker(\varphi_\upsilon) = \ker[o(\varphi/\Lambda)]$$

where

$$\begin{array}{ccc} \mathrm{Ex}_\Lambda(A, \mathbf{k}) & \xrightarrow{\varphi_\upsilon} & \upsilon(R/\Lambda) \\ & \searrow o(\varphi/\Lambda) & \\ & \mathrm{Ex}_\Lambda(R, \mathbf{k}) = o(R/\Lambda) & \end{array}$$

Then there is a natural **k***-linear inclusion*

$$o(R/\Lambda) \subset \upsilon(R/\Lambda)$$

Proof. Choosing φ such that $o(\varphi/\Lambda)$ surjects onto $o(R/\Lambda)$ we obtain an inclusion $o(R/\Lambda) \subset \upsilon(R/\Lambda)$ as asserted. □

In practice this proposition will be applied as follows. Given φ, to give an element of $\mathrm{Ex}_\Lambda(A, \mathbf{k})$ is the same as to give a commutative diagram (2.3). Assume that to each such diagram one associates in a linear way an element of $\upsilon(R/\Lambda)$ which vanishes if and only if there is an extension $\varphi' : R \to B$ making (2.4) commutative. Then the proposition applies.

Taking $\Lambda = \mathbf{k}$ we get the following absolute version of the proposition, where for the last assertion we apply 2.1.2(ii).

Corollary 2.1.8. *Let R be in* ob$(\mathcal{A}^*)$*. Assume that there exists a* **k***-vector space $\upsilon(R)$ such that for every morphism $\varphi : R \to A$ in $\mathcal{A}^*$ with A in* ob$(\mathcal{A})$ *there is a* **k***-linear map*

$$\varphi_\upsilon : o(A) \to \upsilon(R)$$

satisfying

$$\ker(\varphi_\upsilon) = \ker[o(\varphi)]$$

Then there is a natural **k***-linear inclusion*

$$o(R) \subset \upsilon(R)$$

If $\upsilon(R)$ is finite dimensional then

$$\dim_{\mathbf{k}}(t_R) \geq \dim(R) \geq \dim_{\mathbf{k}}(t_R) - \dim_{\mathbf{k}}[\upsilon(R)]$$

A **k**-vector space $\upsilon(R)$ satisfying the conditions of 2.1.8 will sometimes be called *an obstruction space for R.*

2.2 Functors of Artin rings

A *functor of Artin rings* is a covariant functor

$$F : \mathcal{A}_\Lambda \longrightarrow (\text{sets})$$

where $\Lambda \in \text{ob}(\hat{\mathcal{A}})$. Let A be in $\text{ob}(\mathcal{A}_\Lambda)$. An element $\xi \in F(A)$ will be called an *infinitesimal deformation* of $\xi_0 \in F(\mathbf{k})$ if $\xi \mapsto \xi_0$ under the map $F(A) \to F(\mathbf{k})$; if $A = \mathbf{k}[\epsilon]$ then ξ is called a *first-order deformation* of ξ_0.

Examples of functors of Artin rings are obtained by fixing an R in $\text{ob}(\hat{\mathcal{A}}_\Lambda)$ and letting:

$$h_{R/\Lambda}(A) = \text{Hom}_{\hat{\mathcal{A}}_\Lambda}(R, A) \qquad \text{for every } A \text{ in } \text{ob}(\mathcal{A}_\Lambda)$$

Such a functor is clearly nothing but the restriction to $\mathcal{A}_\Lambda$ of a representable functor on $\hat{\mathcal{A}}_\Lambda$. A functor of Artin rings isomorphic to $h_{R/\Lambda}$ for some R in $\text{ob}(\hat{\mathcal{A}}_\Lambda)$ is called *prorepresentable*. In the case $\Lambda = \mathbf{k}$ we write h_R instead of $h_{R/\mathbf{k}}$.

Every *representable* functor $h_{R/\Lambda}$, R in $\text{ob}(\mathcal{A}_\Lambda)$, is a (trivial) example of prorepresentable functor.

Typically, a prorepresentable functor of Artin rings arises as follows. One considers a scheme M and the restriction

$$\Phi : \mathcal{A} \longrightarrow (\text{sets})$$

$$\Phi(A) = \text{Hom}(\text{Spec}(A), M)$$

of the functor of points

$$\text{Hom}(-, M) : (\text{schemes})^\circ \longrightarrow (\text{sets})$$

Φ is a functor of Artin rings; if $\varphi : \text{Spec}(A) \to M$ is an element of $\Phi(A)$ then φ is an infinitesimal deformation of the composition

$$\text{Spec}(\mathbf{k}) \to \text{Spec}(A) \xrightarrow{\varphi} M$$

where the first morphism corresponds to $A \to A/m_A = \mathbf{k}$. For a fixed $\mathbf{k}$-rational point $m \in M$, one may consider the subfunctor

$$F : \mathcal{A} \longrightarrow (\text{sets})$$

of Φ defined as follows:

$$F(A) = \text{Hom}(\text{Spec}(A), M)_m = \left\{ \begin{array}{c} \text{morphisms Spec}(A) \to M \\ \text{whose image is } \{m\} \end{array} \right\} \tag{2.5}$$

We have $F = h_R$, where $R = \hat{\mathcal{O}}_{M,m}$, so F is prorepresentable. We will introduce several functors of Artin rings which are defined by deformation problems. The hope is that they are of the form (2.5) for some $\mathbf{k}$-pointed algebraic scheme (M, m) which has some geometrical meaning with respect to the deformation problem under consideration (a "moduli scheme"). As we will see, this sometimes happens, but most of the time it doesn't. We will thus have to content ourselves to prove that such functors have weaker properties. In this section and in the following one we will study prorepresentability and other properties of functors of Artin rings from an abstract point of view. Later on we will consider specific examples of such functors related to deformation problems.

A prorepresentable functor $F = h_{R/\Lambda}$ satisfies the following conditions:

H_0) $F(\mathbf{k})$ consists of one element (the canonical quotient $R \to R/m_R = \mathbf{k}$)

Let

$$\begin{array}{ccc} A' & & A'' \\ & \searrow \quad \swarrow & \\ & A & \end{array} \tag{2.6}$$

be a diagram in $\mathcal{A}_\Lambda$ and consider the natural map

$$\alpha : F(A' \times_A A'') \to F(A') \times_{F(A)} F(A'')$$

induced by the commutative diagram:

$$\begin{array}{ccc} F(A' \times_A A'') & \to & F(A'') \\ \downarrow & & \downarrow \\ F(A') & \to & F(A) \end{array}$$

Then

H_ℓ) (*left exactness*) For every diagram (2.6) α is bijective (straightforward to check).
H_f) $F(\mathbf{k}[\epsilon])$ has a structure of finite-dimensional $\mathbf{k}$-vector space.

In fact

$$F(\mathbf{k}[\epsilon]) = \mathrm{Hom}_{\Lambda-alg}(R, \mathbf{k}[\epsilon]) = \mathrm{Der}_\Lambda(R, \mathbf{k}) = t_{R/\Lambda}$$

is the relative tangent space of R over Λ (here the Λ-algebra structure of $\mathbf{k}[\epsilon]$ is given by the composition $\Lambda \to \mathbf{k} \to \mathbf{k}[\epsilon]$, (see B.9(vi)).

A property weaker than H_ℓ satisfied by a prorepresentable functor F is the following:

H_ϵ) α is bijective if $A = \mathbf{k}$ and $A'' = \mathbf{k}[\epsilon]$.

It is interesting to observe that if F is prorepresentable the structure of $\mathbf{k}$-vector space on $F(\mathbf{k}[\epsilon])$ can be reconstructed in a purely functorial way using only properties H_0 and H_ϵ, and without using the prorepresentability explicitly.

Indeed it is an easy exercise to check that the homomorphism:

$$\begin{array}{cccc} + : & \mathbf{k}[\epsilon] \times_{\mathbf{k}} \mathbf{k}[\epsilon] & \longrightarrow & \mathbf{k}[\epsilon] \\ & (a + b\epsilon, a + b'\epsilon) & \longmapsto & a + (b + b')\epsilon \end{array}$$

induces the sum operation on $F(\mathbf{k}[\epsilon])$ as the composition

$$F(\mathbf{k}[\epsilon]) \times F(\mathbf{k}[\epsilon]) \to F(\mathbf{k}[\epsilon] \times_{\mathbf{k}} \mathbf{k}[\epsilon]) \stackrel{F(+)}{\longrightarrow} F(\mathbf{k}[\epsilon])$$

where the first map is the inverse of α. Associativity is checked using H_ϵ.

The zero is the image of $F(\mathbf{k}) \to F(\mathbf{k}[\epsilon])$. The multiplication by a scalar $c \in \mathbf{k}$ is induced in $F(\mathbf{k}[\epsilon])$ by the morphism

$$\begin{array}{ccc} \mathbf{k}[\epsilon] & \longrightarrow & \mathbf{k}[\epsilon] \\ a + b\epsilon & \longmapsto & a + (cb)\epsilon \end{array}$$

We can therefore state the following:

Lemma 2.2.1. *If F is a functor of Artin rings having properties H_0 and H_ϵ then the set $F(\mathbf{k}[\epsilon])$ has a structure of* **k***-vector space in a functorial way. This vector space is called the* tangent space of the functor F, *and is denoted by t_F. If $F = h_{R/\Lambda}$ then $t_F = t_{R/\Lambda}$.*

If $f : F \to G$ is a morphism of functors of Artin rings then the induced map $t_F \to t_G$ is called the *differential* of f and it is denoted by df. It is straightforward to check that if F and G satisfy H_0 and H_ϵ then df is **k**-linear.

Every functor of Artin rings F can be extended to a functor

$$\hat{F} : \hat{\mathcal{A}}_\Lambda \longrightarrow (\text{sets})$$

by letting, for every R in $\text{ob}(\hat{\mathcal{A}}_\Lambda)$:

$$\hat{F}(R) = \varprojlim F(R/m_R^{n+1})$$

and for every $\varphi : R \to S$:

$$\hat{F}(\varphi) : \hat{F}(R) \to \hat{F}(S)$$

be the map induced by the maps $F(R/m_R^n) \to F(S/m_S^n)$, $n \geq 1$.

An element $\hat{u} \in \hat{F}(R)$ is called a *formal element* of F. By definition $\hat{u}$ can be represented as a system of elements $\{u_n \in F(R/m_R^{n+1})\}_{n \geq 0}$ such that for every $n \geq 1$ the map

$$F(R/m_R^{n+1}) \to F(R/m_R^n)$$

induced by the projection $R/m_R^{n+1} \to R/m_R^n$ sends

$$u_n \longmapsto u_{n-1} \tag{2.7}$$

$\hat{u}$ is also called a *formal deformation* of u_0. If $f : F \to G$ is a morphism of functors of Artin rings then it can be extended in an obvious way to a morphism of functors $\hat{f} : \hat{F} \to \hat{G}$.

Lemma 2.2.2. *Let R be in* $\text{ob}(\hat{\mathcal{A}}_\Lambda)$. *There is a 1–1 correspondence between $\hat{F}(R)$ and the set of morphisms of functors*

$$h_{R/\Lambda} \longrightarrow F \tag{2.8}$$

Proof. To a formal element $\hat{u} \in \hat{F}(R)$ there is associated a morphism of functors (2.8) in the following way. Each $u_n \in F(R/m_R^{n+1})$ defines a morphism of functors $h_{(R/m^{n+1})/\Lambda} \rightarrow F$. The compatibility conditions (2.7) imply that the following diagram commutes:

$$\begin{array}{ccc} h_{(R/m^n)/\Lambda} & \rightarrow & h_{(R/m^{n+1})/\Lambda} \\ & \searrow & \downarrow \\ & & F \end{array}$$

for every n. Since for each A in $\mathrm{ob}(\mathcal{A}_\Lambda)$

$$h_{(R/m^n)/\Lambda}(A) \rightarrow h_{(R/m^{n+1})/\Lambda}(A)$$

is a bijection for all $n \gg 0$ we may define

$$h_{R/\Lambda}(A) \rightarrow F(A)$$

as

$$\lim_{n\to\infty} [h_{(R/m^{n+1})/\Lambda}(A) \rightarrow F(A)]$$

Conversely, each morphism (2.8) defines a formal element $\hat{u} \in \hat{F}(R)$, where $u_n \in F(R/m_R^{n+1})$ is the image of the canonical projection $R \rightarrow R/m_R^{n+1}$ via the map

$$h_{R/\Lambda}(R/m_R^{n+1}) \rightarrow F(R/m_R^{n+1})$$

□

Definition 2.2.3. *If R is in* $\mathrm{ob}(\hat{\mathcal{A}}_\Lambda)$ *and* $\hat{u} \in \hat{F}(R)$*, we call* $(R, \hat{u})$ *a* formal couple *for F. The differential $t_{R/\Lambda} \rightarrow t_F$ of the morphism $h_{R/\Lambda} \rightarrow F$ defined by $\hat{u}$ is called the* characteristic map *of $\hat{u}$ (or of the formal couple $(R, \hat{u})$) and is denoted by $d\hat{u}$.*
If $(R, \hat{u})$ is such that the induced morphism (2.8) is an isomorphism, then F is prorepresentable, and we also say that F is prorepresented by the formal couple $(R, \hat{u})$. *In this case $\hat{u}$ is called a* universal formal element *for F, and $(R, \hat{u})$ is a* universal formal couple.

An ordinary couple (R, u) with R in $\mathcal{A}_\Lambda$ defines a special case of formal couple, in which $(R_n, u_n) = (R_{n+1}, u_{n+1})$ for all $n \gg 0$. A universal formal couple seldom exists; we will therefore need to introduce some weaker properties of a formal couple. They are based on the following definition.

Definition 2.2.4. *Let $f : F \rightarrow G$ be a morphism of functors of Artin rings. f is called* smooth *if for every surjection $\mu : B \rightarrow A$ in $\mathcal{A}_\Lambda$ the natural map:*

$$F(B) \rightarrow F(A) \times_{G(A)} G(B) \tag{2.9}$$

induced by the diagram:

$$\begin{array}{ccc} F(B) & \to & G(B) \\ \downarrow & & \downarrow \\ F(A) & \to & G(A) \end{array}$$

is surjective. The functor F is called smooth *if the morphism from F to the constant functor*

$$G(A) = \{\textit{one element}\} \qquad \textit{all } A \in \mathrm{ob}(\mathcal{A}_\Lambda)$$

is smooth; equivalently, if

$$F(\mu) : F(B) \to F(A)$$

is surjective for every surjection $\mu : B \to A$ in $\mathcal{A}_\Lambda$.

Note that, since every surjection in $\mathcal{A}_\Lambda$ factors as a finite sequence of small extensions (i.e. surjections with one-dimensional kernel), for f (or F) to be smooth it is necessary and sufficient that the defining condition is satisfied for small extensions in $\mathcal{A}_\Lambda$.

The next proposition states some properties of the notion of smoothness of a morphism of functors of Artin rings.*

Proposition 2.2.5. *(i) Let $f : R \to S$ be a homomorphism in $\hat{\mathcal{A}}_\Lambda$. Then f is formally smooth if and only if the morphism of functors $h_f : h_{S/\Lambda} \to h_{R/\Lambda}$ induced by f is smooth.*

(ii) If $f : F \to G$ is smooth then f is surjective, i.e. $F(B) \to G(B)$ is surjective for every B in $\mathrm{ob}(\mathcal{A}_\Lambda)$. In particular the differential $df : t_F \to t_G$ is surjective.

(iii) If $f : F \to G$ is smooth, then F is smooth if and only if G is smooth.

(iv) If $f : F \to G$ is smooth then the induced morphism of functors $\hat{f} : \hat{F} \to \hat{G}$ is surjective, i.e. $\hat{F}(R) \to \hat{G}(R)$ is surjective for every R in $\mathrm{ob}(\hat{\mathcal{A}}_\Lambda)$.

(v) If $F \to G$ and $G \to H$ are smooth morphisms of functors then the composition $F \to H$ is smooth.

(vi) If $F \to G$ and $H \to G$ are morphisms of functors and $F \to G$ is smooth, then $F \times_G H \to H$ is smooth.

(vii) If F and G are smooth functors then $F \times G$ is smooth.

Proof. (i) Let $B \to A$ be a surjection in $\mathcal{A}_\Lambda$ and let

$$\begin{array}{ccc} A & \leftarrow & S \\ \uparrow & & \uparrow f \\ B & \leftarrow & R \end{array}$$

be a commutative diagram of homomorphisms of Λ-algebras. The formal smoothness of f is equivalent to the existence, for each such diagram, of a morphism $S \to B$ such that the resulting diagram

$$\begin{array}{ccc} A & \leftarrow & S \\ \uparrow & \swarrow & \uparrow f \\ B & \leftarrow & R \end{array}$$

* Here all functors are assumed to satisfy H_0.

is commutative. This is another way to express the condition that the map

$$h_{S/\Lambda}(B) \longrightarrow h_{S/\Lambda}(A) \times_{h_{R/\Lambda}(A)} h_{R/\Lambda}(B)$$

is surjective, i.e. that h_f is smooth.

(ii) follows immediately from the definition.

(iii) Let $\mu : B \to A$ be a surjection in $\mathcal{A}_\Lambda$, and consider the diagram:

$$\begin{array}{ccccc} F(B) & \to & F(A) \times_{G(A)} G(B) & \to & G(B) \\ & & \downarrow & & \downarrow G(\mu) \\ & & F(A) & \underset{f(A)}{\longrightarrow} & G(A) \end{array}$$

Suppose G is smooth and let $\xi \in F(A)$. By the smoothness of G there exists $\eta \in G(B)$ such that $G(\mu)(\eta) = f(A)(\xi)$. By the smoothness of f there is $\zeta \in F(B)$ which is mapped to $(\xi, \eta) \in F(A) \times_{G(A)} G(B)$. It follows that $F(\mu)(\zeta) = \xi$ and F is smooth.

The converse is proved similarly using the surjectivity of $f(A)$.

(iv) Let $\hat{v} = \{v_n\} \in \hat{G}(R)$. Since f is smooth the map

$$F(R/m_R^2) \to F(\mathbf{k}) \times_{G(\mathbf{k})} G(R/m_R^2) = G(R/m_R^2)$$

is surjective. Therefore there exists $w_1 \in F(R/m_R^2)$ such that $f(R/m_R^2)(w_1) = v_1$. Let's assume that for some $n \geq 1$ there exist $w_i \in F(R/m_R^{i+1})$, $i = 1, \ldots, n$ such that:

$$f(R/m_R^{i+1})(w_i) = v_i$$

and $w_i \longmapsto w_{i-1}$ under $F(R/m_R^{i+1}) \longrightarrow F(R/m_R^i)$.

The surjectivity of the map:

$$F(R/m_R^{n+1}) \longrightarrow F(R/m_R^n) \times_{G(R/m^n)} G(R/m_R^{n+1})$$

implies that there exists $w_{n+1} \in F(R/m_R^{n+1})$ whose image is (w_n, v_{n+1}). By induction we conclude that there exists $\hat{w} \in \hat{F}(R)$ whose image under $\hat{f}$ is $\hat{v}$.

The proofs of (v), (vi) and (vii) are straightforward. □

We now introduce the notions of "versality" and "semiuniversality", which are slightly weaker than universality.

Definition 2.2.6. *Let F be a functor of Artin rings. A formal element $\hat{u} \in \hat{F}(R)$, for some R in* ob$(\hat{\mathcal{A}}_\Lambda)$*, is called* versal *if the morphism $h_{R/\Lambda} \to F$ defined by $\hat{u}$ is smooth; $\hat{u}$ is called* semiuniversal *if it is versal and moreover, the differential $t_{R/\Lambda} \to t_F$ is bijective.*
We will correspondingly speak of a versal formal couple *(respectively a* semiuniversal formal couple*).*

It is clear from the definitions that:

$$\hat{u} \ \text{ universal} \Rightarrow \hat{u} \ \text{ semiuniversal} \Rightarrow \hat{u} \ \text{ versal}$$

but none of the inverse implications is true (see Examples 2.6.8). We can also describe these properties as follows.

Assume given $\mu : (B, \xi_B) \to (A, \xi_A)$, a surjection of couples for F (i.e. μ is surjective), and a homomorphism $\varphi : R \to A$ such that $\varphi(\hat{u}) = \xi_A$. Then $\hat{u}$ is *versal* if for all such data there is a lifting $\psi : R \to B$ of φ (i.e. $\mu\psi = \varphi$) such that $\psi(\hat{u}) = \xi_B$:

$$\begin{array}{ccc} & & B \\ & \psi \nearrow & \\ R & & \downarrow \mu \\ & \varphi \searrow & \\ & & A \end{array} \qquad\qquad \begin{array}{ccc} & & \xi_B \\ & \nearrow & \\ \hat{u} & & \downarrow \\ & \searrow & \\ & & \xi_A \end{array}$$

$\hat{u}$ is *universal* if, moreover, the lifting ψ is unique. $\hat{u}$ is *semiuniversal* if it is versal and, moreover, the lifting ψ is unique when $\mu : \mathbf{k}[\epsilon] \to \mathbf{k}$.

Let $(R, \hat{u})$ and $(S, \hat{v})$ be two formal couples for F. A *morphism of formal couples*

$$f : (R, \hat{u}) \to (S, \hat{v})$$

is a morphism $f : R \to S$ in $\hat{\mathcal{A}}$ such that $\hat{F}(f)(\hat{u}) = \hat{v}$. We will call f an *isomorphism of formal couples* if in addition $f : R \to S$ is an isomorphism.

It is obvious that with this definition the formal couples for F and their morphisms form a category containing the category I_F of couples for F as a full subcategory.

Proposition 2.2.7. *Let F be a functor of Artin rings. Then:*

(i) If $(R, \hat{u})$ and $(S, \hat{v})$ are universal formal couples for F there exists a unique isomorphism of formal couples $(R, \hat{u}) \cong (S, \hat{v})$.
(ii) If $(R, \hat{u})$ and $(S, \hat{v})$ are semiuniversal formal couples for F there exists an isomorphism of formal couples $(R, \hat{u}) \cong (S, \hat{v})$, which is not necessarily unique but the induced isomorphism $t_S \cong t_R$ is uniquely determined.
(iii) If $(R, \hat{u})$ is a semiuniversal formal couple for F and $(S, \hat{v})$ is versal, then there is an isomorphism, not necessarily unique:

$$\varphi : R[[X_1, \ldots, X_r]] \to S$$

for some $r \geq 0$, such that $\hat{F}(\varphi j)(\hat{u}) = \hat{v}$, where $j : R \subset R[[X_1, \ldots, X_r]]$ is the inclusion.

Proof. (i) By the universality of $(R, \hat{u})$ for every $n \geq 1$ there exists a unique $f_n \in h_{R/\Lambda}(S/m_S^{n+1})$ such that $f_n \mapsto v_n \in F(S/m_S^{n+1})$ under the isomorphism associated to $\hat{u}$. In this way we obtain

$$f = \lim_{\leftarrow}\{f_n\} : (R, \hat{u}) \to (S, \hat{v})$$

which is uniquely determined. Analogously, we can construct a uniquely determined $g : (S, \hat{v}) \to (R, \hat{u})$. By universality the compositions

$$gf : (R, \hat{u}) \to (R, \hat{u})$$

and

$$fg : (S, \hat{v}) \to (S, \hat{v})$$

are the identity.

(ii) Using versality, and proceeding as above we can construct morphisms of formal couples

$$f : (R, \hat{u}) \to (S, \hat{v})$$

and

$$g : (S, \hat{v}) \to (R, \hat{u})$$

We obtain a commutative diagram:

$$\begin{array}{ccc} t_{R/\Lambda} & & \\ \uparrow\downarrow & \searrow & \\ t_{S/\Lambda} & \longrightarrow & t_F \end{array}$$

where the vertical arrows are the differentials of f and g, and the other arrows are the characteristic maps $d\hat{u}$ and $d\hat{v}$. From this diagram we deduce that $df = (d\hat{u})^{-1}d\hat{v}$ and $dg = (d\hat{v})^{-1}d\hat{u}$ are uniquely determined. Since

$$d(gf) = (dg)(df) = \text{identity of } t_{S/\Lambda}$$

and

$$d(fg) = (df)(dg) = \text{identity of } t_{R/\Lambda}$$

it follows that f and g are bijections inverse of each other.

(iii) By the versality of $(R, \hat{u})$ we can find a morphism of formal couples

$$f : (R, \hat{u}) \to (S, \hat{v})$$

We obtain a commutative diagram:

$$\begin{array}{ccc} t_{R/\Lambda} & & \\ \uparrow & \searrow & \\ t_{S/\Lambda} & \longrightarrow & t_F \end{array}$$

where $d\hat{u} : t_{R/\Lambda} \to t_F$ is bijective because $(R, \hat{u})$ is semiuniversal, and $d\hat{v} : t_{S/\Lambda} \to t_F$ is surjective because $(S, \hat{v})$ is versal. Hence $df : t_{S/\Lambda} \to t_{R/\Lambda}$ is surjective. This means that f induces an inclusion

$$t^{\vee}_{R/\Lambda} \subset t^{\vee}_{S/\Lambda}$$

Let $x_1, \ldots, x_r \in S$ be elements which induce a basis of

$$t^{\vee}_{S/R} = m_S/[m_S^2 + f(m_R)]$$

and define:

$$\varphi : R[[X_1, \ldots, X_r]] \to S$$

by $\varphi(X_i) = x_i$, $i = 1, \ldots, r$. Let $j : R \subset R[[X_1, \ldots, X_r]]$ be the inclusion. Letting

$$\hat{w} = \hat{F}(j)(\hat{u}) \in \hat{F}(R[[X_1, \ldots, X_r]])$$

we have a commutative diagram of formal couples:

$$\begin{array}{ccc} (R[[X_1, \ldots, X_r]], \hat{w}) & \xrightarrow{\varphi} & (S, \hat{v}) \\ \uparrow j & \nearrow f & \\ (R, \hat{u}) & & \end{array}$$

such that $\hat{F}(\varphi j)(\hat{u}) = \hat{v}$. Since

$$\varphi_1 : R[[X_1, \ldots, X_r]]/M^2 \to S/m_S^2$$

is an isomorphism (here $M \subset R[[X_1, \ldots, X_r]]$ denotes the maximal ideal) φ is surjective. Let

$$\psi_1 : S/m_S^2 \to R[[X_1, \ldots, X_r]]/M^2$$

be the inverse of φ_1. We have

$$F(\psi_1)(v_1) = w_1$$

hence, by the versality of $(S, \hat{v})$, it is possible to find a lifting of ψ_1:

$$\psi : S \to R[[X_1, \ldots, X_r]]$$

such that $\hat{F}(\psi)(\hat{v}) = \hat{w}$. Since by construction

$$d(\psi\varphi) = d\psi\, d\varphi = \text{identity of } t_{R[[\underline{X}]]}$$

it follows that $\psi\varphi$ is an automorphism of $R[[X_1, \ldots, X_r]]$. In particular, we deduce that φ is injective. □

The following is useful:

Proposition 2.2.8. *If a functor of Artin rings $F : \mathcal{A} \to$ (sets) satisfying H_0 has a semiuniversal formal element and $t_F = (0)$ then $F = h_{\mathbf{k}}$, the constant functor.*

Proof. The assumptions imply that there is a smooth morphism $h_{\mathbf{k}} \to F$. Since a smooth morphism of functors satisfying H_0 is surjective, the conclusion follows. □

* * * * * *

For the rest of this section we will only consider functors of Artin rings satisfying conditions H_0 and H_ϵ.

Definition 2.2.9. *Let F be a functor of Artin rings. Suppose that $\upsilon(F)$ is a $\mathbf{k}$-vector space such that for every A in $\mathrm{ob}(\mathcal{A}_\Lambda)$ and for every $\xi \in F(A)$ there is a $\mathbf{k}$-linear map*

$$\xi_\upsilon : \mathrm{Ex}_\Lambda(A, \mathbf{k}) \to \upsilon(F)$$

with the following property:

$\ker(\xi_\upsilon)$ consists of the isomorphism classes of extensions $(\tilde{A}, \varphi)$ such that

$$\xi \in \mathrm{Im}[F(\tilde{A}) \to F(A)]$$

Then $\upsilon(F)$ is called an obstruction space *for the functor F. If F has (0) as an obstruction space then it is called* unobstructed.

If F is prorepresented by the formal couple $(R, \hat{u})$, then $o(R/\Lambda)$ is an obstruction space for F, as it follows from the functorial characterization of $o(R/\Lambda)$ given in § 2.1. Conversely, if $\upsilon(F)$ is an obstruction space for F, and F is prorepresented by the formal couple $(R, \hat{u})$, then it follows immediately from the definition that $\upsilon(F)$ is a relative obstruction space for R/Λ.

One can show that under relatively mild conditions a functor F has an obstruction space. We will not discuss this matter here, and we refer the interested reader to [54].

The following are some basic properties of obstruction spaces.

Proposition 2.2.10. *(i) F has (0) as an obstruction space if and only if it is smooth.*
(ii) Let $f : F \to G$ be a smooth morphism of functors of Artin rings. If $\upsilon(G)$ is an obstruction space for G then it is also an obstruction space for F.

Proof. (i) is obvious.

(ii) Let A be in $\mathrm{ob}(\mathcal{A}_\Lambda)$ and $\xi \in F(A)$, and let

$$\xi_\upsilon := f(A)(\xi)_\upsilon : \mathrm{Ex}_\Lambda(A, \mathbf{k}) \to \upsilon(G)$$

the map defined by $f(A)(\xi) \in G(A)$. If $\tilde{A} \to A = \tilde{A}/(t)$ defines an element of $\ker(\xi_\upsilon)$ then, since $\upsilon(G)$ is an obstruction space for G, there is $\eta \in G(\tilde{A})$ such that $\eta \mapsto f(A)(\xi)$ under the map $G(\tilde{A}) \to G(A)$. From the smoothness of f it follows that the map

$$F(\tilde{A}) \to F(A) \times_{G(A)} G(\tilde{A})$$

is surjective, hence there is $\tilde{\xi} \in F(\tilde{A})$ which maps to the formal couple (ξ, η). It follows that $\tilde{\xi} \mapsto \xi$ under $F(\tilde{A}) \to F(A)$.

Conversely, if $\tilde{A}$ is such that there exists $\tilde{\xi} \in F(\tilde{A})$ such that

$$\begin{array}{ccc} F(\tilde{A}) & \to & F(A) \\ \tilde{\xi} & \mapsto & \xi \end{array}$$

then from the diagram:

$$\begin{array}{ccc} F(\tilde{A}) & \to & F(A) \\ \downarrow f(\tilde{A}) & & \downarrow f(A) \\ G(\tilde{A}) & \to & G(A) \end{array}$$

we see that $f(\tilde{A})(\tilde{\xi}) \mapsto f(A)(\xi)$ hence $f(A)(\xi) \in \mathrm{Im}[G(\tilde{A}) \to G(A)]$. Therefore the extension $\tilde{A} \to A$ defines an element of $\ker[f(A)(\xi)_\upsilon]$. This proves that ξ_υ satisfies the conditions of Definition 2.2.9. □

Corollary 2.2.11. *Let $F : \mathcal{A} \to$ (sets) be a functor of Artin rings, and suppose that $(R, \hat{u})$ is a versal formal couple for F. If F has a finite-dimensional obstruction space $\upsilon(F)$ then*

$$\dim_{\mathbf{k}}(t_R) \geq \dim(R) \geq \dim_{\mathbf{k}}(t_R) - \dim_{\mathbf{k}}[\upsilon(F)]$$

Proof. From the definition of versal formal couple and from Proposition 2.2.10 we deduce that $\upsilon(F)$ is an obstruction space for h_R, hence for R. The conclusion follows from 2.1.6. □

Corollary 2.2.11 is very important in deformation theory since it gives basic geometrical information about a deformation problem. If F is induced by a **k**-pointed scheme (M, m), i.e. it is of the form (2.5), then 2.2.11 gives information about the dimension of M at m. We refer the reader to Proposition 2.3.6 for a result about obstruction spaces which is often used in applications.

NOTES

1. For an interesting discussion on the geometric motivations behind the consideration of functors of Artin rings we refer the reader to [72], part C.1.

2. Corollary 2.2.11 is due to Mumford (unpublished, see [189], p. 126). A proof is outlined in [157], p. 153. The proof given here is based on such outline, and has appeared already in [163], Corollary (B2.4).

3. Let R, P be in $\hat{A}$, $\varphi : P \to R$ a homomorphism and $F : h_R \to h_P$ be the corresponding morphism between the prorepresentable functors

$$h_R, h_P : \mathcal{A} \to \text{(sets)}$$

Then $F(A) : h_R(A) \to h_P(A)$ is injective for all A in ob($\mathcal{A}$) if and only if φ is surjective, i.e. if and only if $R = P/J$ for an ideal $J \subset P$. (*Proof:* left to the reader.)

2.3 The theorem of Schlessinger

In this section we will prove a well-known theorem of Schlessinger which gives necessary and sufficient conditions, easy to verify in practice, for the existence of a (semi)universal element for a functor of Artin rings. Before stating the theorem we want to make some introductory remarks, which can be useful in what follows. Let's fix $\Lambda \in \mathrm{ob}(\hat{\mathcal{A}})$ throughout this section. We start with a characterization of prorepresentable functors:

Proposition 2.3.1. *Let $F : \mathcal{A}_\Lambda \to$ (sets) be a functor of Artin rings satisfying condition H_0. Then F is prorepresentable if and only if it is left exact and has finite-dimensional tangent space, i.e. it has properties H_ℓ and H_f.*

Proof. The "only if" implication is obvious (see also § 2.2). So let's assume that F satisfies H_ℓ and H_f. Then, by Proposition E.9, the category I_F of couples for F is cofiltered and

$$F = \lim_{\substack{\to \\ (X,\xi)}} h_X$$

Let $(A_i, \xi_i) \in \text{ob}(I_F)$ and consider all the subrings of A_i images of morphisms $(A_j, \xi_j) \to (A_i, \xi_i)$. By the descending chain condition there is $(\bar{A}_i, \bar{\xi}_i)$ such that

$$\bar{A}_i = \bigcap \text{Im}[A_j \to A_i]$$

By construction the couples $(\bar{A}_i, \bar{\xi}_i)$ form a full subcategory of I_F in which all maps are surjective. Moreover, the corresponding category of representable functors $h_{\bar{A}_i/\Lambda}$ is clearly cofinal in $(I_F)^\circ$, and therefore

$$F = \lim_{\to} h_{\bar{A}_i/\Lambda}$$

Therefore replacing I_F by this subcategory we can assume that all homomorphisms are surjective, and therefore we have:

$$F = \bigcup h_{A_i/\Lambda}$$

Moreover, since H_f holds, $F(\mathbf{k}[\epsilon]) = h_{A_i}(\mathbf{k}[\epsilon])$ when $i \gg 0$. We can discard all those A_i for which this is not true and we get again a full subcategory. Therefore

$$F = \lim_{\to} h_{A_i/\Lambda}$$

and $F(\mathbf{k}[\epsilon]) = h_{A_i/\Lambda}(\mathbf{k}[\epsilon])$ for all i. Then for each i we can find a surjection

$$\Lambda[[X_1, \ldots, X_r]] \to A_i$$

with $r = \dim(F(\mathbf{k}[\epsilon]))$, and these surjections are compatible, i.e. the diagrams

$$\begin{array}{ccc} \Lambda[[X_1, \ldots, X_r]] & \to & A_i \\ & \searrow & \downarrow \\ & & A_j \end{array}$$

are commutative. Define $B_1 = \Lambda[[X_1, \ldots, X_r]]/(\underline{X})^2 = A_i/m_i^2$ for all i. Then fix $\nu \geq 2$, and set $A_{i,\nu} = A_i/m_i^{\nu+1}$. All $A_{i,\nu}$'s are quotients of $\Lambda[[X_1, \ldots, X_r]]/(\underline{X})^{\nu+1}$ and form a projective system:

$$\cdots \to A_{j,\nu} \to A_{i,\nu} \to \cdots$$

Let

$$B_\nu = \varprojlim A_{i,\nu} = A_{i,\nu} \quad i \gg 0$$

Then by construction

$$\cdots B_{\nu+1} \to B_\nu \to \cdots$$

form a projective system and

$$F = \varinjlim h_{B_\nu/\Lambda}$$

Then

$$\hat{B} := \varprojlim B_\nu$$

is in $\mathrm{ob}(\hat{\mathcal{A}}_\Lambda)$ and prorepresents F. □

Unfortunately this characterization of prorepresentable functors is not very useful in practice because given homomorphisms

$$\begin{array}{ccc} A' & & A'' \\ & \searrow \swarrow & \\ & A & \end{array}$$

in $\mathcal{A}_\Lambda$ we have

$$\mathrm{Spec}(A' \times_A A'') = \mathrm{Spec}(A') \bigcup_{\mathrm{Spec}(A)} \mathrm{Spec}(A'')$$

and this is not easy to visualize. That's why left exactness of a functor of Artin rings F is hard to check. On the other hand, if at least one of the above homomorphisms is surjective then $\mathrm{Spec}(A' \times_A A'')$ is easier to describe. The theorem of Schlessinger reduces the prorepresentability to the verification of the condition of left exactness only in cases when at least one of the above maps is surjective. An analogous condition is given for the existence of a semiuniversal element. The result is the following.

Theorem 2.3.2 (Schlessinger [155]). *Let $F : \mathcal{A}_\Lambda \to$ (sets) be a functor of Artin rings satisfying condition H_0. Let $A' \to A$ and $A'' \to A$ be homomorphisms in $\mathcal{A}_\Lambda$ and let*

$$\alpha : F(A' \times_A A'') \longrightarrow F(A') \times_{F(A)} F(A'') \tag{2.10}$$

be the natural map. Then:

(i) F has a semiuniversal formal element if and only if it satisfies the following conditions:
  $\bar{H}$) if $A'' \to A$ is a small extension then the map (2.10) is surjective.
  H_ϵ) If $A = \mathbf{k}$ and $A'' = \mathbf{k}[\epsilon]$ then the map (2.10) is bijective.
  H_f) $\dim_{\mathbf{k}}(t_F) < \infty$

(ii) F has a universal element if and only if it also satisfies the following additional condition:
  H) the natural map

$$F(A' \times_A A') \longrightarrow F(A') \times_{F(A)} F(A')$$

  is bijective for every small extension $A' \to A$ in $\mathcal{A}_\Lambda$.

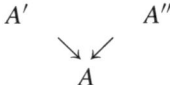

Remark 2.3.3. It will be helpful to explain the meaning of the conditions of the theorem before giving the proof. Consider a small extension in $\mathcal{A}_\Lambda$:

$$0 \to (t) \to A' \xrightarrow{\mu} A \to 0$$

Assume that $F \cong h_{R/\Lambda}$ is prorepresentable. Then two Λ-homomorphisms $f, g : R \to A'$ have the same image in $\mathrm{Hom}_{\Lambda-alg}(R, A)$ if and only if there exists a Λ-derivation $d : R \to \mathbf{k}$ (which is uniquely determined) such that

$$g(r) = f(r) + d(r)t$$

equivalently if and only if g and f differ by an element of

$$\mathrm{Der}_\Lambda(R, \mathbf{k}) = t_{R/\Lambda}$$

Therefore the fibres of

$$F(\mu) : \mathrm{Hom}_{\Lambda-alg}(R, A') \to \mathrm{Hom}_{\Lambda-alg}(R, A)$$

that are nonempty are principal homogeneous spaces under the above action of $t_{R/\Lambda} = t_F$.

Assume now that F is just a functor of Artin rings having properties H_0 and H_ϵ, so that it has tangent space t_F. We can define an action of t_F on $F(A')$ by means of the composition

$$\tau : t_F \times F(A') \xrightarrow{\alpha^{-1}} F(\mathbf{k}[\epsilon] \times_{\mathbf{k}} A') \xrightarrow{F(b)} F(A')$$

where α^{-1} exists by property H_ϵ and $F(b)$ is induced by the morphism

$$\begin{array}{rccc} b: & \mathbf{k}[\epsilon] \times_{\mathbf{k}} A' & \longrightarrow & A' \\ & (x + y\epsilon, a') & \longmapsto & a' + yt \end{array}$$

The action τ maps the fibres of $F(\mu)$ into themselves. Indeed the isomorphism

$$\begin{array}{rccc} \gamma: & \mathbf{k}[\epsilon] \times_{\mathbf{k}} A' & \longrightarrow & A' \times_A A' \\ & (x + y\epsilon, a') & \longmapsto & (a' + yt, a') \end{array}$$

induces a map

$$\beta : t_F \times F(A') \xrightarrow{\alpha^{-1}} F(\mathbf{k}[\epsilon] \times_{\mathbf{k}} A') \xrightarrow{F(\gamma)} F(A' \times_A A') \longrightarrow F(A') \times_{F(A)} F(A')$$

which coincides with

$$\begin{array}{ccc} t_F \times F(A') & \longrightarrow & F(A') \times_{F(A)} F(A') \\ (v, \xi) & \longmapsto & (\tau(v, \xi), \xi) \end{array}$$

In case F is prorepresentable we have just given another description of the action of t_F on the fibres of $F(\mu)$ introduced before. In general the map β is neither injective (i.e. in general the action τ is not free on the fibres of $F(\mu)$) nor surjective (i.e. τ is not transitive on the fibres of $F(\mu)$). This depends on the properties of the map

$$\alpha' : F(A' \times_A A') \longrightarrow F(A') \times_{F(A)} F(A')$$

If F is left exact then α' is bijective, hence β is bijective, and the action of t_F on the fibres of $F(\mu)$ is free and transitive, as expected, since F is prorepresentable by Proposition 2.3.1.

Conversely, what this analysis shows is that for τ to be free and transitive on the fibres of $F(\mu)$ we only need α' bijective, i.e. the condition H, weaker than H_ℓ. $\bar{H}$ only guarantees the transitivity of such action: the failure from prorepresentability is therefore related to the existence of fixed points of this action. In applications this is usually due to the existence of automorphisms of geometric objects associated to an element $\xi \in F(A)$ which don't lift to automorphisms of objects associated to an element $\xi' \in F(\mu)^{-1}(\xi)$ (see § 2.6).

Since Theorem 2.3.2 guarantees that F has a semiuniversal element if it satisfies H_0, H_ϵ and $\bar{H}$, we can summarize this discussion in the following statement.

Proposition 2.3.4. *Let F be a functor of Artin rings satisfying conditions H_0 and H_ϵ and let*

$$0 \to (t) \to A' \to A \to 0$$

be a small extension in $\mathcal{A}$. Then there is an action τ of t_F on $F(A')$ such that:

(a) if F has a semiuniversal formal element then t_F acts transitively on the non-empty fibres of $F(A') \to F(A)$;
(b) if F is prorepresentable then t_F acts freely and transitively on the non-empty fibres of $F(A') \to F(A)$.

The following definition introduces a convenient terminology.

Definition 2.3.5. *Let $f : F \to G$ be a morphism of functors of Artin rings having a semiuniversal formal element and obstruction spaces $\upsilon(F)$ and $\upsilon(G)$ respectively. An* obstruction map *for f is a linear map*

$$o(f) : \upsilon(F) \to \upsilon(G)$$

such that for each A in ob($\mathcal{A}$) *and for each $\xi \in F(A)$ the diagram:*

$$\begin{array}{ccccc} & & \mathrm{Ex}_{\mathbf{k}}(A, \mathbf{k}) & & \\ & \swarrow \xi_\upsilon & & \searrow \eta_\upsilon & \\ \upsilon(F) & & \xrightarrow{o(f)} & & \upsilon(G) \end{array} \qquad (2.11)$$

is commutative, where η denotes $f(A)(\xi)$. If f has an injective obstruction map we say that F is less obstructed than G.

With this terminology we can state a result often used in concrete situations:

Proposition 2.3.6. *Let $f : F \to G$ be a morphism of functors of Artin rings having a semiuniversal formal element and obstruction spaces $\upsilon(F)$ and $\upsilon(G)$ respectively. Consider the following conditions:*

(a) df is surjective
(b) F is less obstructed than G.

If they are both satisfied then f is smooth. If condition (b) is satisfied (but not necessarily (a)) and G is smooth then F is smooth.

Proof. Let $\varphi : \tilde{A} \to A$ be a small extension and assume (a) and (b) hold. Consider the map

$$F(\tilde{A}) \to F(A) \times_{G(A)} G(\tilde{A}) \tag{2.12}$$

and let $(\xi, \tilde{\eta}) \in F(A) \times_{G(A)} G(\tilde{A})$. Since $\tilde{\eta} \mapsto \eta = f(A)(\xi)$ we have $\eta_\upsilon(\varphi) = 0$. By (b) and the commutativity of (2.11) we also have $\xi_\upsilon(\varphi) = 0$ and therefore there exists $\tilde{\xi} \in F(\tilde{A})$ such that $\tilde{\xi} \mapsto \xi$. Let $\tilde{\eta}' = f(\tilde{A})(\tilde{\xi}) \in G(\tilde{A})$. We have

$$\begin{array}{ccc} \tilde{\eta} & & \tilde{\eta}' \\ & \searrow \swarrow & \\ & \eta & \end{array}$$

Since by Theorem 2.3.2 the functor G satisfies condition $\bar{H}$, there is $w \in t_G$ such that $\tau(w, \tilde{\eta}') = \tilde{\eta}$. From the surjectivity of df it follows that there is $\upsilon \in t_F$ such that $\upsilon \mapsto w$. Now $\tau(\upsilon, \tilde{\xi}) \in F(\tilde{A})$ satisfies

$$\tau(\upsilon, \tilde{\xi}) \mapsto (\xi, \tilde{\eta})$$

because the action τ is functorial (as it easily follows from its definition). This shows that (2.12) is surjective and f is smooth.

Now assume that only (b) holds and that G is smooth. We must show that

$$F(\tilde{A}) \to F(A)$$

is surjective or equivalently that $\xi_\upsilon(\varphi) = 0$. Let $\xi \in F(A)$ and let $\eta := f(A)(\xi) \in G(A)$. Since G is smooth we have $\eta_\upsilon(\varphi) = 0$. By (b) and the commutativity of (2.11) we also have $\xi_\upsilon(\varphi) = 0$. □

Corollary 2.3.7. *Let $f : F \to G$ be a morphism of functors of Artin rings having a semiuniversal formal element. Assume that F is smooth, G has an obstruction space $\upsilon(G)$ and df is surjective. Then f and G are smooth.*

Proof. Since F is smooth it has (0) as an obstruction space and the obvious map $(0) \to \upsilon(G)$ is an obstruction map. Therefore f is smooth by Proposition 2.3.6 and we conclude by Proposition 2.2.5(iii). □

Remark 2.3.8. When F and G are prorepresentable the first part of Proposition 2.3.6 is a reformulation of Theorem 2.1.5(iii) and Corollary 2.3.7 is a reformulation of Corollary 2.1.6(i). If in Corollary 2.3.7 we replace the condition "df surjective" by "df bijective" we cannot conclude that f is an isomorphism unless both functors are prorepresentable: if they are then f is an isomorphism by Corollary 2.1.6(ii) because a formally etale homomorphism of **k**-algebras in $\hat{\mathcal{A}}$ is an isomorphism.

Proof of Theorem 2.3.2

(i) Let's assume that F has a semiuniversal formal element $(R, \hat{u})$. Consider a homomorphism $f : A' \to A$ and a small extension $\pi : A'' \to A$ both in $\mathcal{A}_\Lambda$, and let

$$(\xi', \xi'') \in F(A') \times_{F(A)} F(A'')$$

with

$$F(f)(\xi') = F(\pi)(\xi'') =: \xi \in F(A)$$

By the versality of $(R, \hat{u})$ the following maps are surjective:

$$\mathrm{Hom}_{\Lambda-alg}(R, A') \longrightarrow F(A')$$

$$\mathrm{Hom}_{\Lambda-alg}(R, A'') \longrightarrow \mathrm{Hom}_{\Lambda-alg}(R, A) \times_{F(A)} F(A'') \tag{2.13}$$

therefore there are

$$g' \in \mathrm{Hom}_{\Lambda-alg}(R, A'), \quad g'' \in \mathrm{Hom}_{\Lambda-alg}(R, A'')$$

such that

$$\hat{F}(g')(\hat{u}) = \xi'$$

and $g'' \mapsto (fg', \xi'')$ under the map (2.13). This last condition gives:

$$\pi g'' = fg', \quad \hat{F}(g'')(\hat{u}) = \xi''$$

and consequently, $\hat{F}(\pi g'')(\hat{u}) = \xi$. Using the morphism $g' \times g'' : R \to A' \times_A A''$ we obtain an element

$$\zeta := \hat{F}(g' \times g'')(\hat{u}) \in F(A' \times_A A'')$$

which, by construction, is mapped to (ξ', ξ'') by (2.13). This proves that $(R, \hat{u})$ satisfies condition $\bar{H}$.

If $A'' = \mathbf{k}[\epsilon]$ and $A = \mathbf{k}$ the map (2.13) reduces to the bijection

$$\mathrm{Hom}_{\Lambda-alg}(R, \mathbf{k}[\epsilon]) \longrightarrow t_F \tag{2.14}$$

In this case if $\zeta_1, \zeta_2 \in F(A' \times_A \mathbf{k}[\epsilon])$ are such that $\alpha(\zeta_1) = \alpha(\zeta_2) = (\xi', \xi'')$ choose $g' \in \mathrm{Hom}_{\Lambda-alg}(R, A')$ as before. By the versality the map

$$\begin{array}{c}\mathrm{Hom}_{\Lambda-alg}(R, A' \times_{\mathbf{k}} \mathbf{k}[\epsilon]) \longrightarrow F(A' \times_{\mathbf{k}} \mathbf{k}[\epsilon]) \times_{F(\mathbf{k}[\epsilon])} \mathrm{Hom}_{\Lambda-alg}(R, \mathbf{k}[\epsilon]) \\ \| \\ F(A' \times_{\mathbf{k}} \mathbf{k}[\epsilon])\end{array}$$

induced by the projection $A' \times_{\mathbf{k}} \mathbf{k}[\epsilon] \to \mathbf{k}[\epsilon]$ is surjective. Hence we obtain two morphisms:

$$g' \times g_i : R \longrightarrow A' \times_{\mathbf{k}} \mathbf{k}[\epsilon]$$

such that

$$\hat{F}(g' \times g_i)(\hat{u}) = \zeta_i$$

$i = 1, 2$. But then $\hat{F}(g_i)(\hat{u}) = \xi''$, $i = 1, 2$, hence, by the bijectivity of (2.14), $g_1 = g_2$, i.e. $\zeta_1 = \zeta_2$. This proves that F satisfies condition H_ϵ. Condition H_f is satisfied because the differential $t_{R/\Lambda} \to t_F$ is linear and is a bijection by definition of a semiuniversal formal couple.

Conversely, let's assume that F satisfies conditions $\bar{H}$, H_ϵ and H_f. We will find a semiuniversal formal couple $(R, \hat{u})$ by constructing a projective system $\{R_n; p_{n+1} : R_{n+1} \to R_n\}_{n \geq 0}$ of Λ-algebras in $\mathcal{A}_\Lambda$ and a sequence $\{u_n \in F(R_n)\}_{n \geq 0}$ such that $F(p_n)(u_n) = u_{n-1}$, $n \geq 1$.

We take $R_0 = \mathbf{k}$ and $u_0 \in F(\mathbf{k})$ its unique element. Let $r = \dim_{\mathbf{k}}(t_F)$, $\{e_1, \ldots, e_r\}$ a basis of t_F and letting $\Lambda[[\underline{X}]] = \Lambda[[X_1, \ldots, X_r]]$ we set

$$R_1 = \Lambda[[\underline{X}]]/((\underline{X})^2 + m_\Lambda \Lambda[[\underline{X}]])$$

Since we have

$$R_1 = \mathbf{k}[\epsilon] \times_{\mathbf{k}} \cdots \times_{\mathbf{k}} \mathbf{k}[\epsilon] \qquad (r \quad \text{times})$$

from H_ϵ we deduce that $F(R_1) = t_F \times \cdots \times t_F$.

Let's take $u_1 = (e_1, \ldots, e_r) \in F(R_1)$. Note that the map induced by u_1:

$$\mathbf{k}^r \cong ((\underline{X})/(\underline{X})^2)^\vee = \mathrm{Hom}_{\Lambda-alg}(R_1, \mathbf{k}[\epsilon]) \longrightarrow t_F$$

is the isomorphism

$$(\lambda_1, \ldots, \lambda_r) \longmapsto \sum_j \lambda_j e_j$$

Let's proceed by induction on n: assume that couples

$$(R_0, u_0), (R_1, u_1), \ldots, (R_{n-1}, u_{n-1})$$

such that

$$R_h = \Lambda[[\underline{X}]]/J_h, \quad u_h \in F(R_h), \quad u_h \mapsto u_{h-1}$$

$h = 1, \ldots, n-1$ have been already constructed. In order to construct (R_n, u_n) we consider the family $\mathcal{I}$ of all ideals $J \subset \Lambda[[\underline{X}]]$ having the following properties:

(a) $J_{n-1} \supset J \supset (\underline{X})J_{n-1}$;
(b) there exists $u \in F(\Lambda[[\underline{X}]]/J)$ such that the map

$$F(\Lambda[[\underline{X}]]/J) \longrightarrow F(R_{n-1})$$

sends $u \mapsto u_{n-1}$.

$\mathcal{I} \neq \emptyset$ because $J_{n-1} \in \mathcal{I}$. Moreover, $\mathcal{I}$ has a minimal element J_n. This will be proved if we show that $\mathcal{I}$ is closed with respect to finite intersections. Let $I, J \in \mathcal{I}$ and $K = I \cap J$. It is obvious that K satisfies condition (a). We may assume that $I + J = J_{n-1}$ (making J larger without changing K if necessary). This implies that the natural homomorphism

$$\Lambda[[\underline{X}]]/K \longrightarrow \Lambda[[\underline{X}]]/I \times_{R_{n-1}} \Lambda[[\underline{X}]]/J$$

is an isomorphism. By $\bar{H}$ the map

$$\alpha : F(\Lambda[[\underline{X}]]/K) \longrightarrow F(\Lambda[[\underline{X}]]/I) \times_{F(R_{n-1})} F(\Lambda[[\underline{X}]]/J)$$

is surjective, therefore there exists $u \in F(\Lambda[[\underline{X}]]/K)$ whose image in $F(R_{n-1})$ is u_{n-1}; this means that K satisfies condition (b) as well, hence it is in $\mathcal{I}$.

We take $R_n = \Lambda[[\underline{X}]]/J_n$ and $u_n \in F(R_n)$ an element which is mapped to u_{n-1}. By induction we have constructed a formal couple $(R, \hat{u})$. We now show that it is a semiuniversal formal couple for F.

As already remarked, $\hat{u}$ induces an isomorphism of tangent spaces $t_{R/\Lambda} \cong t_F$. Therefore we only have to prove versality, namely that the map

$$\hat{u}_\pi : \mathrm{Hom}_{\Lambda-alg}(R, A') \longrightarrow \mathrm{Hom}_{\Lambda-alg}(R, A) \times_{F(A)} F(A')$$

is surjective for every small extension $\pi : A' \to A$.

Let $(f, \xi') \in \mathrm{Hom}_{\Lambda-alg}(R, A) \times_{F(A)} F(A')$, i.e. $\hat{F}(f)(\hat{u}) = F(\pi)(\xi')$. We must find $f' \in \mathrm{Hom}_{\Lambda-alg}(R, A')$ such that $\hat{u}_\pi(f') = (f, \xi')$, i.e. such that

$$(1) \quad \pi f' = f; \qquad (2) \quad \hat{F}(f')(\hat{u}) = \xi'$$

Let's consider the commutative diagram:

$$\begin{array}{ccc} \mathrm{Hom}_{\Lambda-alg}(R, \mathbf{k}[\epsilon]) \times \mathrm{Hom}_{\Lambda-alg}(R, A') & \to & t_F \times F(A') \\ \downarrow \beta_1 & & \downarrow \beta_2 \\ \mathrm{Hom}_{\Lambda-alg}(R, A') \times_{\mathrm{Hom}(R,A)} \mathrm{Hom}_{\Lambda-alg}(R, A') & \to & F(A') \times_{F(A)} F(A') \end{array} \tag{2.15}$$

where the horizontal arrows are induced by $\hat{u}$. The map β_1 is a bijection because the action of $\mathrm{Hom}_{\Lambda-alg}(R, \mathbf{k}[\epsilon])$ on the fibres of

$$\mathrm{Hom}_{\Lambda-alg}(R, A') \to \mathrm{Hom}_{\Lambda-alg}(R, A)$$

is free and transitive (see Remark 2.3.3). From $\bar{H}$ it follows that β_2 is surjective (Remark 2.3.3 again). Therefore if $f' \in \mathrm{Hom}_{\Lambda-alg}(R, A')$ satisfies (1) then, letting

$$\eta' := F(f')(\hat{u}) \in F(A')$$

there exist $v \in \mathrm{Hom}_{\Lambda-alg}(R, \mathbf{k}[\epsilon]) = t_F$, $\quad f'' \in \mathrm{Hom}_{\Lambda-alg}(R, A')$ such that in diagram (2.15) we have:

$$\begin{array}{ccc} (v, f') & \longmapsto & (v, \eta') \\ \downarrow & & \downarrow \\ (f'', f') & \longmapsto & (\xi', \eta') \end{array}$$

This means that $\hat{u}_\pi(f'') = (f, \xi')$. It follows that it suffices to find f' satisfying condition (1).

Let $n > 0$ be such that f factors as

$$R \to R_{n-1} \xrightarrow{f_{n-1}} A$$

Then f' exists if and only if there exists φ which makes the following diagram commutative:

$$\begin{array}{ccc} R_n & \xrightarrow{\varphi} & A' \\ \downarrow p_n & & \downarrow \pi \\ R_{n-1} & \longrightarrow & A \end{array}$$

equivalently, if and only if the extension

$$\begin{array}{lccccccccc} f_n^*(A', \pi): & 0 \to & \ker(\pi) & \to & R_n \times_A A' & \xrightarrow{\pi'} & R_n & \to 0 \\ & & \| & & \downarrow f_n' & & \downarrow f_n & \\ & 0 \to & \ker(\pi) & \to & A' & \xrightarrow{\pi} & A & \to 0 \end{array}$$

is trivial.

Suppose not. This means that π' induces an isomorphism of tangent spaces (see 1.1.2(ii)), i.e. that there exists an ideal $I \subset \Lambda[[\underline{X}]]$ such that

$$R_n \times_A A' = \Lambda[[\underline{X}]]/I$$

By construction

$$J_{n-1} \supset I \supset (\underline{X})J_{n-1}$$

Moreover, since by $\bar{H}$ the map

$$F(R_n \times_A A') \longrightarrow F(R_n) \times_{F(A)} F(A')$$

is surjective, there exists $u \in F(R_n \times_A A')$ inducing $u_n \in F(R_n)$, hence inducing $u_{n-1} \in F(R_{n-1})$. It follows that I satisfies conditions (a) and (b) and, by the minimality of J_n in $\mathcal{I}$, it follows that $J_n \subset I$. But this is a contradiction because from the fact that π is a surjection with nontrivial kernel it follows that I is properly contained in J_n. This proves that $f_n^*(A', \pi)$ is trivial and concludes the proof of the fact that $(R, \hat{u})$ is semiuniversal and of part (i) of the theorem.

(ii) If F is prorepresentable then it trivially satisfies conditions H, H_ϵ and H_f, as already remarked.

Conversely, suppose that F satisfies conditions $\bar{H}$, H, H_ϵ and H_f. We have just proved that F has a semiuniversal formal couple $(R, \hat{u})$. We will prove that this is a universal formal couple by showing that for every A in $\mathcal{A}_\Lambda$ the map

$$\hat{u}(A) : \mathrm{Hom}_{\Lambda-alg}(R, A) \longrightarrow F(A)$$

induced by $\hat{u}$ is bijective.

This is clearly true if $A = \mathbf{k}$. We will proceed by induction on $\dim_{\mathbf{k}}(A)$. Let $\pi : A' \to A$ be a small extension in $\mathcal{A}_\Lambda$. By the inductive hypothesis

$$\mathrm{Hom}_{\Lambda-alg}(R, A) \longrightarrow F(A)$$

is bijective and, by the versality, the map

$$\hat{u}_\pi : \mathrm{Hom}_{\Lambda-alg}(R, A') \to \mathrm{Hom}_{\Lambda-alg}(R, A) \times_{F(A)} F(A') \cong F(A')$$

is surjective. The map β_2 in diagram (2.15) is bijective by condition H, and this implies that $\hat{u}_\pi$ is bijective. □

NOTES

1. Theorem 2.3.2 has been published in [155]. It had also appeared in [154]. See also [119].

2. From Theorem 2.3.2 it follows that if F has a semiuniversal element then it has a tangent space which is of finite dimension, because F satisfies H_0, H_ϵ and H_f. This property was not explicitly stated in the definition.

2.4 The local moduli functors

2.4.1 Generalities

If X is an algebraic scheme then for every A in $\mathrm{ob}(\mathcal{A})$ we let

$$\mathrm{Def}_X(A) = \{\text{deformations of } X \text{ over } A\}/\text{isomorphism}$$

By the functoriality properties already observed in § 1.2 this defines a functor of Artin rings

$$\mathrm{Def}_X : \mathcal{A} \to (\text{sets})$$

which is called the *local moduli functor* of X. If $X = \mathrm{Spec}(B_0)$ is affine, we will often write Def_{B_0} instead of Def_X. We can define the subfunctor

$$\mathrm{Def}'_X : \mathcal{A} \to (\text{sets})$$

by

$$\mathrm{Def}'_X(A) = \{\text{locally trivial deformations of } X \text{ over } A\}/\text{isomorphism}$$

called the *locally trivial moduli functor* of X.

Theorem 2.4.1. *(i) For any algebraic scheme X the functors Def_X and Def'_X satisfy conditions H_0, $\bar{H}$, H_ϵ of Schlessinger's theorem. Therefore, if $\mathrm{Def}_X(\mathbf{k}[\epsilon])$ (resp. $\mathrm{Def}'_X(\mathbf{k}[\epsilon])$) is finite dimensional, then Def_X (resp. Def'_X) has a semiuniversal element.*

(ii) There is a canonical identification of $\mathbf{k}$*-vector spaces*

$$\mathrm{Def}'_X(\mathbf{k}[\epsilon]) = H^1(X, T_X) \tag{2.16}$$

In particular, if X *is nonsingular then*

$$\mathrm{Def}_X(\mathbf{k}[\epsilon]) = \mathrm{Def}'_X(\mathbf{k}[\epsilon]) = H^1(X, T_X)$$

(iii) If X *is an arbitrary algebraic scheme then we have a natural identification*

$$\mathrm{Def}_X(\mathbf{k}[\epsilon]) = \mathrm{Ex}_{\mathbf{k}}(X, \mathcal{O}_X)$$

and an exact sequence:

$$0 \to H^1(X, T_X) \xrightarrow{\tau} \mathrm{Def}_X(\mathbf{k}[\epsilon]) \xrightarrow{\ell} H^0(X, T^1_X) \xrightarrow{\rho} H^2(X, T_X) \tag{2.17}$$

In particular,

$$\mathrm{Def}_{B_0}(\mathbf{k}[\epsilon]) = T^1_{B_0}$$

if $X = \mathrm{Spec}(B_0)$ *is affine.*

(iv) If X *is a reduced algebraic scheme then there is an isomorphism*

$$\mathrm{Def}_X(\mathbf{k}[\epsilon]) \cong \mathrm{Ext}^1_{\mathcal{O}_X}(\Omega^1_X, \mathcal{O}_X)$$

and the exact sequence (2.17) is isomorphic to the local-to-global exact sequence for Exts:

$$\begin{aligned} 0 \to H^1(X, T_X) &\to \mathrm{Ext}^1_{\mathcal{O}_X}(\Omega^1_X, \mathcal{O}_X) \\ &\to H^0(X, \mathcal{E}xt^1_{\mathcal{O}_X}(\Omega^1_X, \mathcal{O}_X)) \to H^2(X, T_X) \end{aligned} \tag{2.18}$$

Proof. Obviously Def_X and Def'_X satisfy condition H_0. To verify the other conditions we assume first that $X = \mathrm{Spec}(B_0)$ is affine.

Let's prove that Def_{B_0} *satisfies* $\bar{H}$. Let

$$\begin{array}{ccc} A' & & A'' \\ & \searrow \swarrow & \\ & A & \end{array}$$

be homomorphisms in $\mathcal{A}$, with $A'' \to A$ a small extension. Letting $\bar{A} = A' \times_A A''$ we have a commutative diagram with exact rows:

$$\begin{array}{ccccccccc} 0 & \to & (\epsilon) & \to & \bar{A} & \to & A' & \to & 0 \\ & & \| & & \downarrow & & \downarrow & & \\ 0 & \to & (\epsilon) & \to & A'' & \to & A & \to & 0 \end{array} \tag{2.19}$$

Consider an element of

$$\mathrm{Def}_{B_0}(A') \times_{\mathrm{Def}_{B_0}(A)} \mathrm{Def}_{B_0}(A'')$$

which is represented by a pair of deformations $f' : A' \to B'$ and $f'' : A'' \to B''$ of B_0 such that $A \to B' \otimes_{A'} A$ and $A \to B'' \otimes_{A''} A$ are isomorphic deformations. Assume that the isomorphism is given by A-isomorphisms $B' \otimes_{A'} A \cong B \cong B'' \otimes_{A''} A$, where $A \to B$ is a deformation. In order to check $\bar{H}$ it suffices to find a deformation $\bar{f} : \bar{A} \to \bar{B}$ inducing (f', f''). Let

$$\bar{B} = B' \times_B B''$$

endowed with the obvious homomorphism $\bar{f} : \bar{A} \to \bar{B}$. It is elementary to check that there are an A'-isomorphism $\bar{B} \otimes_{\bar{A}} A' \cong B'$ and an A''-isomorphism $\bar{B} \otimes_{\bar{A}} A'' \cong B''$.

Therefore we only need to check that $\bar{f}$ is flat. Tensoring diagram (2.19) with $\otimes_{\bar{A}} \bar{B}$ we obtain the following diagram with exact rows:

$$\begin{array}{ccccccccc} & & (\epsilon) \otimes_{\bar{A}} \bar{B} & \to & \bar{B} & \to & B' & \to & 0 \\ & & \| & & \downarrow & & \downarrow & & \\ 0 & \to & B_0 & \to & B'' & \to & B & \to & 0 \end{array}$$

where the second row is given by Lemma A.9. This diagram shows that

$$(\epsilon) \otimes_{\bar{A}} \bar{B} \to \bar{B}$$

is injective: the flatness of $\bar{f}$ follows from Lemma A.9.

Let's prove that Def_{B_0} *satisfies* H_ϵ. Assume in the above situation that $A'' = \mathbf{k}[\epsilon]$ and $A = \mathbf{k}$, and let $\tilde{f} : \bar{A} \to \tilde{B}$ be a deformation such that $\alpha(\tilde{f}) = (f', f'')$. Then the diagram

$$\begin{array}{ccc} \tilde{B} & \to & \tilde{B} \otimes_{\bar{A}} A'' \cong B'' \\ \downarrow & & \downarrow \\ \tilde{B} \otimes_{\bar{A}} A' \cong B' & \to & B_0 \end{array}$$

commutes: the universal property of the fibre product implies that we have a homomorphism $\gamma : \tilde{B} \to \bar{B}$ of deformations, hence an isomorphism by Lemma A.4. This proves that the fibres of α contain only one element, i.e. α is bijective. Therefore Def_{B_0} satisfies condition H_ϵ.

Let's prove (i) for X arbitrary. Consider a diagram in $\mathcal{A}$:

$$\begin{array}{ccc} A' & & A'' \\ & \searrow\swarrow & \\ & A & \end{array}$$

with $A'' \to A$ a small extension and let $\bar{A} = A' \times_A A''$. Consider an element

$$([\mathcal{X}'], [\mathcal{X}'']) \in \mathrm{Def}_X(A') \times_{\mathrm{Def}_X(A)} \mathrm{Def}_X(A'')$$

Therefore we have a diagram of deformations:

$$\begin{array}{ccccc} \mathcal{X}' & & & & \mathcal{X}'' \\ & \nwarrow f' & & \nearrow f'' & \\ \downarrow & & \mathcal{X} & & \downarrow \\ \mathrm{Spec}(A') & & \downarrow & & \mathrm{Spec}(A'') \\ & \nwarrow & & \nearrow & \\ & & \mathrm{Spec}(A) & & \end{array}$$

where the morphisms f' and f'' induce isomorphisms of deformations

$$\mathcal{X}' \times_{\mathrm{Spec}(A')} \mathrm{Spec}(A) \cong \mathcal{X} \cong \mathcal{X}'' \times_{\mathrm{Spec}(A'')} \mathrm{Spec}(A)$$

Consider the sheaf of $\bar{A}$-algebras $\mathcal{O}_{\mathcal{X}'} \times_{\mathcal{O}_{\mathcal{X}}} \mathcal{O}_{\mathcal{X}''}$ on X. Then $\bar{\mathcal{X}} := (|X|, \mathcal{O}_{\mathcal{X}'} \times_{\mathcal{O}_{\mathcal{X}}} \mathcal{O}_{\mathcal{X}''})$ is a scheme over $\mathrm{Spec}(\bar{A})$ (by the proof of the affine case). Reducing to the affine case one shows that $\bar{\mathcal{X}}$ is flat over $\mathrm{Spec}(\bar{A})$. Therefore $\bar{\mathcal{X}}$ is a deformation of X over $\mathrm{Spec}(\bar{A})$ inducing the pair $([\mathcal{X}'], [\mathcal{X}''])$. This shows that the map

$$\mathrm{Def}_X(\bar{A}) \to \mathrm{Def}_X(A') \times_{\mathrm{Def}_X(A)} \mathrm{Def}_X(A'')$$

is surjective, proving $\bar{H}$ for Def_X. Moreover, if the deformations $\mathcal{X}'$ and $\mathcal{X}''$ are locally trivial then so is $\bar{\mathcal{X}}$, and therefore $\bar{H}$ holds for the functor Def'_X as well.

Now assume that $A'' = \mathbf{k}[\epsilon]$ and $A = \mathbf{k}$. Then the previous diagram becomes

$$\begin{array}{ccccc} \mathcal{X}' & & & & \mathcal{X}'' \\ & \nwarrow f' & & \nearrow f'' & \\ & & X & & \end{array}$$

In this case any $\tilde{\mathcal{X}} \to \mathrm{Spec}(\bar{A})$ inducing the pair

$$([\mathcal{X}'], [\mathcal{X}'']) \in \mathrm{Def}_X(A') \times_{\mathrm{Def}_X(A)} \mathrm{Def}_X(A'')$$

is such that the isomorphisms

$$\tilde{\mathcal{X}} \times_{\mathrm{Spec}(\bar{A})} \mathrm{Spec}(A') \cong \mathcal{X}', \quad \tilde{\mathcal{X}} \times_{\mathrm{Spec}(\bar{A})} \mathrm{Spec}(A'') \cong \mathcal{X}''$$

induce the identity on $X = \tilde{\mathcal{X}} \times_{\mathrm{Spec}(\bar{A})} \mathrm{Spec}(\mathbf{k})$. Therefore $\tilde{\mathcal{X}}$ fits into a commutative diagram

$$\begin{array}{ccccc} & & \tilde{\mathcal{X}} & & \\ & \nearrow & & \nwarrow & \\ \mathcal{X}' & & & & \mathcal{X}'' \\ & \nwarrow f' & & \nearrow f'' & \\ & & X & & \end{array}$$

By the universal property of the fibred sum of schemes we then get a morphism of deformations $\bar{\mathcal{X}} \to \tilde{\mathcal{X}}$, which is necessarily an isomorphism. This proves property H_ϵ for Def_X. The proof for Def'_X is similar.

(ii) The identification (2.16) has been proved in 1.2.9. The verification that it is $\mathbf{k}$-linear is elementary and will be left to the reader.

(iii) If

$$\begin{array}{ccc} X & \subset & \mathcal{X} \\ \downarrow & & \downarrow \\ \mathrm{Spec}(\mathbf{k}) & \to & \mathrm{Spec}(\mathbf{k}[\epsilon]) \end{array}$$

is a first-order deformation of X then $X \subset \mathcal{X}$ is an extension of X by $\mathcal{O}_X$ because by the flatness of $\mathcal{X}$ over $\mathrm{Spec}(\mathbf{k}[\epsilon])$ we have $\epsilon\mathcal{O}_{\mathcal{X}} \cong \mathcal{O}_X$ (Lemma A.9). Conversely, given such an extension $X \subset \mathcal{X}$ we have an exact sequence

$$0 \to \mathcal{O}_X \xrightarrow{j} \mathcal{O}_{\mathcal{X}} \to \mathcal{O}_X \to 0$$

$\mathcal{O}_{\mathcal{X}}$ has a natural structure of $\mathbf{k}[\epsilon]$-algebra by sending, for any open $U \subset X$, $\epsilon \mapsto j(1)$. It follows from Lemma A.9 that $\mathcal{X}$ is flat over $\mathrm{Spec}(\mathbf{k}[\epsilon])$.

Let's assume first that $X = \mathrm{Spec}(B_0)$ is affine. Then

$$\mathrm{Ex}_{\mathbf{k}}(X, \mathcal{O}_X) = H^0(X, T^1_{B_0}) = T^1_{B_0}$$

and the exact sequence (2.17) reduces to the isomorphism

$$\mathrm{Def}_{B_0}(\mathbf{k}[\epsilon]) \cong T^1_{B_0}$$

which holds by what we have just remarked.

Let's assume that X is general. The map τ corresponds to the inclusion $\mathrm{Def}'_X(\mathbf{k}[\epsilon]) \subset \mathrm{Def}_X(\mathbf{k}[\epsilon])$ in view of (2.16). The map ℓ associates to a first-order deformation ξ of X the section of T^1_X defined by the restrictions $\{\xi_{|U_i}\}$ for some affine open cover $\{U_i\}$. It is clear that $\mathrm{Im}(\tau) = \ker(\ell)$.

We now define ρ. Let $h \in H^0(X, T^1_X)$ be represented, in a suitable affine open cover $\mathcal{U} = \{U_i = \mathrm{Spec}(B_i)\}$ of X, by a collection of $\mathbf{k}$-extensions $\mathcal{E}_i$ of B_i by B_i. Since h is a global section there exist isomorphisms $\sigma_{ij} : \mathcal{E}_{j|U_i\cap U_j} \cong \mathcal{E}_{i|U_i\cap U_j}$. These isomorphisms patch together to give an extension $\mathcal{E}$ if and only if $h \in \mathrm{Im}(\ell)$ if and only if we can find new isomorphisms σ'_{ij} such that

$$\sigma'_{ij}\sigma'_{jk} = \sigma'_{ik} \tag{2.20}$$

on $U_i \cap U_j \cap U_k$. Such isomorphisms are of the form

$$\sigma'_{ij} = \sigma_{ij}\theta_{ij}$$

where θ_{ij} is an automorphism of the extension $\mathcal{E}_{j|U_i\cap U_j}$. The collection of automorphisms (θ_{ij}) corresponds, via Lemma 1.2.6, to a 1-cochain $(t_{ij}) \in C^1(\mathcal{U}, T_X)$; conversely, every 1-cochain (t_{ij}) defines a system of isomorphisms (σ'_{ij}); and the condition (2.20) is satisfied if and only if (t_{ij}) is a 1-cocycle. Therefore we define $\rho(h)$ to be the class of the 2-cocycle $(t_{ij} + t_{jk} - t_{ik})$. With this definition we clearly have $\ker(\rho) = \mathrm{Im}(\ell)$. We leave to the reader to verify that the definition of ρ does not depend on the choices made.

(iv) Since we have a natural identification $\mathrm{Def}_X(\mathbf{k}[\epsilon]) = \mathrm{Ex}_{\mathbf{k}}(X, \mathcal{O}_X)$ we conclude by Theorem 1.1.10. The exact sequences (2.17) and (2.18) are isomorphic in view of the isomorphism

$$T^1_X \cong \mathcal{E}xt^1_{\mathcal{O}_X}(\Omega^1_X, \mathcal{O}_X)$$

given by Corollary 1.1.11, page 18. □

Corollary 2.4.2. *If X is a projective scheme or an affine scheme with at most isolated singularities then* Def_X *has a semiuniversal formal element.*

Proof. The condition implies that $H^1(X, T_X)$ and $H^0(X, T^1_X)$ are finite-dimensional vector spaces. Therefore the conclusion follows from Theorem 2.4.1 and Schlessinger's theorem. □

The stronger property of being prorepresentable is not satisfied in general by Def_X. We will discuss this matter in § 2.6. The case X affine will be taken up again and analysed in detail in Section 3.1.

Definition 2.4.3. *If $(R, \hat{u})$ is a semiuniversal couple for* Def_X *then the Krull dimension of R (i.e. the maximum of the dimensions of the irreducible components of* $\mathrm{Spec}(R)$*) is called the* number of moduli *of X and it is denoted by $\mu(X)$.*

Remark 2.4.4. Let X be a reduced scheme and let $\xi : \mathcal{X} \to \mathrm{Spec}(\mathbf{k}[\epsilon])$ be a first-order deformation of X. Then the conormal sequence of $X \subset \mathcal{X}$

$$0 \to \mathcal{O}_X \to \Omega^1_{\mathcal{X}|X} \to \Omega^1_X \to 0 \tag{2.21}$$

is exact and defines the element of $\mathrm{Ext}^1_{\mathcal{O}_X}(\Omega^1_X, \mathcal{O}_X)$ which corresponds to ξ in the identification $\mathrm{Def}_X(\mathbf{k}[\epsilon]) = \mathrm{Ext}^1_{\mathcal{O}_X}(\Omega^1_X, \mathcal{O}_X)$ of Theorem 2.4.1(iv) (see also Theorem 1.1.10).

Given an infinitesimal deformation $\xi : \mathcal{X} \to \mathrm{Spec}(A)$ of X we have a *Kodaira–Spencer map* $\kappa_\xi : t_A \to \mathrm{Ext}^1_{\mathcal{O}_X}(\Omega^1_X, \mathcal{O}_X)$ which associates to a tangent vector $\theta \in t_A$ the conormal sequence 2.21 of the pullback of ξ to $\mathrm{Spec}(\mathbf{k}[\epsilon])$ defined by θ.

Example 2.4.5. Let X be a projective curve. Then $H^2(X, T_X) = (0)$ and the exact sequence (2.17) shows that if $h^0(X, T^1_X) \neq 0$ then $\mathrm{Def}_X(\mathbf{k}[\epsilon]) \neq (0)$ and X is not rigid. In particular, if X is reduced and has at least one nonrigid singular point then X is not rigid. By the way, it is not known whether rigid curve singularities exist at all.

2.4.2 Obstruction spaces

The elementary analysis of obstructions to lifting infinitesimal deformations carried out in Chapter 1 can be interpreted as the description of obstruction spaces for the corresponding deformation functors. More precisely, we have the following:

Proposition 2.4.6. *Let X be a nonsingular algebraic variety. Then $H^2(X, T_X)$ is an obstruction space for the functor* Def_X. *If X is an arbitrary algebraic scheme then $H^2(X, T_X)$ is an obstruction space for the functor* Def'_X.

Proof. The proposition is just a rephrasing of 1.2.12. □

Corollary 2.4.7. *Let X be a nonsingular projective algebraic variety. Then*

$$h^1(X, T_X) \geq \mu(R) \geq h^1(X, T_X) - h^2(X, T_X)$$

The first equality holds if and only if X is unobstructed.

Proof. It is an immediate consequence of 2.4.6 and of Corollary 2.2.11. □

If X is a singular algebraic scheme the previous results give no information about obstructions of the functor Def_X. The following proposition addresses the case of a reduced local complete intersection (l.c.i.) scheme.

Proposition 2.4.8. *Let X be a reduced l.c.i. algebraic scheme X, and assume $char(\mathbf{k}) = 0$. Then $\mathrm{Ext}^2_{\mathcal{O}_X}(\Omega^1_X, \mathcal{O}_X)$ is an obstruction space for the functor Def_X.*

Proof. Let A be in $\mathrm{ob}(\mathcal{A})$ and let

$$\begin{array}{cccc} & X & \rightarrow & \mathcal{X} \\ \xi : & \downarrow & & \downarrow f \\ & \mathrm{Spec}(\mathbf{k}) & \rightarrow & \mathrm{Spec}(A) \end{array}$$

be a deformation of X over A. We need to define a $\mathbf{k}$-linear map

$$o_\xi : \mathrm{Ex}_{\mathbf{k}}(A, \mathbf{k}) \rightarrow \mathrm{Ext}^2_{\mathcal{O}_X}(\Omega^1_X, \mathcal{O}_X)$$

having the properties of an obstruction map according to Definition 2.2.9. Consider an element of $\mathrm{Ex}_{\mathbf{k}}(A, \mathbf{k})$ represented by an extension

$$e: \quad 0 \rightarrow (t) \rightarrow \tilde{A} \rightarrow A \rightarrow 0$$

and the conormal sequence of e:

$$0 \rightarrow (t) \rightarrow \Omega_{\tilde{A}/\mathbf{k}} \otimes_{\tilde{A}} A \rightarrow \Omega_{A/\mathbf{k}} \rightarrow 0 \tag{2.22}$$

which is exact by Lemma B.10. Since f is flat, pulling back (2.22) to $\mathcal{X}$ we obtain the exact sequence

$$0 \rightarrow \mathcal{O}_X \rightarrow f^*(\Omega^1_{\mathrm{Spec}(\tilde{A})|\mathrm{Spec}(A)}) \rightarrow f^*(\Omega^1_{\mathrm{Spec}(A)}) \rightarrow 0 \tag{2.23}$$

Since X is a reduced l.c.i. the morphism f satisfies the hypothesis of Theorem D.2.8. Therefore the relative cotangent sequence of f:

$$0 \rightarrow f^*(\Omega^1_{\mathrm{Spec}(A)}) \rightarrow \Omega^1_{\mathcal{X}} \rightarrow \Omega^1_{\mathcal{X}/\mathrm{Spec}(A)} \rightarrow 0 \tag{2.24}$$

is exact. Composing (2.23) and (2.24) we obtain the 2-term extension

$$0 \rightarrow \quad \mathcal{O}_X \quad \rightarrow \quad f^*(\Omega^1_{\mathrm{Spec}(\tilde{A})|\mathrm{Spec}(A)}) \quad \rightarrow \quad \Omega^1_{\mathcal{X}} \quad \rightarrow \quad \Omega^1_{\mathcal{X}/\mathrm{Spec}(A)} \quad \rightarrow 0$$

which defines an element

$$o_\xi(e) \in \mathrm{Ext}^2_{\mathcal{O}_{\mathcal{X}}}(\Omega^1_{\mathcal{X}/\mathrm{Spec}(A)}, \mathcal{O}_X) = \mathrm{Ext}^2_{\mathcal{O}_X}(\Omega^1_X, \mathcal{O}_X)$$

This defines the map o_ξ. The linearity of o_ξ is a consequence of the linearity of the map $\mathrm{Ex}_{\mathbf{k}}(A, \mathbf{k}) \rightarrow \mathrm{Ext}^1_A(\Omega_{A/\mathbf{k}}, \mathbf{k})$ associating to an extension e the conormal sequence (2.22).

Assume that there is a lifting of ξ to $\tilde{A}$, i.e. that we have a diagram:

$$\begin{array}{ccc} \mathcal{X} & \subset & \tilde{\mathcal{X}} \\ \downarrow f & & \downarrow \tilde{f} \\ \mathrm{Spec}(A) & \subset & \mathrm{Spec}(\tilde{A}) \end{array}$$

Then we have a commutative and exact diagram as follows:

$$\begin{array}{ccccccc} & & & & 0 & & \\ & & & & \downarrow & & \\ 0 \to & \mathcal{O}_X & \to & f^*(\Omega^1_{\mathrm{Spec}(\tilde{A})|\mathrm{Spec}(A)}) & \to & f^*(\Omega^1_{\mathrm{Spec}(A)}) & \to 0 \\ & \| & & \downarrow & & \downarrow & \\ 0 \to & \mathcal{O}_X & \to & \Omega^1_{\tilde{\mathcal{X}}|\mathcal{X}} & \to & \Omega^1_{\mathcal{X}} & \to 0 \\ & & & \downarrow & & \downarrow & \\ & & & (\Omega^1_{\tilde{\mathcal{X}}/\mathrm{Spec}(\tilde{A})})_{|\mathcal{X}} & = & \Omega^1_{\mathcal{X}/\mathrm{Spec}(A)} & \\ & & & \downarrow & & \downarrow & \\ & & & 0 & & 0 & \end{array}$$

In this diagram the first row is the pullback of the second row and this implies that $o_\xi(e) = 0$.

Conversely, assume that $o_\xi(e) = 0$. Then we have a commutative and exact diagram as follows:

$$\begin{array}{ccccccc} & & & & 0 & & \\ & & & & \downarrow & & \\ 0 \to & \mathcal{O}_X & \to & f^*(\Omega^1_{\mathrm{Spec}(\tilde{A})|\mathrm{Spec}(A)}) & \to & f^*(\Omega^1_{\mathrm{Spec}(A)}) & \to 0 \\ & \| & & \downarrow & & \downarrow & \\ 0 \to & \mathcal{O}_X & \to & \mathcal{E} & \to & \Omega^1_{\mathcal{X}} & \to 0 \\ & & & & & \downarrow & \\ & & & & & \Omega^1_{\mathcal{X}/\mathrm{Spec}(A)} & \\ & & & & & \downarrow & \\ & & & & & 0 & \end{array}$$

for some coherent sheaf $\mathcal{E}$ on $\mathcal{X}$. By the construction of Theorem 1.1.10 one finds a sheaf of A-algebras $\mathcal{O}_{\tilde{\mathcal{X}}}$ and an extension of sheaves of A-algebras

$$0 \to \mathcal{O}_X \to \mathcal{O}_{\tilde{\mathcal{X}}} \to \mathcal{O}_{\mathcal{X}} \to 0$$

such that $\mathcal{E} = \Omega^1_{\tilde{\mathcal{X}}|\mathcal{X}}$. It remains to be shown that $\mathcal{O}_{\tilde{\mathcal{X}}}$ can be given a structure of sheaf of flat $\tilde{A}$-algebras. The shortest way to see this is to use Corollary 3.1.13(ii), to be proved in the next chapter: since X is a l.c.i. it implies that there are no obstructions to lifting $\mathcal{X}$ locally to $\tilde{A}$. This means that the restriction of $\mathcal{O}_{\tilde{\mathcal{X}}}$ to every affine open subset $U \subset X$ defines a lifting of $\mathcal{X}_{|U}$ to $\tilde{A}$, and therefore it is a sheaf of flat $\tilde{A}$-algebras. □

Example 2.4.9. If X is a reduced l.c.i. curve then $Ext^1_{\mathcal{O}_X}(\Omega^1_X, \mathcal{O}_X)$ is a torsion sheaf and $Ext^2_{\mathcal{O}_X}(\Omega^1_X, \mathcal{O}_X) = 0$; therefore

$$H^i(X, Ext^j_{\mathcal{O}_X}(\Omega^1_X, \mathcal{O}_X)) = (0)$$

for all i, j with $i + j = 2$. It follows that $\mathrm{Ext}^2_{\mathcal{O}_X}(\Omega^1_X, \mathcal{O}_X) = 0$ and X is unobstructed by Proposition 2.4.8.

For a discussion of obstructions in the affine case we refer the reader to Subsection 3.1.2.

2.4.3 Algebraic surfaces

In this subsection we will assume $\mathrm{char}(\mathbf{k}) = 0$. We will denote by S a projective nonsingular connected algebraic surface. Let $(R, \hat{u})$ be a semiuniversal formal deformation of S, and denote by

$$\mu(S) := \dim(R)$$

the number of moduli of S.

Proposition 2.4.10.

$$10\chi(\mathcal{O}_S) - 2(K^2) + h^0(S, T_S) \leq \mu(S) \leq h^1(S, T_S) \tag{2.25}$$

If $h^2(S, T_S) = 0$ then both inequalities are equalities.

Proof. A direct application of the Riemann–Roch formula gives

$$h^1(S, T_S) - h^2(S, T_S) = 10\chi(\mathcal{O}_S) - 2(K^2) + h^0(S, T_S)$$

By applying Corollary 2.4.7 we obtain the conclusion. □

The first inequality was proved by Enriques (see [52]).

Examples 2.4.11. (i) If S is a minimal ruled surface and $S \to C$ is the ruling over a projective nonsingular curve C, then letting $S(x) \cong \mathbb{P}^1$ be the fibre of any $x \in C$ we have

$$h^1(S(x), T_{S|S(x)}) = 0$$

as an immediate consequence of the exact sequence

$$\begin{array}{ccccccc} 0 \to T_{S(x)} & \to & T_{S|S(x)} & \to & N_{S(x)/S} & \to & 0 \\ & & & & \| & & \\ & & & & \mathcal{O}_{S(x)} & & \end{array}$$

Therefore $R^1 p_* T_S = 0$ by Corollary 4.2.6, and the Leray spectral sequence implies that $h^2(S, T_S) = 0$. Therefore S is unobstructed and $\mu(S) = h^1(S, T_S)$. This computation for the rational ruled surfaces F_m is also done in Example B.11(iii), page 288.

(ii) Assume that S is a K3-surface. Then

$$h^2(S, T_S) = h^0(S, \Omega^1_S) = h^1(S, \mathcal{O}_S) = 0$$

Therefore S is unobstructed. Moreover,

$$h^0(S, T_S) = h^2(S, \Omega^1_S) = h^1(S, \omega_S) = 0$$

and Def_S is prorepresentable (Corollary 2.6.4). Formula (2.25) gives in this case

$$\mu(S) = h^1(S, T_S) = 20$$

(iii) Let $\pi : X \to S$ be the blow-up of S at a point s and $E = \pi^{-1}(s)$ the exceptional curve. Then we have a standard exact sequence

$$0 \to T_X \xrightarrow{d\pi} \pi^* T_S \to \mathcal{O}_E(-E) \to 0$$

(to see it one can use the exact sequence (3.50), page 172, and note that $\mathrm{coker}(d\pi) = N_\pi$, the normal sheaf of π, which is defined on page 162). We thus see that

$$H^2(X, T_X) \cong H^2(S, T_S)$$

This implies, for example, that non-minimal rational or ruled surfaces and blow-ups of K3-surfaces are unobstructed.

(iv) When the Kodaira dimension of S is ≥ 1 then in general

$$h^2(S, T_S) = h^0(S, \Omega^1 \otimes \omega_S) \neq 0$$

and in fact such surfaces can be obstructed. Examples are given in Theorem 3.4.26, page 185. If we assume that $h^0(S, T_S) = 0$ the estimate for $\mu(S)$ given by Proposition 2.4.10 becomes

$$10\chi(\mathcal{O}_S) - 2(K^2) \leq \mu(S) \leq h^1(S, T_S) = 10\chi(\mathcal{O}_S) - 2(K^2) + h^0(S, \Omega^1 \otimes \omega_S)$$

and to give an upper bound for $\mu(S)$ amounts to giving one for $h^0(S, \Omega^1 \otimes \omega_S)$ in terms of p_a, K^2, q. We refer the reader to [27] for a more detailed discussion of this point. For examples of obstructed surfaces see also [24], [87], [96], [28].

(v) If $h^0(S, \omega_S) > 0$ and $q > 0$ then certainly

$$h^0(S, \Omega^1 \otimes \omega_S) > 0$$

This is because there is a bilinear pairing

$$H^0(S, \Omega^1_S) \times H^0(S, \omega_S) \to H^0(S, \Omega^1 \otimes \omega_S)$$

which is non-degenerate on each factor.

For example, take $S = C_1 \times C_2$ where C_1 and C_2 are projective nonsingular connected curves of genera $g_1 \geq 2$ and $g_2 \geq 2$ respectively. Then $h^0(S, \omega_S) = g_1 g_2$, $q = g_1 + g_2$ and

$$T_S = p_1^* \omega_{C_1}^{-1} \oplus p_2^* \omega_{C_2}^{-1}$$

where $p_i : S \to C_i$ is the i-th projection, $i = 1, 2$. Therefore

$$h^0(T_S) = 0, \quad h^1(T_S) = 3(g_1 + g_2) - 6, \quad h^2(T_S) = g_2(3g_1 - 3) + g_1(3g_2 - 3)$$

On the other hand, S is unobstructed. In fact there is a natural morphism of functors

$$f : \mathrm{Def}_{C_1} \times \mathrm{Def}_{C_2} \to \mathrm{Def}_S$$

which is clearly an isomorphism on tangent spaces: in fact

$$(\mathrm{Def}_{C_1} \times \mathrm{Def}_{C_2})(\mathbf{k}[\epsilon]) := \mathrm{Def}_{C_1}(\mathbf{k}[\epsilon]) \times \mathrm{Def}_{C_2}(\mathbf{k}[\epsilon]) =$$

$$= H^1(C_1, \omega_{C_1}^{-1}) \oplus H^1(C_2, \omega_{C_2}^{-1}) = H^1(S, T_S)$$

Therefore, since $\mathrm{Def}_{C_1} \times \mathrm{Def}_{C_2}$ is smooth because both factors are (Proposition 2.2.5(vii)), Def_S is smooth by Corollary 2.3.7. Actually, f is an isomorphism because both functors are prorepresentable (see Remark 2.3.8, page 60). This example can be generalized to any finite product of nonsingular projective connected curves of genera ≥ 2.

If S is an abelian surface then

$$h^0(S, \Omega^1 \otimes \omega_S) = 2 = h^0(S, T_S)$$

Formula (2.25) gives

$$2 \leq \mu(S) \leq h^1(S, T_S) = 4$$

In fact the second equality holds because abelian surfaces are unobstructed despite the fact that $h^2(S, T_S) \neq 0$. This is a property common to all abelian varieties (see [137] and [135]).

(vi) One should keep in mind that $\mu(S)$ is defined as the number of moduli of S in a formal sense. This is because the semiuniversal formal deformation $(R, \hat{u})$ can be non-algebraizable (see § 2.5 for the notion of algebraizability). For example $\mu(S) = 20$ for a K3-surface, but every algebraic family of K3-surfaces with injective Kodaira–Spencer map at every point has dimension ≤ 19 (see § 3.3). Similarly an abelian surface has $\mu(S) = 4$ but every algebraic family of abelian surfaces with injective Kodaira–Spencer map at every point has dimension ≤ 3 (see Example 3.3.13). In order to give an algebraic meaning to the number of moduli one should take the maximum dimension of a semiuniversal formal deformation of a pair (S, L) where L is an ample invertible sheaf on S. See § 3.3 and the appendix by Mumford to Chapter V of [189].

2.5 Formal versus algebraic deformations

We have already mentioned (see Examples 1.2.5 and 1.2.11) that infinitesimal deformations do not explain faithfully some of the phenomena which can occur when one considers deformations parametrized by algebraic schemes or by the spectrum of an arbitrary noetherian, or even e.f.t., local ring. In this section we will make such statements precise. We start with a few definitions and some terminology.

Let (S,s) be a pointed scheme. An *etale neighbourhood* of s in S is an etale morphism of pointed schemes $f:(T,t)\to(S,s)$ such that the following diagram commutes:

$$\begin{array}{ccc} & & T \\ & t\nearrow & \downarrow f \\ \mathrm{Spec}(\mathbf{k}(s)) & \xrightarrow{s} & S \end{array}$$

The definition implies that $\mathbf{k}(s)\cong\mathbf{k}(t)$, i.e. f induces a trivial extension of the residue fields at s and t; therefore $\hat{\mathcal{O}}_{S,s}\cong\hat{\mathcal{O}}_{T,t}$ by C.7. Affine neighbourhoods of s are particular etale neighbourhoods.

Given two etale neighbourhoods (T,t) and (U,u) of $s\in S$ a *morphism* $(T,t)\to(U,u)$ is given by a commutative diagram of pointed schemes:

$$\begin{array}{ccc} (T,t) & \to & (U,u) \\ & \searrow \quad \swarrow & \\ & (S,s) & \end{array}$$

Lemma 2.5.1. *Let $f:X\to Y$ be an etale morphism and $g:Y\to X$ a section of f. Then g is etale.*

Proof. Use C.2(v). □

Proposition 2.5.2. *Let S be a scheme. The etale neighbourhoods of a given $s\in S$ form a filtered system of pointed schemes.*

Proof. Given two etale neighbourhoods (S',s') and (S'',s''), they are dominated by a third, namely:

$$\begin{array}{ccc} S'\times_S S'' & \to & S'' \\ \downarrow & & \downarrow \\ S' & \to & S \end{array}$$

Now let $f_1,f_2:(S'',s'')\to(S',s')$ be two morphisms between etale neighbourhoods. Then there exists a third etale neighbourhood (S''',s''') and a morphism $(S''',s''')\to(S'',s'')$ which equalizes them. In fact, consider the diagram:

$$\begin{array}{ccc} S''\times_S S' & \xrightarrow{pr'} & S' \\ \downarrow pr'' & & \downarrow \\ S'' & \to & S \end{array}$$

We can shrink S' and S'' so that S' is affine and S'' is connected. Then the graphs Γ_1 and Γ_2 of f_1 and f_2 are closed, because S' is affine, and open, because they are images of sections of the etale morphism pr'', which are etale. Therefore they are connected components of $S'' \times_S S'$. But $(s'', s') \in \Gamma_1 \cap \Gamma_2$ and therefore $\Gamma_1 = \Gamma_2$. It follows that $f_1 = f_2$ on $S'' = \Gamma_1 = \Gamma_2$. □

Definition 2.5.3. *Given a scheme S and a point $s \in S$ we define* the local ring of S in s in the etale topology *to be*

$$\tilde{\mathcal{O}}_{S,s} := \lim_{\substack{\rightarrow \\ (S',s')}} \mathcal{O}_{S',s'}$$

where the limit is taken for (S', s') varying through all the etale neighbourhoods of s. The ring $\tilde{\mathcal{O}}_{S,s}$ is also called the henselization *of $\mathcal{O}_{S,s}$.* (Note that $\tilde{\mathcal{O}}_{S,s}$ is a local ring, because it is a limit of a filtering system of local rings and local homomorphisms.)

A local ring A is called henselian *if for the closed point s of $S = \mathrm{Spec}(A)$ one has*

$$\tilde{A} := \tilde{\mathcal{O}}_{S,s} = \mathcal{O}_{S,s} = A$$

The henselization of an e.f.t. local **k***-algebra is called an* algebraic local ring.

Therefore the local ring in the etale topology of a point of an algebraic scheme is an algebraic local ring.

For a given scheme S and point $s \in S$ there is a canonical homomorphism $\mathcal{O}_{S,s} \to \tilde{\mathcal{O}}_{S,s}$ which is flat and induces an isomorphism of the completions

$$\hat{\mathcal{O}}_{S,s} \cong \widehat{\tilde{\mathcal{O}}_{S,s}}$$

because every $\mathcal{O}_{S,s} \to \mathcal{O}_{S',s'}$ does. Moreover,

$$\mathcal{O}_{S,s} \subset \tilde{\mathcal{O}}_{S,s} \subset \hat{\mathcal{O}}_{S,s}$$

because $\mathcal{O}_{S,s} \to \tilde{\mathcal{O}}_{S,s}$ is faithfully flat and $\mathcal{O}_{S,s}$ is separated for the m-adic topology. In particular, we see that $\mathcal{O}_{S,s} = \tilde{\mathcal{O}}_{S,s}$ if $\mathcal{O}_{S,s} = \hat{\mathcal{O}}_{S,s}$, i.e. *a local* **k***-algebra in $\hat{\mathcal{A}}$ (i.e. complete noetherian with residue field* **k***) is henselian.*

Theorem 2.5.4 (Nagata). *If A is a noetherian local ring then $\tilde{A}$ is noetherian.*

Proof. We have $A \subset \tilde{A} \subset \hat{A}$ and $\hat{A} = \widehat{\tilde{A}}$. Moreover,

$$\tilde{A} = \lim_{\rightarrow} A'$$

with A' local algebras etale over A and inducing trivial residue field extension. To prove that $\tilde{A}$ is noetherian it suffices to prove that every ascending chain of finitely generated ideals of $\tilde{A}$

$$\underline{a}_1 \subseteq \underline{a}_2 \subseteq \cdots \subseteq \underline{a}_n \subseteq$$

stabilizes. The chain $\{\underline{a}_n \hat{A}\}$ stabilizes because $\hat{A}$ is noetherian. Therefore it suffices to prove that if $\underline{a}, \underline{b} \subset \tilde{A}$ are finitely generated ideals such that $\underline{a}\hat{A} = \underline{b}\hat{A}$ then

$\underline{a} = \underline{b}$. Since $\underline{a}$ and $\underline{b}$ are finitely generated one can find $A' \supset A$ as above and finitely generated ideals $\underline{a}', \underline{b}' \subset A'$ such that $\underline{a} = \underline{a}'\tilde{A}$, $\underline{b} = \underline{b}'\tilde{A}$. It follows that $\underline{a}'\hat{A} = \underline{b}'\hat{A}$. But since A' is noetherian it follows that $\underline{a}' = \underline{b}'$ and therefore $\underline{a} = \underline{b}$. □

The following proposition gives a geometrical characterization of the henselization.

Proposition 2.5.5. *Let A be a local ring, $S = \mathrm{Spec}(A)$, $s \in S$ the closed point. A is henselian if and only if every morphism $f : Z \to S$ such that there is a point $z \in Z$ with $f(z) = s$, $\mathbf{k}(s) = \mathbf{k}(z)$ and f etale at z, admits a section.*

Proof. Assume the condition is satisfied. If $A \to A'$ is an etale homomorphism inducing an isomorphism of the residue fields then the induced morphism $f : \mathrm{Spec}(A') \to S$ admits a section, which defines an isomorphism $A' \cong A$; therefore A is henselian. Conversely, assume A is henselian and let $f : Z \to S$ be a morphism satisfying the stated conditions. Then f induces an isomorphism $A \cong \mathcal{O}_{Z,z}$ because A is henselian. The section is the composition

$$S = \mathrm{Spec}(A) \cong \mathrm{Spec}(\mathcal{O}_{Z,z}) \subset Z$$

□

* * * * * *

We will need the notion of "formal deformation of an algebraic scheme", already introduced in § 2.2 for a general functor of Artin rings. Let X be an algebraic scheme and let $\bar{A}$ be in $\mathrm{ob}(\hat{\mathcal{A}})$. Then a formal deformation of X over $\bar{A}$ is just a formal element $\hat{\eta} \in \widehat{\mathrm{Def}}_X(\bar{A})$ for Def_X: by definition it is a sequence $\{\eta_n\}$ of infinitesimal deformations of X

$$\eta_n : \begin{array}{ccc} X & \xrightarrow{f_n} & \mathcal{X}_n \\ \downarrow & & \downarrow \pi_n \\ \mathrm{Spec}(\mathbf{k}) & \to & \mathrm{Spec}(\bar{A}_n) \end{array}$$

where $\bar{A}_n = \bar{A}/m_{\bar{A}}^{n+1}$, such that for all $n \geq 1$ η_n induces η_{n-1} by pullback under the natural inclusion $\mathrm{Spec}(\bar{A}_{n-1}) \to \mathrm{Spec}(\bar{A}_n)$, i.e. we have:

$$\eta_{n-1} : \begin{array}{ccc} X & \xrightarrow{f_{n-1}} & \mathcal{X}_n \otimes_{\bar{A}_n} \bar{A}_{n-1} = \mathcal{X}_{n-1} \\ \downarrow & & \downarrow \pi_{n-1} \\ \mathrm{Spec}(\mathbf{k}) & \to & \mathrm{Spec}(\bar{A}_{n-1}) \end{array}$$

We will denote such a formal deformation by $(\bar{A}, \{\eta_n\})$ or by $(\bar{A}, \hat{\eta})$. It can be also viewed as the morphism of formal schemes

$$\bar{\pi} : \mathcal{X} \to \mathrm{Specf}(\bar{A})$$

where

$$\mathcal{X} = (X, \varprojlim \mathcal{O}_{\mathcal{X}_n}), \quad \bar{\pi} = \varprojlim \pi_n$$

(see [84] and [16] for the definition and main properties of formal schemes).

Note that $\bar{\pi}$ is a flat morphism of formal schemes, namely for every $x \in X$ the local ring $\mathcal{O}_{\mathcal{X},x}$ is flat over $\bar{A} = \mathcal{O}_{\text{Specf}(\bar{A}),\bar{\pi}(x)}$. This is an almost immediate consequence of the general version of the local criterion of flatness ([3], Exposé IV, Corollaire 5.8).

The formal deformation $(\bar{A}, \{\eta_n\})$ will be called *trivial* (resp. *locally trivial*) if each η_n is trivial (resp. locally trivial).

A formal deformation $(\bar{A}, \{\eta_n\})$ is not to be confused with a deformation of X over $\text{Spec}(\bar{A})$, just as a formal scheme over $\text{Specf}(\bar{A})$ is not in general a formal completion of a scheme over $\text{Spec}(\bar{A})$ (see Definition 2.5.10 below and the discussion following it).

Let X be a projective scheme and consider a flat family of deformations of X parametrized by an affine scheme $S = \text{Spec}(B)$, with B in ($\mathbf{k}$-algebras)

$$\begin{array}{cccc} & X & \xrightarrow{f} & \mathcal{X} \\ \eta: & \downarrow & & \downarrow \pi \\ & \text{Spec}(\mathbf{k}) & \xrightarrow{s} & S \end{array}$$

namely a cartesian diagram with π projective and flat. The deformation η is called *algebraic* if B is a $\mathbf{k}$-algebra of finite type. If B is in $\text{ob}(\mathcal{A}^*)$ then η is called a *local deformation* of X. If B is in $\text{ob}(\mathcal{A})$ we obtain an infinitesimal deformation of X; it is simultaneously local and algebraic. We will identify the deformation η with the couple (S, η) or (B, η) and we will also denote it by (S, s, η) or (B, s, η).

Given such a deformation (S, s, η) let η_n be the infinitesimal deformation induced by pulling back η under the natural closed embedding

$$\text{Spec}(\mathcal{O}_{S,s}/m^{n+1}) \to S$$

We have $\mathcal{O}_{S,s}/m^{n+1} = \hat{\mathcal{O}}_{S,s}/\hat{m}^{n+1}$ and therefore it follows that $(\hat{\mathcal{O}}_{S,s}, \{\eta_n\})$ is a formal deformation of X. It will be called *the formal deformation defined by* (or *associated to*) η.

(S, s, η) is called *formally trivial* (resp. *formally locally trivial*) if the formal deformation defined by η is trivial (resp. locally trivial).

Lemma 2.5.6. *Let (S, s, η) be a deformation of an algebraic scheme X, $f : (\tilde{S}, \tilde{s}) \to (S, s)$ an etale neighbourhood of s in S and $(\tilde{S}, \tilde{s}, \tilde{\eta})$ the deformation of X obtained by pulling back η by f. Then the formal deformations of X defined by η and by $\tilde{\eta}$ are isomorphic.*

Proof. We have a cartesian diagram of formal schemes:

$$\begin{array}{ccc} \tilde{\mathcal{X}} & \to & \mathcal{X} \\ \downarrow \bar{\bar{\pi}} & & \downarrow \bar{\pi} \\ \text{Specf}(\hat{\mathcal{O}}_{\tilde{S},\tilde{s}}) & \to & \text{Specf}(\hat{\mathcal{O}}_{S,s}) \end{array}$$

where $\bar{\bar{\pi}}$ and $\bar{\pi}$ are the formal deformations defined by $\tilde{\eta}$ and by η respectively. Since f is etale it induces an isomorphism $\hat{\mathcal{O}}_{S,s} \cong \hat{\mathcal{O}}_{\tilde{S},\tilde{s}}$ and the conclusion follows. □

Definition 2.5.7. *A deformation* (S, s, η) *of* X *is called* formally universal *(resp.* formally semiuniversal, formally versal*) if the formal deformation* $(\hat{\mathcal{O}}_{S,s}, \{\eta_n\}_{n\geq 0})$ *associated to* η *is universal (resp. semiuniversal, versal). An algebraic formally versal deformation of* X *is also called* with general moduli.

A flat family $\pi : \mathcal{X} \to S$ *is called* formally universal *(resp.* formally semiuniversal, formally versal, with general moduli) at a **k**-rational point $s \in S$ *if*

$$\eta : \begin{array}{ccc} \mathcal{X}(s) & \subset & \mathcal{X} \\ \downarrow & & \downarrow \pi \\ \mathrm{Spec}(\mathbf{k}) & \xrightarrow{s} & S \end{array}$$

is a formally universal, (resp. formally semiuniversal, formally versal, with general moduli) deformation of $\mathcal{X}(s)$.

The expression "general moduli" goes back to the classical geometers. Informally, it means that the family parametrizes all possible "sufficiently small" deformations of $\mathcal{X}(s)$; when the family π parametrizes varieties for which there is a moduli scheme, or just a moduli stack (whatever this means for the reader since we have not introduced these notions), π with general moduli means that the functorial morphism from S to the moduli scheme or stack is open at s. An expression like "consider a variety X with general moduli" is used to mean: "choose X as a general fibre in a family with general moduli".

The early literature on deformation theory of complex analytic manifolds (in the approach of Kodaira and Spencer) considered only families parametrized by complex analytic manifolds. In that context the expression "effectively parametrized" was used to mean "with injective Kodaira–Spencer map", and the word "complete" was used to mean the analogue of the notion of formal versality, in the category of germs of complex analytic manifolds.

Proposition 2.5.8. *Let* (S, s, η) *be a deformation of* X. *Then:*

(i) If η *is formally versal (resp. formally semiuniversal or formally universal) then the Kodaira–Spencer map*

$$\kappa_{\pi,s} : T_s S \to \mathrm{Def}_X(\mathbf{k}[\epsilon])$$

is surjective (resp. an isomorphism).

(ii) If S *is nonsingular at* s *and the Kodaira–Spencer map* $\kappa_{\pi,s}$ *is surjective (resp. an isomorphism) then* η *is formally versal (resp. formally semiuniversal) and* X *is unobstructed, i.e. the functor* Def_X *is smooth.*

Proof. (i) is obvious in view of the definitions of versality and semiuniversality of a formal couple applied to $h_{\hat{\mathcal{O}}_{S,s}} \to \mathrm{Def}_X$.

(ii) follows from Proposition 2.2.5(iii) applied to $f : h_{\hat{\mathcal{O}}_{S,s}} \to \mathrm{Def}_X$. □

The proposition applies in particular to an algebraic deformation, giving a criterion for it to have general moduli. A classical result (see [107]) states the completeness of a complex analytic family of compact complex manifolds if the map $\kappa_{\pi,s}$ is surjective. Part (ii) of Proposition 2.5.8 is the algebraic version of this result. It turns out to be very useful because it reduces the verification of formal versality and of unobstructedness to the computation of the Kodaira–Spencer map.

Definition 2.5.9. *A formal deformation* $(\bar{A}, \{\eta_n\})$ *of* X *is called* algebraizable *if there exists an algebraic deformation* (S, s, ξ) *of* X *and an isomorphism* $\hat{\mathcal{O}}_{S,s} \cong \bar{A}$ *sending* η_n *to* ξ_n *for all* n *(i.e.* $(\bar{A}, \{\eta_n\})$ *is isomorphic to the formal deformation defined by* ξ*). The deformation* (S, s, ξ) *is called an* algebraization *of* $(\bar{A}, \{\eta_n\})$.

It goes without saying that an algebraization of a formal versal (resp. semiuniversal, universal) deformation is formally versal (resp. formally semiuniversal, formally universal). The existence of algebraizations is a highly nontrivial problem. It can be considered as the counterpart of the convergence step in the construction of local families of deformations in the Kodaira–Spencer theory of deformations of compact complex manifolds. But the algebraic case presents some characteristic features which make the two theories radically different in methods and in results. In particular, a projective algebraic variety X defined over the field of complex numbers need not have an algebraic formally versal deformation, even in the case when Def_X is prorepresentable and unobstructed (see Example 2.5.12 below); but, according to the KNSK theory (see the Introduction), such a variety has a complete deformation in the complex analytic sense.

* * * * * *

For the rest of this section we will need to assume some familiarity with formal schemes: for some definitions and results not contained in [84] we will refer to [1]. The following definition introduces an important notion weaker than algebraizability.

Definition 2.5.10. *Let* X *be an algebraic scheme and let* $\bar{A}$ *be in* $\mathrm{ob}(\hat{A})$. *A formal deformation* $(\bar{A}, \hat{\eta})$ *of* X *is called* effective *if there exists a deformation*

$$\bar{\eta}: \quad \begin{array}{ccc} X & \to & \mathcal{X} \\ \downarrow & & \downarrow \pi \\ \mathrm{Spec}(\mathbf{k}) & \to & \mathrm{Spec}(\bar{A}) \end{array}$$

of X *over* $\bar{A}$ *such that* $\hat{\eta}$ *is the formal deformation associated to* $\bar{\eta}$. *Equivalently, letting* $\bar{\pi} : \boldsymbol{\mathcal{X}} \to \mathrm{Specf}(\bar{A})$ *be the morphism of formal schemes associated to the formal deformation* $(\bar{A}, \hat{\eta})$*, its effectivity means that there is a deformation* $\bar{\eta}$ *such that* $\boldsymbol{\mathcal{X}}$ *is the formal completion of* $\mathcal{X}$ *along* X *and* $\bar{\pi}$ *is the morphism of formal schemes induced by* π*; in symbols:*

$$\boldsymbol{\mathcal{X}} \cong \mathrm{Specf}(\bar{A}) \times_{\mathrm{Spec}(\bar{A})} \mathcal{X}$$

It is obvious that the *trivial formal deformation* $\boldsymbol{\mathcal{X}} \to \mathrm{Specf}(\bar{A})$, where

$$\boldsymbol{\mathcal{X}} = X \times \mathrm{Specf}(\bar{A}) = (X, \varprojlim \mathcal{O}_X \otimes_{\mathbf{k}} \bar{A}/m^n)$$

is effective: it is the formal completion of the trivial deformation

$$X \times \mathrm{Spec}(\bar{A}) \to \mathrm{Spec}(\bar{A})$$

In particular, for any $\bar{A}$ in $\mathrm{ob}(\hat{\mathcal{A}})$ the *formal projective space* over $\bar{A}$

$$\mathcal{P}^r_{\bar{A}} := (I\!P^r, \lim_{\leftarrow} \mathcal{O}_{I\!P^r_{\bar{A}_n}})$$

is effective, being the formal completion of $I\!P^r \times \mathrm{Spec}(\bar{A})$ along the closed fibre $I\!P^r = I\!P^r \times \mathrm{Spec}(\mathbf{k})$. A consequence of Grothendieck's theorem of formal functions is that a formal deformation of a proper algebraic scheme can be effective in at most one way; more precisely we have:

Theorem 2.5.11. *Let X be a proper algebraic scheme. If*

$$\bar{\eta}: \begin{array}{ccc} X & \to & \mathcal{X} \\ \downarrow & & \downarrow \pi \\ \mathrm{Spec}(\mathbf{k}) & \to & \mathrm{Spec}(\bar{A}) \end{array} \qquad \bar{\xi}: \begin{array}{ccc} X & \to & \mathcal{Y} \\ \downarrow & & \downarrow q \\ \mathrm{Spec}(\mathbf{k}) & \to & \mathrm{Spec}(\bar{A}) \end{array}$$

are two deformations of X with $\bar{A}$ in $\mathrm{ob}(\hat{\mathcal{A}})$ such that the associated formal deformations $(\bar{A}, \hat{\eta})$, $(\bar{A}, \hat{\xi})$ are isomorphic, then $\bar{\eta}$ and $\bar{\xi}$ are isomorphic deformations.

Proof. Let $S = \mathrm{Spec}(\bar{A})$ and $\hat{S} = \mathrm{Specf}(\bar{A})$. For a given proper S-scheme $f : Z \to S$ we denote by $\hat{Z}$ the formal completion of Z along $Z(s) = f^{-1}(s)$, where $s \in S$ is the closed point, and by $\hat{\mathcal{F}}$ the $\mathcal{O}_{\hat{Z}}$-sheaf obtained as the formal completion of a quasi-coherent $\mathcal{O}_Z$-sheaf $\mathcal{F}$. It suffices to show that the correspondence $\varphi \mapsto \hat{\varphi}$ defines a one-to-one correspondence

$$\mathrm{Hom}_S(\mathcal{X}, \mathcal{Y}) \to \mathrm{Hom}_{\hat{S}}(\hat{\mathcal{X}}, \hat{\mathcal{Y}})$$

because it will follow that $\hat{\mathcal{X}} \cong \hat{\mathcal{Y}}$, i.e. $(\bar{A}, \hat{\eta}) \cong (\bar{A}, \hat{\xi})$, implies $\mathcal{X} \cong \mathcal{Y}$.

The proof goes as follows. For any proper S-scheme $f : Z \to S$ and coherent sheaves $\mathcal{F}$ and $\mathcal{G}$ on Z the theorem of formal functions ([84], p. 277) implies that there is an isomorphism

$$\mathrm{Hom}(\mathcal{F}, \mathcal{G})^\wedge \cong \lim_{\leftarrow} \mathrm{Hom}_{\mathcal{O}_{Z_n}}(\mathcal{F}_n, \mathcal{G}_n) = \mathrm{Hom}_{\hat{Z}}(\hat{\mathcal{F}}, \hat{\mathcal{G}})$$

Moreover, since $\mathrm{Hom}(\mathcal{F}, \mathcal{G})$ is a finitely generated $\bar{A}$-module, it coincides with its $m_{\bar{A}}$-adic completion ([127], Theorem 8.7, p. 60); thus we have an isomorphism

$$\mathrm{Hom}(\mathcal{F}, \mathcal{G}) \cong \mathrm{Hom}_{\hat{Z}}(\hat{\mathcal{F}}, \hat{\mathcal{G}})$$

It maps a homomorphism $u : \mathcal{F} \to \mathcal{G}$ to its completion $\hat{u} : \hat{\mathcal{F}} \to \hat{\mathcal{G}}$. It is easy to see that $\hat{u}$ is injective (resp. surjective) if and only if u is injective (resp. surjective) ([1], ch. III, Cor. 5.1.3). In particular, we have a one-to-one correspondence between sheaves of ideals of $\mathcal{O}_Z$ and of $\mathcal{O}_{\hat{Z}}$, equivalently between closed subschemes of Z and closed formal subschemes of $\hat{Z}$.

Now take $Z = \mathcal{X} \times_S \mathcal{Y}$ and view it as an S-scheme by any of the projections. Then we have a natural identification $\hat{Z} = \hat{\mathcal{X}} \times_{\hat{S}} \hat{\mathcal{Y}}$ ([1], ch I, Prop. 10.9.7). We deduce a one-to-one correspondence between closed subschemes of $\mathcal{X} \times_S \mathcal{Y}$ and of $\hat{\mathcal{X}} \times_{\hat{S}} \hat{\mathcal{Y}}$, in particular between graphs of S-morphisms $\mathcal{X} \to \mathcal{Y}$ and graphs of morphisms $\hat{\mathcal{X}} \to \hat{\mathcal{Y}}$ of $\hat{S}$-formal schemes. □

In general, not every formal deformation of a given algebraic scheme need to be effective, just as deformations of algebraic varieties defined over the field of complex numbers need not be algebraic. Here is a typical example.

Example 2.5.12 (K3-surfaces). For this example we will need a result to be proved in Chapter 3. Let X be a projective $K3$-surface, and assume for simplicity that $\mathrm{Pic}(X) = \mathbb{Z}[H]$ for some divisor H. Let $(\bar{A}, \hat{\eta})$ be the formal universal deformation of X and $\boldsymbol{\mathcal{X}}$ the corresponding formal scheme over $\hat{S} = \mathrm{Specf}(\bar{A})$. By Example 2.4.11(ii) we know that

$$\bar{A} \cong \mathbf{k}[[X_1, \ldots, X_{20}]]$$

where $X_1, \ldots, X_{20}$ are indeterminates.

Claim: $(\bar{A}, \hat{\eta})$ is not effective.

Otherwise there is a proper smooth morphism $f : \mathcal{X} \to S = \mathrm{Spec}(\bar{A})$ such that $\hat{\mathcal{X}} \cong \boldsymbol{\mathcal{X}}$, where $\hat{\mathcal{X}}$ is the formal completion of $\mathcal{X}$ along $X = f^{-1}(s)$, $s \in S$ the closed point. Since $\mathcal{X}$ is of finite type over the integral scheme S, it has a nontrivial line bundle $\mathcal{L}$ which we may assume to correspond to a Cartier divisor whose support does not contain X and has a nonempty intersection with X; therefore $\mathcal{L}$ has a nontrivial restriction to the closed fibre X, say $\mathcal{L} \otimes \mathcal{O}_X \cong nH$, $n \neq 0$. By [1], Ch. III, Cor. 5.1.6, $\mathcal{L}$ corresponds to a line bundle $\hat{\mathcal{L}}$ on $\boldsymbol{\mathcal{X}}$ which extends $\mathcal{L} \otimes \mathcal{O}_X \cong nH$. But nH cannot be extended to all of $\boldsymbol{\mathcal{X}}$ because it extends to a 19-dimensional space of first-order deformations of X (see page 150). Therefore we have a contradiction and the claim is proved.

The following is a basic result of effectivity of formal deformations.

Theorem 2.5.13 (Grothendieck [71]). *Let X be a projective scheme. Then:*

(i) Let $\bar{\pi} : \boldsymbol{\mathcal{X}} \to \mathrm{Specf}(\bar{A})$, $\bar{A}$ in $\mathrm{ob}(\hat{\mathcal{A}})$, be a formal deformation of X. Assume that there is a closed embedding of formal schemes $j : \boldsymbol{\mathcal{X}} \subset \mathcal{P}^r_{\bar{A}}$ such that $\bar{\pi} = pj$ where $p : \mathcal{P}^r_{\bar{A}} \to \mathrm{Specf}(\bar{A})$ is the projection. Then $\bar{\pi}$ is effective.

(ii) Assume that $h^2(X, \mathcal{O}_X) = 0$. Then every formal deformation of X is effective.

Proof. (i) The structure sheaf $\mathcal{O}_{\boldsymbol{\mathcal{X}}}$ and the ideal sheaf $\mathcal{I}_{\boldsymbol{\mathcal{X}}}$ are coherent sheaves of $\mathcal{O}_{\mathcal{P}^r_{\bar{A}}}$-modules. Repeating verbatim the proof of the classical result of Serre (see [84], Theorem II.5.15, p. 121), one shows that there is an exact sequence

$$\bigoplus_{i=1}^{N} \mathcal{O}_{\mathcal{P}^r_{\bar{A}}}(-n_i) \xrightarrow{\bar{g}} \mathcal{O}_{\mathcal{P}^r_{\bar{A}}} \to \mathcal{O}_{\boldsymbol{\mathcal{X}}} \to 0$$

for some positive integers $n_1, \ldots, n_N$. We have

$$\bar{g} \in \bigoplus_{i=1}^{N} H^0(\mathcal{P}^r_{\bar{A}}, \mathcal{O}_{\mathcal{P}^r_{\bar{A}}}(n_i))$$

and, by the theorem on formal functions ([84], ch. III, Theorem 11.1) we have

$$H^0(\mathcal{P}^r_{\bar{A}}, \mathcal{O}_{\mathcal{P}^r_{\bar{A}}}(n_i)) = H^0(\mathbb{P}^r_{\bar{A}}, \mathcal{O}_{\mathbb{P}^r_{\bar{A}}}(n_i))^{\wedge}$$

Since $H^0(\mathbb{P}^r_{\bar{A}}, \mathcal{O}_{\mathbb{P}^r_{\bar{A}}}(n_i))$ is an $\bar{A}$-module of finite type we have

$$H^0(\mathbb{P}^r_{\bar{A}}, \mathcal{O}_{\mathbb{P}^r_{\bar{A}}}(n_i))^{\wedge} = H^0(\mathbb{P}^r_{\bar{A}}, \mathcal{O}_{\mathbb{P}^r_{\bar{A}}}(n_i))$$

([127], Theorem 8.7, p. 60) and therefore

$$H^0(\mathcal{P}^r_{\bar{A}}, \mathcal{O}_{\mathcal{P}^r_{\bar{A}}}(n_i)) = H^0(\mathbb{P}^r_{\bar{A}}, \mathcal{O}_{\mathbb{P}^r_{\bar{A}}}(n_i))$$

This implies that the homomorphism $\bar{g}$ is induced by a homomorphism

$$g : \bigoplus_{i=1}^{N} \mathcal{O}_{\mathbb{P}^r_{\bar{A}}}(-n_i) \to \mathcal{O}_{\mathbb{P}^r_{\bar{A}}}$$

whose cokernel is the structure sheaf of a closed subscheme $\mathcal{X} \subset \mathbb{P}^r_{\bar{A}}$ having $\boldsymbol{\mathcal{X}}$ as formal completion along $\mathcal{X}(o)$. The flatness of $\mathcal{X}$ over $\mathrm{Spec}(\bar{A})$ follows from Theorem A.5.

(ii) Let $(\bar{A}, \{\eta_n\})$ be a formal deformation of X, where:

$$\eta_n : \begin{array}{ccc} X & \xrightarrow{f_n} & \mathcal{X}_n \\ \downarrow & & \downarrow \pi_n \\ \mathrm{Spec}(\mathbf{k}) & \to & \mathrm{Spec}(\bar{A}_n) \end{array}$$

For each $n \geq 1$ we have an exact sequence:

$$0 \to \frac{m^n_{\bar{A}}}{m^{n+1}_{\bar{A}}} \otimes_{\mathbf{k}} \mathcal{O}_X \xrightarrow{\exp} \mathcal{O}^*_{\mathcal{X}_n} \xrightarrow{\rho} \mathcal{O}^*_{\mathcal{X}_{n-1}} \to 0$$

where ρ is the natural restriction, $\exp(\sigma) = 1 + \sigma$ and where we have identified

$$\frac{m^n_{\bar{A}}}{m^{n+1}_{\bar{A}}} \otimes_{\mathbf{k}} \mathcal{O}_X = \frac{m^n_{\bar{A}}}{m^{n+1}_{\bar{A}}} \otimes_{\bar{A}_n} \mathcal{O}_{X_n}$$

using the flatness of π_n. From the hypothesis $h^2(X, \mathcal{O}_X) = 0$ we deduce that the homomorphism of Picard groups $\mathrm{Pic}(\mathcal{X}_n) \to \mathrm{Pic}(\mathcal{X}_{n-1})$ is surjective. This implies that if we fix a very ample line bundle L on X such that $H^1(X, L) = 0$ then for each $n \geq 1$ we can find a line bundle L_n on $\mathcal{X}_n$ such that $L_{n|\mathcal{X}_{n-1}} = L_{n-1}$. By Note 3 of § 4.2, L_n is very ample and defines an embedding $\mathcal{X}_n \subset \mathbb{P}^N_{\bar{A}_n}$ where $N + 1 = h^0(X, L)$. This means that the formal scheme $\boldsymbol{\mathcal{X}}$ is a formal closed subscheme of $\mathcal{P}^N_{\bar{A}}$. Now we conclude by part (i). □

After these preliminaries we can state the following special case of Artin's algebraization theorem.

Theorem 2.5.14 (Artin [12]). *Let X be a projective scheme and let $(\bar{A}, \hat{\eta})$ be an effective formal versal deformation of X. Then $(\bar{A}, \hat{\eta})$ is algebraizable.*

The universal formal deformation $(\bar{A}, \hat{\eta})$ of a K3-surface is not effective (Example 2.5.12) and in fact it is not algebraizable: this is well known if $\mathbf{k} = \mathbf{C}$ because K3-surfaces have arbitrarily small deformations which are non-algebraic complex manifolds.

* * * * * *

To give a complete proof of Theorem 2.5.14 is beyond our goals. Nevertheless, we want to introduce some relevant definitions and preliminary results which are needed for its general statement (see below) and proof.

Consider a covariant functor

$$F : (\mathbf{k}\text{-algebras}) \to (\text{sets})$$

Suppose given R in $\text{ob}(\mathcal{A}^*)$ and an element $\bar{u} \in F(\hat{R})$. An important question is whether it is possible to approximate $\bar{u}$ in some way by an element $u \in F(R)$. Indeed, every algebraization problem can be reduced to such a question for an appropriate functor F. In this context the following is a natural definition of approximation:

Definition 2.5.15. *Suppose given R in* ob$(\mathcal{A}^*)$, *$\bar{u} \in F(\hat{R})$ and $u \in F(R)$. If $c > 0$ is an integer we say that $\bar{u}$ and u are* congruent modulo m^c, *in symbols*

$$\bar{u} \equiv u \quad (\textit{modulo } m^c)$$

if $\bar{u}$ and u induce the same element in $F(\hat{R}/m^c_{\hat{R}}) = F(R/m^c_R)$.

In order to make the problem tractable it turns out to be natural to impose the following finiteness condition on the functor F.

Definition 2.5.16. *A functor*

$$F : (\mathbf{k}\text{-}algebras) \to (sets)$$

is said to be locally of finite presentation *if for every filtering inductive system of* $\mathbf{k}$*-algebras $\{B_i\}$ the canonical map*

$$\varinjlim F(B_i) \to F(\varinjlim B_i)$$

is bijective. A functor locally of finite presentation is sometimes also called limit preserving.

This is a natural finiteness property, first introduced in [1], ch. IV, Prop. 8.14.2, which is usually satisfied by the functors arising in algebraic geometry.

An illustration of the meaning of Definition 2.5.16 is given by the following proposition and by its corollary.

Proposition 2.5.17. *Let R be a noetherian ring and B an R-algebra. Then B is an R-algebra of finite type if and only if for every filtering inductive system of R-algebras $\{C_i\}$ the canonical map*

$$\varinjlim \left[\mathrm{Hom}_{R-alg}(B, C_i)\right] \to \mathrm{Hom}_{R-alg}(B, \varinjlim C_i)$$

is bijective.

Proof. Assume that the condition of the statement is satisfied, and write

$$B = \varinjlim B_i$$

where the $B_i \subset B$ are R-subalgebras of finite type. Then the hypothesis implies that the identity $B \to B$ factors as $B \to B_i \to B$ for some i, and this implies $B = B_i$. Therefore B is of finite type.

Conversely, assume that B is of finite type, and let $\{t_1, \dots, t_n\}$ be a system of generators of B as an R-algebra. Let $\{C_i\}$ be a filtering inductive system of R-algebras, $C = \lim C_i$ and let

$$\varinjlim \left[\mathrm{Hom}_{R-alg}(B, C_i)\right] \to \mathrm{Hom}_{R-alg}(B, C) \tag{2.26}$$

be the natural map. Consider two elements of $\lim \left[\mathrm{Hom}_{R-alg}(B, C_i)\right]$, given by two compatible systems of homomorphisms (θ_i) and (θ_i'), defining the same homomorphism $\theta : B \to C$. Then, letting $\varphi_i : C_i \to C$ and $\varphi_{ji} : C_i \to C_j$ be the canonical homomorphisms of the inductive system, for each $s = 1, \dots, n$ there is an index i_s such that

$$\theta(t_s) = \varphi_{i_s}(\theta_{i_s}(t_s)) = \varphi_{i_s}(\theta'_{i_s}(t_s))$$

We may assume that all the i_s are equal, say to i_0. In the same way we can find an index $j_0 \geq i_0$ such that

$$\varphi_{j_0 i_0}(\theta_{i_0}(t_s)) = \varphi_{j_0 i_0}(\theta'_{i_0}(t_s))$$

for all $s = 1, \dots, n$; this means that $\theta_{j_0}(t_s) = \theta'_{j_0}(t_s)$, i.e. $\theta_{j_0} = \theta'_{j_0}$. Therefore the map (2.26) is injective.

Now let $\theta : B \to C$ be a homomorphism, and let $c_s = \theta(t_s)$, $s = 1, \dots, n$. Write $B = R[T_1, \dots, T_n]/J$, and let $\{P_k\}_{k=1,\dots,m}$ be a system of generators of the ideal J. Then we have

$$P_k(c_1, \dots, c_n) = \theta(P_k(t_1, \dots, t_n)) = 0, \qquad k = 1, \dots, m$$

Let i_0 be an index such that there exist elements $x_1, \dots, x_n \in C_{i_0}$ such that $\varphi_{i_0}(x_s) = c_s$, $i = 1, \dots, n$. One has

$$\varphi_{i_0}(P_k(x_1, \dots, x_n)) = P_k(c_1, \dots, c_n) = 0 \qquad k = 1, \dots, m$$

Then there is an index $j_0 \geq i_0$ such that

$$0 = \varphi_{j_0 i_0}((P_k(x_1, \ldots, x_n)) = P_k(\varphi_{j_0 i_0}(x_1), \ldots, \varphi_{j_0 i_0}(x_n)) \qquad k = 1, \ldots, m$$

One deduces the existence of a **k**-algebra homomorphism $\theta_{j_0} : B \to C_{j_0}$ such that $\theta_{j_0}(t_s) = \varphi_{j_0 i_0}(x_s)$, $s = 1, \ldots, n$. This implies the existence of a homomorphism $\theta_j = \varphi_{j j_0}\theta_{j_0} : B \to C_j$ for every $j \geq j_0$. It follows that θ is the inductive limit of the system (θ_j), and therefore the map (2.26) is also surjective. □

The following is an immediate consequence.

Corollary 2.5.18. *Let B be a* **k***-algebra and*

$$F = \mathrm{Hom}_{\mathbf{k}-alg}(B, -) : (\mathbf{k}\text{-}algebras) \to (sets)$$

Then F is locally of finite presentation if and only if B is a **k***-algebra of finite type.*

Therefore we see that the condition of being locally of finite presentation for a representable functor coincides with the condition of being represented by an algebra of finite type.

For the statement of the algebraization theorem we need some terminology which generalizes notions we have already introduced.

Consider a covariant functor

$$F : (\mathbf{k}\text{-algebras}) \to (\text{sets})$$

Let $u_0 \in F(\mathbf{k})$. A couple (A, u) where A is in ob$(\mathcal{A})$ and $u \in F(A)$ is called an *infinitesimal deformation* of u_0 if $u \mapsto u_0$ under the map

$$F(A) \to F(A/m_A) = F(\mathbf{k})$$

We will denote by

$$F_{u_0} : \mathcal{A} \to (\text{sets})$$

the *functor of infinitesimal deformations* of u_0, i.e.:

$$F_{u_0}(A) = \{u \in F(A) : u \mapsto u_0 \text{ under } F(A) \to F(\mathbf{k})\}$$

This is a functor of Artin rings and, according to the definitions given in § 2.2, we have the notions of *formal deformation* of u_0, and of *universal* (resp. *semiuniversal*, resp. *versal*) *formal deformation* of u_0.

Definition 2.5.19. *A* local deformation *of $u_0 \in F(\mathbf{k})$ is a couple (A, u) where $u \in F(A)$, A is in $\mathcal{A}^*$ and $u \mapsto u_0$ under the map*

$$F(A) \to F(A/m_A) = F(\mathbf{k})$$

An algebraic deformation *of u_0 is a triple (B, x, u) where B is a* **k***-algebra of finite type, $x \in$ Spec(B) is a* **k***-rational point and $u \in F(B)$ is such that $u \mapsto u_0$ under the map $F(B) \to F(\mathbf{k})$ induced by the composition*

$$B \to \mathcal{O}_{\mathrm{Spec}(B),x} \to \mathbf{k}(x) = \mathbf{k}$$

To every local deformation (A, u) of u_0 one associates the formal deformation $(\hat{A}, \{u_n\})$, where $\hat{A}$ is the m_A-adic completion of A, and $u_n \in F(A/m_A^{n+1}) = F(\hat{A}/m_{\hat{A}}^{n+1})$ is the image of u under the map $F(A) \to F(A/m_A^{n+1})$. We call $(\hat{A}, \{u_n\})$ *the formal deformation of* u_0 *associated to* (or *defined by*) (A, u). Similarly, *the formal deformation of* u_0 *defined by an algebraic deformation* (B, x, u) of u_0 is the formal deformation $(\hat{\mathcal{O}}_{\mathrm{Spec}(B),x}, \{u_n\})$ where $u \mapsto u_n$ under the map

$$F(B) \to F(\hat{\mathcal{O}}_{\mathrm{Spec}(B),x}/m^{n+1})$$

induced by the natural homomorphism

$$B \to \mathcal{O}_{\mathrm{Spec}(B),x}/m^{n+1} = \hat{\mathcal{O}}_{\mathrm{Spec}(B),x}/m^{n+1}$$

A local deformation is called *formally universal* (resp. *formally semiuniversal*, resp. *formally versal*) if the associated formal deformation has the corresponding property. A similar definition is given for algebraic deformations.

Definition 2.5.20. *A formal deformation* $(\bar{A}, \{u_n\})$ *of* $u_0 \in F(\mathbf{k})$ *is called* algebraizable *if it is associated to some algebraic deformation* (B, x, u). *In this case the deformation* (B, x, u) *is called an algebraization of* $(\bar{A}, \{u_n\})$.

A formal deformation $(\bar{A}, \{u_n\})$ *of* u_0 *is called* effective *if there exists* $\bar{u} \in F(\bar{A})$ *which induces* $\{u_n\}$. *In this case we will call* $(\bar{A}, \bar{u})$ *an* effective formal deformation *as well.*

It is clear from the definition that an algebraic deformation (B, x, u) of u_0 is an algebraization of the effective formal deformation $(\bar{A}, \bar{u})$ if and only if $\bar{A} = \hat{\mathcal{O}}_{X,x}$ and $u \equiv \bar{u} \bmod m^n$ for all $n \geq 0$.

Note that if the formal deformation $(\bar{A}, \{u_n\})$ of u_0 is associated to a local deformation (A, u) then it is effective: in fact it follows that $\{u_n\}$ is also associated to the deformation $\bar{u} \in F(\hat{A})$ which is the image of u under the map $F(A) \to F(\hat{A})$. Similarly, an algebraizable deformation is effective.

The algebrization theorem states that the converse is true for versal deformations. Namely:

Theorem 2.5.21 (algebraization theorem [12]). *Let*

$$F : (\mathbf{k}\text{-}algebras) \to (sets)$$

be a functor locally of finite presentation and let $u_0 \in F(\mathbf{k})$. *Then every effective versal formal deformation of* u_0 *is algebraizable.*

Theorem 2.5.21 is a generalization of Theorem 2.5.14 because the deformation functor Def_X of a projective scheme X can be extended to a functor locally of finite presentation defined on the category ($\mathbf{k}$-algebras).

The main technical ingredient in the proof of the algebraization theorem is the following:

Theorem 2.5.22 (approximation theorem [11]). *Let*

$$F : (\mathbf{k}\text{-}algebras) \to (sets)$$

be a functor locally of finite presentation, A an e.f.t. local **k**-*algebra and $\bar{u} \in F(\hat{A})$. Then for every positive integer c there is an element $u \in F(\tilde{A})$ such that $\bar{u} \equiv u$ modulo m^c.*

Outline of the proof of Theorem 2.5.21 (in a special case, when it follows almost directly from the approximation theorem.)

Let $(\bar{A}, \bar{u})$ be an effective formal versal deformation of u_0. We will assume that $\bar{A} = \hat{A}$, where A is an e.f.t. local **k**-algebra in $\mathcal{A}^*$. Suppose that we can find $\tilde{u} \in F(\tilde{A})$ whose associated formal deformation is $\{\bar{u}_n\}$. Then, since

$$\tilde{A} = \lim_{\to} \mathcal{O}_{S,s}$$

where (S, s) runs through all the etale neighbourhoods of $(\mathrm{Spec}(A), \{m_A\})$, and since F is locally of finite presentation there is an etale neighbourhood $(\mathrm{Spec}(B), x)$ of $(\mathrm{Spec}(A), \{m_A\})$ and $u \in F(B)$ such that $u \mapsto \tilde{u}$ under $F(B) \to F(\tilde{A})$. Again, because F is locally of finite presentation we may assume that $\mathrm{Spec}(B)$ is an affine algebraic scheme: therefore (B, x, u) is an algebraization of $(\bar{A}, \bar{u})$. Therefore we only need to find $\tilde{u}$ as above. By the approximation theorem there exists $\tilde{u} \in F(\tilde{A})$ such that $\bar{u} \equiv \tilde{u} \bmod m^2$. Therefore the homomorphism $\psi_1 : \bar{A} \to \tilde{A}/m^2$ sends $\bar{u} \mapsto \tilde{u}_1 = u_1$. By versality there is a compatible sequence of homomorphisms $\psi_n : \bar{A} \to \tilde{A}/m^{n+1}$ lifting ψ_1 and such that $F(\psi_n)(\bar{u}) = \tilde{u}_n$ for all n. We obtain an induced local homomorphism $\psi : \bar{A} \to \widehat{\tilde{A}} = \bar{A}$ such that $F(\psi)(\bar{u}) \equiv \tilde{u}_n \bmod m^{n+1}$ for all n. It suffices to prove that ψ is an isomorphism. By construction ψ is the identity mod m^2: the conclusion now follows from Lemma C.5. □

NOTES

1. The terminology introduced in Definition 2.5.10 is the most commonly used today but it differs from Grothendieck's: in [71] he calls a formal deformation "algebraizable" if it is effective. The same terminology is used in [84], p. 195.

2. Other references for Theorem 2.5.13 are [1], ch. III, Th. 5.4.5, and [3], Exp. III, Prop. 7.2.

2.6 Automorphisms and prorepresentability

... chaque fois que ... une variété de modules ... ne peut exister, malgré de bonnes hypothéses de platitude, propreté, et non singularité eventuellement, la raison en est seulement l'existence d'automorphismes de la structure ...
(Grothendieck [40], p. 94.)

2.6.1 The automorphism functor

The following theorem gives a criterion on an algebraic scheme X to decide whether Def_X, resp. Def'_X, has a universal formal element and not merely a semiuniversal one.

Theorem 2.6.1. *Assume that X is an algebraic scheme such that* Def_X *has a semiuniversal element (e.g. X affine with isolated singularities or X projective). Then the following conditions are equivalent:*

(i) Def_X *is prorepresentable*
(ii) for each small extension $A' \to A$ in $\mathcal{A}$, and for each deformation $\mathcal{X}'$ of X over $\mathrm{Spec}(A')$, *every automorphism of the deformation*

$$\mathcal{X}' \times_{\mathrm{Spec}(A')} \mathrm{Spec}(A) \to \mathrm{Spec}(A)$$

is induced by an automorphism of $\mathcal{X}'$.

A similar statement holds for the functor Def'_X.

Proof. (i) $\Rightarrow$ (ii) Let $\mathcal{X} = \mathcal{X}' \times_{\mathrm{Spec}(A')} \mathrm{Spec}(A)$ and let $f : \mathcal{X} \to \mathcal{X}'$ be the induced morphism; assume that θ is an automorphism of $\mathcal{X}$. Letting $\bar{A} = A' \times_A A'$, one can construct two deformations $\mathcal{Z}$ and $\mathcal{W}$ of X over $\bar{A}$ as we did in the proof of Theorem 2.4.1 as fibred sums fitting into the two diagrams:

$$\begin{array}{ccccc} & & \mathcal{Z} & & \\ & \nearrow & & \nwarrow & \\ \mathcal{X}' & & & & \mathcal{X}' \\ & \nwarrow f\theta & & \nearrow f & \\ & & \mathcal{X} & & \end{array} \qquad \begin{array}{ccccc} & & \mathcal{W} & & \\ & \nearrow & & \nwarrow & \\ \mathcal{X}' & & & & \mathcal{X}' \\ & \nwarrow f & & \nearrow f & \\ & & \mathcal{X} & & \end{array}$$

Since $[\mathcal{Z}], [\mathcal{W}] \in \mathrm{Def}_X(\bar{A})$ have the same image $([\mathcal{X}'], [\mathcal{X}'])$ under the map

$$\mathrm{Def}_X(\bar{A}) \to \mathrm{Def}_X(A') \times_{\mathrm{Def}_X(A)} \mathrm{Def}_X(A')$$

and since this map is bijective by (i), we have an isomorphism of deformations $\rho : \mathcal{Z} \cong \mathcal{W}$. The isomorphism ρ induces automorphisms θ_1 and θ_2 of $\mathcal{X}'$ and an automorphism ψ of $\mathcal{X}$ such that:

$$\theta_1 f\theta = f\psi, \quad \theta_2 f = f\psi$$

This equality implies $f\theta = \theta_1^{-1}\theta_2 f$:

$$\begin{array}{ccc} \mathcal{X}' & \stackrel{\theta_1^{-1}\theta_2}{\longrightarrow} & \mathcal{X}' \\ \uparrow f & & \uparrow f \\ \mathcal{X} & \stackrel{\theta}{\longrightarrow} & \mathcal{X} \end{array}$$

and therefore $\theta_1^{-1}\theta_2$ induces θ.

(ii) $\Rightarrow$ (i) Since Def_X has a semiuniversal element, it suffices to show that it satisfies condition H of Theorem 2.3.2. Let $A' \to A$ be a small extension in $\mathcal{A}$: letting $\bar{A} = A' \times_A A'$ we must show that the map

$$\alpha : \mathrm{Def}_X(\bar{A}) \to \mathrm{Def}_X(A') \times_{\mathrm{Def}_X(A)} \mathrm{Def}_X(A')$$

is bijective. Given deformations $\mathcal{X}'$ and $\tilde{\mathcal{X}}'$ of X over A' inducing the deformation $\mathcal{X}$ over A, we have the "fibred sum" deformation $\bar{\mathcal{X}}$ over $\bar{A}$, which fits into the diagram:

$$\begin{array}{ccccc} & & \bar{\mathcal{X}} & & \\ & \nearrow & & \nwarrow & \\ \mathcal{X}' & & & & \tilde{\mathcal{X}}' \\ & \nwarrow f & & \nearrow \tilde{f} & \\ & & \mathcal{X} & & \end{array}$$

and satisfies $\alpha([\bar{\mathcal{X}}]) = ([\mathcal{X}'], [\mathcal{X}'])$. Suppose that $\mathcal{Z}$ is another deformation of X over $\bar{A}$ such that $\alpha([\mathcal{Z}]) = ([\mathcal{X}'], [\tilde{\mathcal{X}}'])$. We have isomorphisms of deformations induced by the two projections:

$$\mathcal{X}' \cong \mathcal{Z} \times_{\mathrm{Spec}(\bar{A})} \mathrm{Spec}(A') \cong \tilde{\mathcal{X}}'$$

There remains induced an automorphism θ of $\mathcal{X}$ as the composition:

$$\mathcal{X} \cong \mathcal{X}' \times_{\mathrm{Spec}(A')} \mathrm{Spec}(A) \cong \mathcal{Z} \times_{\mathrm{Spec}(\bar{A})} \mathrm{Spec}(A) \cong \tilde{\mathcal{X}}' \times_{\mathrm{Spec}(A')} \mathrm{Spec}(A) \cong \mathcal{X}$$

and θ fits into the commutative diagram:

$$\begin{array}{ccccc} & & \mathcal{Z} & & \\ & \nearrow & & \nwarrow & \\ \mathcal{X}' & & & & \tilde{\mathcal{X}}' \\ & \nwarrow f & & \nearrow \tilde{f} & \\ & & \mathcal{X} \stackrel{\theta}{\longrightarrow} \mathcal{X} & & \end{array}$$

By (ii) we can lift θ to an automorphism $\sigma : \mathcal{X}' \cong \mathcal{X}'$. Replacing the lower left map f by σf we obtain the commutative diagram

$$\begin{array}{ccccc} & & \mathcal{Z} & & \\ & \nearrow & & \nwarrow & \\ \mathcal{X}' & & & & \tilde{\mathcal{X}}' \\ & \nwarrow \sigma f & & \nearrow \tilde{f} & \\ & & \mathcal{X} & & \end{array}$$

By the universal property of the fibred sum we obtain an isomorphism $\bar{\mathcal{X}} \cong \mathcal{Z}$ which is an isomorphism of deformations. Therefore $[\mathcal{Z}] = [\bar{\mathcal{X}}]$ and α is bijective.

In the case of Def'_X the proof is similar. □

When X is a projective scheme, condition (ii) of Theorem 2.6.1 can be stated in a different way by means of the *automorphism functor*, which we now introduce.

Assume that X is an algebraic scheme such that Def_X has a semiuniversal couple $(R, \hat{u})$. Consider the functor of Artin rings

$$\begin{aligned} \mathrm{Aut}_{\hat{u}} : \mathcal{A}_R &\to (\text{ sets}) \\ \mathrm{Aut}_{\hat{u}}(A) &= \text{the group of automorphisms of the deformation } \mathcal{X}_A \end{aligned}$$

where $\mathcal{X}_A$ is the deformation induced by $\hat{u}$ under the morphism $R \to A$. Then we have the following:

Proposition 2.6.2. *If X is projective then* $\mathrm{Aut}_{\hat{u}}$ *has $H^0(X, T_X)$ as tangent space and is prorepresented by a complete local R-algebra S. Moreover, the deformation functor* Def_X *is prorepresentable if and only if S is a formally smooth R-algebra, i.e. if it is a power series ring over R.*

Proof. Obviously, $\mathrm{Aut}_{\hat{u}}$ satisfies condition H_0 because by definition the only automorphism of the deformation $\mathcal{X}_{\mathbf{k}} = X$ is the identity. Now consider a diagram in $\mathcal{A}_R$:

$$\begin{array}{ccc} A' & & A'' \\ & \searrow\swarrow & \\ & A & \end{array}$$

with $A'' \to A$ a small extension and let $\bar{A} = A' \times_A A''$. There is induced a diagram of deformations:

$$\begin{array}{ccc} & \mathcal{X}_{\bar{A}} & \\ & \nearrow\nwarrow & \\ \mathcal{X}_{A'} & & \mathcal{X}_{A''} \\ & \nwarrow\nearrow & \\ & \mathcal{X}_A & \end{array}$$

and therefore a natural homomorphism:

$$\mathcal{O}_{X_{\bar{A}}} \to \mathcal{O}_{\mathcal{X}_{A'}} \times_{\mathcal{O}_{\mathcal{X}_A}} \mathcal{O}_{\mathcal{X}_{A''}} \tag{2.27}$$

Since $\mathcal{O}_{\mathcal{X}_{A'}} = \mathcal{O}_{\mathcal{X}_{\bar{A}}} \otimes_{\bar{A}} A'$ it follows from Lemma A.4 that (2.27) is an isomorphism; in particular, we obtain an induced isomorphism

$$\mathcal{O}^*_{\mathcal{X}_{\bar{A}}} \cong \mathcal{O}^*_{\mathcal{X}_{A'}} \times_{\mathcal{O}^*_{\mathcal{X}_A}} \mathcal{O}^*_{\mathcal{X}_{A''}}$$

and therefore

$$H^0(\mathcal{O}^*_{\mathcal{X}_{\bar{A}}}) \cong H^0(\mathcal{O}^*_{\mathcal{X}_{A'}}) \times_{H^0(\mathcal{O}^*_{\mathcal{X}_A})} H^0(\mathcal{O}^*_{\mathcal{X}_{A''}}) \tag{2.28}$$

Now note that for every A in $\mathrm{ob}(\mathcal{A}_R)$ the elements of $\mathrm{Aut}_{\hat{u}}(A)$ are identified with those elements of $H^0(\mathcal{O}^*_{\mathcal{X}_A})$ which restrict to $1 \in \mathcal{O}^*_X$. Hence we see that (2.28) immediately implies the bijection

$$\mathrm{Aut}_{\hat{u}}(\bar{A}) \cong \mathrm{Aut}_{\hat{u}}(A') \times_{\mathrm{Aut}_{\hat{u}}(A)} \mathrm{Aut}_{\hat{u}}(A'')$$

Therefore the functor $\mathrm{Aut}_{\hat{u}}$ also satisfies conditions H and H_ϵ.

From Lemma 1.2.6 it follows that

$$\mathrm{Aut}_{\hat{u}}(\mathbf{k}[\epsilon]) \cong H^0(X, T_X) \tag{2.29}$$

which has finite dimension since X is projective, and also H_f is satisfied. This concludes the proof of the first part.

Condition (ii) of Theorem 2.6.1 can be rephrased by saying that the functor $\mathrm{Aut}_{\hat{u}}$ is smooth over Def_X. □

$H^0(X, T_X)$ is usually called the *space of infinitesimal automorphisms* of X. As a corollary we obtain:

Corollary 2.6.3. *If X is a projective scheme then the following conditions are equivalent:*

(i) $h^0(X, T_X) = 0$, i.e. X has no infinitesimal automorphisms;
(ii) $\mathrm{Aut}_{\hat{u}} \cong \mathrm{Def}_X$;
(iii) every infinitesimal deformation of X has no nontrivial automorphisms.

Proof. It is an immediate consequence of the proposition. It can be also proved directly without using Proposition 2.6.2: just observe that if $H^0(X, T_X) = 0$ then using Lemma 1.2.6 one shows by induction that every infinitesimal deformation of X has no automorphisms. □

An important application of the above proposition is the following result, which can be considered as the scheme-theoretic version of a classical theorem due to Kodaira–Nirenberg–Spencer [106]:

Corollary 2.6.4. *If X is a projective scheme such that $h^0(X, T_X) = 0$ then Def_X is prorepresentable. If, moreover, X is nonsingular and $h^2(X, T_X) = 0$ then Def_X is prorepresented by a formal power series ring.*

Proof. From (2.29) it follows that $S = R$ if $h^0(X, T_X) = 0$ and in particular S is a formally smooth R-algebra. Then the first part follows from Proposition 2.6.2. The last assertion is a consequence of Corollary 2.4.7. □

The condition $h^0(X, T_X) = 0$ (no infinitesimal automorphisms) implies that the group $\mathrm{Aut}(X)$ is finite. We thus see that the existence of a nontrivial automorphism group does not prevent Def_X from being prorepresentable provided there are no infinitesimal automorphisms. On the other hand, the existence of any automorphisms is a source of difficulties when one considers *local* deformations (see Subsection 2.6.2).

The prorepresentability criterion for Def_X given by Proposition 2.6.2 is not easy to apply when $h^0(X, T_X) > 0$. Note that the condition $h^0(X, T_X) = 0$ is not necessary for the prorepresentability of Def_X. An example is given by $X = \mathbb{P}^r$, $r \geq 1$: in this case Def_X is trivially prorepresentable because X is rigid, but $h^0(X, T_X) = (r+1)^2 - 1 > 0$. For another example see the following proposition.

Corollary 2.6.4 can be generalized in a straightforward way to conclude that any functor of Artin rings F classifying isomorphism classes of deformations of a scheme with some additional structure or of any other algebro-geometric object Ξ (a morphism, etc.) is prorepresentable provided F has a semiuniversal element and Ξ has no infinitesimal automorphisms. This remark will be applied in the proof of the following:

Proposition 2.6.5. *Let X be a projective irreducible and nonsingular curve of genus 1. Then* Def_X *is prorepresentable.*

Proof. Fix a closed point $p \in X$. For each A in $\mathrm{ob}(\mathcal{A})$ we define a deformation of the pointed curve (X, p) to be a pair (ξ, σ) where

$$\xi : \begin{array}{ccc} X & \to & \mathcal{X} \\ \downarrow & & \downarrow \pi \\ \mathrm{Spec}(\mathbf{k}) & \to & \mathrm{Spec}(A) \end{array}$$

is an infinitesimal deformation of X over A and $\sigma : \mathrm{Spec}(A) \to \mathcal{X}$ is a section of π such that $\mathrm{Im}(\sigma) = \{p\}$. We have an obvious definition of isomorphism of two deformations of (X, p) over A, and we define a functor of Artin rings

$$\mathrm{Def}_{(X,p)} \to (\mathrm{sets})$$

by

$$\mathrm{Def}_{(X,p)}(A) = \{\text{isom. classes of deformations of } (X, p) \text{ over } A\}$$

We have a morphism of functors:

$$\phi : \mathrm{Def}_{(X,p)} \to \mathrm{Def}_X$$

induced by the correspondence

$$(\xi, \sigma) \mapsto \xi$$

which forgets the section σ. The proposition is a consequence of the following two facts:

(a) ϕ is an isomorphism of functors.
(b) $\mathrm{Def}_{(X,p)}$ is prorepresentable.

To prove (a) let $A \in \mathrm{ob}(\mathcal{A})$ and consider an infinitesimal deformation ξ of X over A. The point p defines a morphism $\mathrm{Spec}(\mathbf{k}) \to \mathcal{X}$ making the following diagram commutative:

$$\begin{array}{ccc} & & \mathcal{X} \\ & p \nearrow & \downarrow \pi \\ \mathrm{Spec}(\mathbf{k}) & \to & \mathrm{Spec}(A) \end{array}$$

By the smoothness of $\mathcal{X}$ over $\mathrm{Spec}(A)$ there is an extension of p to a section σ : $\mathrm{Spec}(A) \to \mathcal{X}$ of π; this proves that $\mathrm{Def}_{(X,p)}(A) \to \mathrm{Def}_X(A)$ is surjective. Now let (ξ, σ) and (η, τ) be two deformations with section of X over A, where

$$\eta : \begin{array}{ccc} X & \to & \mathcal{Y} \\ \downarrow & & \downarrow q \\ \mathrm{Spec}(\mathbf{k}) & \to & \mathrm{Spec}(A) \end{array}$$

and suppose that there is an isomorphism of deformations

$$\begin{array}{ccccc} & & X & & \\ & \swarrow & & \searrow & \\ \mathcal{X} & & \xrightarrow{\psi} & & \mathcal{Y} \\ & \searrow & & \swarrow & \\ & & \mathrm{Spec}(A) & & \end{array}$$

Then $\psi\sigma, \tau : \mathrm{Spec}(A) \to \mathcal{Y}$ are two sections of q. We now use the fact that $\mathcal{Y}$ has a structure of group scheme over $\mathrm{Spec}(A)$ with identity τ (in outline this can be seen as follows: X is a group scheme with identity p and the group structure is given by a multiplication morphism $\mu : X \times X \to X$; the group operation on $\mathcal{Y}$ is defined by a morphism $\mu_A : \mathcal{Y} \times_A \mathcal{Y} \to \mathcal{Y}$ which extends μ and which exists because we have a commutative diagram:

$$\begin{array}{ccccc} X \times X & \xrightarrow{\mu} & X & \subset & \mathcal{Y} \\ \cap & & & & \downarrow \\ \mathcal{Y} \times_A \mathcal{Y} & & \to & & \mathrm{Spec}(A) \end{array}$$

and $\mathcal{Y}$ is smooth over $\mathrm{Spec}(A)$). Replacing ψ by by $\psi' = (\psi\sigma)^{-1}\psi$ we obtain an isomorphism of deformations ψ' such that $\psi'\sigma = \tau$ and therefore ψ' defines an isomorphism of (ξ, σ) with (η, τ); this proves that $\mathrm{Def}_{(X,p)}(A) \to \mathrm{Def}_X(A)$ is injective as well, and (a) is proved.

In particular, it follows that $\mathrm{Def}_{(X,p)}$ has a semiuniversal element because Def_X does. Now observe that the vector space of automorphisms of the trivial deformation of (X, p) can be identified with the vector subspace of $H^0(X, T_X) = H^0(X, \mathcal{O}_X)$ consisting of the derivations $D : \mathcal{O}_X \to \mathcal{O}_X$ vanishing at p, and this is equal to $H^0(X, \mathcal{O}_X(-p)) = (0)$. Now the remark following Corollary 2.6.4 applies to conclude that $\mathrm{Def}_{(X,p)}$ is prorepresentable, i.e. (b) holds. □

The following corollary computes in particular the number of moduli of projective nonsingular curves.

Corollary 2.6.6. *If X is a projective nonsingular connected curve of genus g then* Def_X *is prorepresentable. More precisely,* $\mathrm{Def}_X = h_R$ *where*

$$R = \begin{cases} \mathbf{k} & \textit{if } g = 0 \\ \mathbf{k}[[X]] & \textit{if } g = 1 \\ \mathbf{k}[[X_1, \ldots, X_{3g-3}]] & \textit{if } g \geq 2 \end{cases}$$

.

Proof. X is unobstructed by Proposition 2.4.6 and

$$h^1(X, T_X) = \begin{cases} 0 & \text{if } g = 0 \\ 1 & \text{if } g = 1 \\ 3g - 3 & \text{if } g \geq 2 \end{cases}$$

So it remains to be shown that Def_X is prorepresentable. In the case $g = 0$ this is because $\mathbb{P}^1$ is rigid; in the case $g \geq 2$, since $\deg(T_X) = 2 - 2g < 0$ we have $H^0(X, T_X) = 0$ and therefore 2.6.4 applies. If $g = 1$ we use 2.6.5. □

Recalling Theorems 2.5.13 and 2.5.14 we deduce the following:

Theorem 2.6.7. *Let X be a projective nonsingular connected curve. Then X has an algebraic formally universal deformation.*

Examples 2.6.8. (i) ([154]) Let $C = \mathrm{Spec}(B)$, where $B = \mathbf{k}[x, y]/(xy)$, be a reducible affine plane conic. Then Def_C is not prorepresentable although C has a semiuniversal deformation by Corollary 2.4.2. In fact, consider the deformation of C over $\mathbf{k}[\epsilon]$ given by $xy + \epsilon = 0$ and its automorphism:

$$\begin{array}{ccc} x & \mapsto & x + x\epsilon \\ y & \mapsto & y \end{array}$$

This automorphism does not extend to an automorphism of $xy+\bar{t} = 0$ over $\mathbf{k}[t]/(t^3)$; if it did it would be of the form

$$\begin{array}{ccc} x & \mapsto & x + x\bar{t} + a\bar{t}^2 \\ y & \mapsto & y + b\bar{t}^2 \end{array}$$

for some $a, b \in \mathbf{k}[x, y]$. But this implies that $bx + ay = -1$ in $\mathbf{k}[x, y]$, which is impossible. From Theorem 2.6.1 we deduce that Def_C is not prorepresentable.

This holds more generally for the union of the coordinate axes in $\mathbf{A}^n$, $n \geq 2$ (see [154]).

(ii) The condition of Corollary 2.6.4 is not satisfied by the rational ruled surfaces F_m, $m \geq 0$ (see Example B.11(iii)). Since $h^1(T_{F_0}) = 0 = h^1(T_{F_1})$ we find that F_0 and F_1 are rigid; in particular, Def_{F_0} and Def_{F_1} are prorepresentable. On the other hand, Def_{F_m} is unobstructed when $m \geq 2$ (since $h^2(T_{F_m}) = 0$) and has a semiuniversal element but it is not prorepresentable. To see it we can argue as follows. We can identify F_m with the hypersurface Σ_m of $\mathbb{P}^2 \times \mathbb{P}^1$ of equation

$$x^m v - y^m u = 0$$

where $(u, v, w; x, y)$ are bihomogeneous coordinates in $\mathbb{P}^2 \times \mathbb{P}^1$ (for the elementary proof of this fact see [5], p. 55). For simplicity let's consider the case $m = 2$. The linear pencil

$$\mathcal{V} \subset \mathbb{P}^2 \times \mathbb{P}^1 \times \mathbf{A}^1$$

of equation:

$$x^2 v - y^2 u - txyw = 0$$

defines a flat family $\mathcal{V} \to \mathbf{A}^1$ such that $\mathcal{V}(0) = \Sigma_2$ and $\mathcal{V}(t) \cong \Sigma_0$ for all $t \neq 0$. In fact, we have an isomorphism $\mathcal{V}\backslash\mathcal{V}(0) \to \mathcal{V}(1) \times \mathbf{A}^1\backslash\{0\}$ over $\mathbf{A}^1\backslash\{0\}$ sending

$$(x, y; u, v, w; t) \mapsto (x, y; u, v, tw; t);$$

on the other hand, $\Sigma_0 \cong \mathcal{V}(1)$ by the isomorphism

$$(x, y; u, v, w) \mapsto (x, y; -x^2uw - xyw^2, xyu^2 + y^2vw, x^2u^2 + 2xyuw + y^2w^2)$$

($\mathcal{V}$ is essentially the family considered in Example 1.2.11(iii) for $m = 2$). The pullback

$$\mathcal{V}_\epsilon = V(x^2 v - y^2 u - \epsilon xyw) \subset I\!P^2 \times I\!P^1 \times \mathrm{Spec}(\mathbf{k}[\epsilon])$$

has the automorphism defined by sending $w \mapsto w + \epsilon u$ and leaving all the other coordinates unchanged. We leave the reader to check that this automorphism does not extend to the pullback of $\mathcal{V}$ over $\mathrm{Spec}(\mathbf{k}[t]/(t^3))$. From Theorem 2.6.1 we deduce that Def_{F_2} is not prorepresentable.

2.6.2 Isotriviality

The notions of triviality and of formal triviality of an algebraic deformation are related in a quite subtle way, as shown by Example 1.2.11(ii). This example is a special case of an important phenomenon, called *isotriviality*, considered for the first time in the literature in [165].

Definition 2.6.9. *Let $\pi : \mathcal{X} \to S$ be a flat family of schemes. Then:*

(i) π is called isotrivial *if there is an etale cover (i.e. a finite surjective etale morphism) $f : S' \to S$ such that the family $\pi_{S'} : S' \times_S \mathcal{X} \to S'$ is trivial. If $S' \times_S \mathcal{X} \cong X \times S'$ we say that π is* isotrivial with fibre X.

(ii) If $s \in S$ is a **k***-rational point then π is called* locally isotrivial at s *if there is an etale neighbourhood $f : (S', s') \to (S, s)$ such that the pullback $\pi_{S'} : S' \times_S \mathcal{X} \to S'$ of π is trivial. π is called* locally isotrivial *if it is locally isotrivial at every* **k***-rational point of S.*

Every trivial family is isotrivial. A rational ruled surface $\pi : F_m \to I\!P^1$ with $m \geq 2$ is locally isotrivial because locally around each point of $I\!P^1$ it is the product family with fibre $I\!P^1$; on the other hand, π is not isotrivial because it is not trivial and the identity $I\!P^1 \to I\!P^1$ is the only connected etale cover of $I\!P^1$. In particular, the trivial family $I\!P^1 \times I\!P^1 \to I\!P^1$ cannot be obtained by pulling back π by an etale cover of $I\!P^1$. Therefore the notions of isotriviality and of local isotriviality are different.

If $\pi : \mathcal{X} \to S$ is isotrivial then all the fibres over the **k**-rational points are isomorphic. The next proposition considers the opposite implication.

Proposition 2.6.10. *Let $\pi : \mathcal{X} \to S$ be a flat family of algebraic schemes, and let $s \in S$ be a closed point. If π is locally isotrivial at s then the formal deformation of $\mathcal{X}(s)$ associated to π is trivial. If the morphism π is projective then the converse is also true.*

Proof. The first part follows immediately from Lemma 2.5.6. Conversely, let $X = \mathcal{X}(s)$ and $A = \mathcal{O}_{S,s}$, and assume that π is projective and that the formal deformation $(\hat{A}, \hat{\eta})$ of X associated to the family π is trivial. Let $(\hat{A}, \bar{\eta})$ be the deformation of X over $\mathrm{Spec}(\hat{A})$ induced by π under the morphism $\mathrm{Spec}(\hat{A}) \to S$. By Theorem 2.5.11 this deformation is uniquely determined by the formal deformation $(\hat{A}, \hat{\eta})$ and therefore it is trivial.

Extend Def_X to a functor F locally of finite presentation defined on (**k**-algebras). We may assume that $S = \mathrm{Spec}(B)$ is affine. We have a commutative diagram of functors defined on the category (**k**-algebras):

$$\begin{array}{ccccc} & & h_{\hat{A}} & & \\ & \swarrow & & \searrow & \\ h_B & & & & h_B \\ & \searrow & & \swarrow & \\ & & F & & \end{array}$$

The lower arrows are induced by the two algebraic deformations $\pi : \mathcal{X} \to S$ and $p : X \times S \to S$ of X (by Yoneda's lemma), while the upper arrows are the same and correspond to the natural homomorphism $B \to \hat{A}$. There remains induced a morphism of functors

$$h_{\hat{A}} \to h_B \times_F h_B$$

which corresponds to an effective formal element for the functor $h_B \times_F h_B$. Since F is locally of finite presentation and B is of finite type the functor $h_B \times_F h_B$ is locally of finite presentation as well (a fibred product of functors locally of finite presentation is again locally of finite presentation: see [12], p. 33) so that we can apply the approximation theorem 2.5.22. Therefore we can find an etale neighbourhood $\mathrm{Spec}(B') \to \mathrm{Spec}(B)$ such that both deformations π and p pull back to the same deformation over $\mathrm{Spec}(B')$; in particular, π pulls back to a trivial deformation over $\mathrm{Spec}(B')$. □

The inverse implication of Proposition 2.6.10 is false in general: families of nonsingular affine schemes are formally trivial but not isotrivial in general, as shown by Example 1.2.5, page 26. The following is immediate.

Corollary 2.6.11. *If $\pi : \mathcal{X} \to S$ is a smooth projective family of algebraic schemes which is locally isotrivial at a* **k***-rational point $s \in S$ then the Kodaira–Spencer map*

$$\kappa_{\pi,s} : T_s S \to H^1(\mathcal{X}(s), T_{\mathcal{X}(s)})$$

is 0.

Example 2.6.12. The converse of Corollary 2.6.11 is false. For example, consider a smooth projective family $\pi : \mathcal{X} \to \mathrm{Spec}(\mathbf{k}[t])$ such that the induced deformations over $\mathbf{k}[t]/(t^n)$ are nontrivial for $n \gg 0$; in particular, π is not locally isotrivial at the point $t = 0$. Let $q^\sharp : \mathrm{Spec}(\mathbf{k}[u]) \to \mathrm{Spec}(\mathbf{k}[t])$ be the morphism defined by $q : \mathbf{k}[t] \to \mathbf{k}[u]$ sending $t \mapsto u^2$. Then $q^\sharp$ has zero differential at 0 and therefore the pulled back family

$$\mathcal{X} \times_{\mathrm{Spec}(\mathbf{k}[t])} \mathrm{Spec}(\mathbf{k}[u]) \to \mathrm{Spec}(\mathbf{k}[u])$$

has a vanishing Kodaira–Spencer map at 0. But the restriction of this family to $\mathbf{k}[u]/(u^n)$ is nontrivial for $n \gg 0$ and therefore the family is not locally isotrivial at s by Proposition 2.6.10.

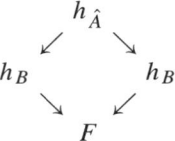

The existence of nontrivial isotrivial deformations of a scheme X is closely related to the existence of nontrivial automorphisms of X. Before investigating this fact we give some examples.

Examples 2.6.13. (i) Let X be a quasi-projective scheme such that there is a finite nontrivial subgroup $G \subset \mathrm{Aut}(X)$. Let S' be a quasi-projective scheme on which G acts freely and let $S := S'/G$ be the quotient. Then G acts on $X \times S'$ componentwise and the action is easily seen to be free. The quotient $\mathcal{X} := (X \times S')/G$ exists and is an algebraic scheme (see [166] or [3], exp. V). Since the projection $X \times S' \to S'$ is G-equivariant it induces a morphism

$$\pi : \mathcal{X} \to S$$

and we have a commutative diagram:

$$\begin{array}{ccc} X \times S' & \rightarrow & \mathcal{X} \\ \downarrow & & \downarrow \pi \\ S' & \rightarrow & S \end{array} \tag{2.30}$$

where the horizontal arrows are etale morphisms and all the fibres of π over the closed points are isomorphic to X. Moreover, (2.30) is a cartesian diagram (there is an S'-morphism $X \times S' \to S' \times_S \mathcal{X}$ which is easily seen to be an isomorphism) and π is flat (use A.5).

Claim: The family π is not trivial.

In fact, since (2.30) is cartesian, if π were trivial the action of G on $X \times S'$ would be trivial on the first factor, a contradiction.

It follows that π is an isotrivial nontrivial family.

(ii) Let G be a nontrivial finite group scheme and Z a quasi-projective scheme on which G acts freely. Then the quotient scheme Z/G exists (see [166] or [3], exp. V). Let $\pi : Z \to Z/G$ be the canonical morphism. Then π is an etale cover; in particular, it is flat. Moreover, we have a commutative diagram

$$\begin{array}{ccc} Z \times_{Z/G} Z & \rightarrow & Z \\ \downarrow \pi' & & \downarrow \pi \\ Z & \rightarrow & Z/G \end{array}$$

and since the action is free we have an isomorphism $G \times Z \to Z \times_{Z/G} Z$ induced by the map $(g, z) \mapsto (z, gz)$. Therefore π' is the trivial family and π is isotrivial. If Z is integral then π has no sections and it follows that π is not trivial. The morphism π is called a *principal G-bundle*.

(iii) Perhaps the simplest examples of isotrivial nontrivial families are those described in Example 1.2.11(ii), page 31. We leave to the reader to verify that they are isotrivial.

We need the following well-known lemma ([165], n. 1.5).

Lemma 2.6.14. *Let* $f : Z \to Y$ *be an etale cover of algebraic schemes. Then there is an etale cover* $\varphi : P \to Z$ *such that the composition*

$$f\varphi : P \to Z \to Y$$

is a principal G*-bundle with respect to a finite group* G*. In particular,* f *is isotrivial and, if* Z *is integral and* $\deg(f) > 1$*, it is nontrivial.*

Proof. Let $n = \deg(f)$. In the n-fold fibre product $Z \times_Y Z \times_Y \cdots \times_Y Z$ consider the set P of points $(z_1, \ldots, z_n)$ such that $z_i \neq z_j$ for all $i \neq j$. Then P is a union of connected components of $Z \times_Y Z \times_Y \cdots \times_Y Z$ which is stable under the natural action of the symmetric group S_n. The natural morphism $\phi : P \to Y$ is an etale cover of degree $n!$ and therefore it induces an isomorphism $P/S_n \cong Y$. Therefore ϕ is a principal S_n-bundle. Moreover, the first projection $\varphi : P \to Z$ is etale of degree $(n-1)!$ and satisfies $f\varphi = \phi$.

In order to prove the last assertion recall that by Example 2.6.13(ii) ϕ is isotrivial. More precisely, we have a commutative diagram:

$$\begin{array}{ccccc} P \times_Y P \cong & S_n \times P & \to & P \\ & \downarrow & & \downarrow \phi \\ & P & \to & Y \end{array}$$

The left vertical morphism is the projection and, because of the factorization $f\varphi = \phi$, it factors as

$$S_n \times P \to P \times_Y Z \to P$$

It follows that $P \times_Y Z \to P$ is trivial as well and this implies that f is isotrivial. If Z is integral and $\deg(f) > 1$ then f cannot be trivial because it has no sections. □

Theorem 2.6.15. *The following conditions are equivalent on a quasi-projective scheme* X*:*

(a) There exists a nontrivial isotrivial algebraic family with fibre X*.*
(b) $\mathrm{Aut}(X)$ *contains a nontrivial finite subgroup.*

Proof. (b) ⇒ (a) has been proved in Example 2.6.13(i).

(a) ⇒ (b). Let $\pi : \mathcal{X} \to S$ be a family as in (a). By hypothesis there is an etale cover $\phi : P \to S$ such that $\pi_P : P \times_S \mathcal{X} \to P$ is trivial with fibre X; let's identify $X \times P = P \times_S \mathcal{X}$ and let $\psi : X \times P \to \mathcal{X}$ be the projection:

$$\begin{array}{ccc} X \times P & \xrightarrow{\psi} & \mathcal{X} \\ \downarrow & & \downarrow \pi \\ P & \xrightarrow{\phi} & S \end{array}$$

For each $p \in P$ denote by $\psi_p : X \to \mathcal{X}$ the morphism defined by

$$\psi_p(x) = \psi(x, p)$$

By Lemma 2.6.14 we may assume that ϕ is a principal G-bundle with respect to some finite group G. We define an action of G on $X \times P$ by the following rule:

$$g(x, p) = (\psi_{gp}^{-1}\psi(x, p), gp)$$

$(X \times P)/G$ exists and, since the action is clearly free and transitive on the fibres of ψ, there is an induced morphism $(X \times P)/G \to \mathcal{X}$ which is an etale cover of degree one, i.e. it is an isomorphism. Fix $p \in P$ and define

$$G \times X \to X$$

by $gx = \psi_{gp}^{-1}\psi(x, p)$. Then one checks immediately that this is an action of G on X. Since π is nontrivial it is quite clear that this action cannot be trivial for all $p \in P$. Therefore for some $p \in P$ the above action defines a homomorphism $G \to \mathrm{Aut}(X)$ whose image is $\neq \{1\}$. □

If a scheme X has an isotrivial local deformation η which is nontrivial then the local moduli functor

$$\mathrm{Def}_X : \mathcal{A}^* \to (\text{sets})$$

considered in the Introduction cannot be representable, i.e. the local deformation υ considered there cannot exist. Assume by contradiction that υ exists; then, since η is nontrivial it must be pulled back from υ by a nonconstant morphism g : $\mathrm{Spec}(A) \to \mathrm{Spec}(\mathcal{O})$. On the other hand, since η is isotrivial its pullback to $\mathrm{Spec}(\tilde{A})$ is trivial and is therefore obtained by pulling back υ in two different ways: by the constant morphism and by the composition

$$\mathrm{Spec}(\tilde{A}) \xrightarrow{\varphi} \mathrm{Spec}(A) \xrightarrow{g} \mathcal{O}$$

which is nonconstant because φ is faithfully flat hence surjective; this contradicts the universality of υ.

These remarks explain why we cannot expect to be able to construct families representing functors defined on $\mathcal{A}^*$ or on (**k**-algebras) or on (schemes), which classify isomorphism classes of schemes having nontrivial isotrivial deformations; and the existence of such deformations is closely related to the existence of nontrivial automorphisms of such schemes, as we saw in Theorem 2.6.15.

This discussion suggests that while the consideration of isomorphism classes of deformations is not a drawback when one is studying infinitesimal deformations, it becomes inadequate for the classification of algebraic deformations and for global moduli problems. In other words, because of the presence of nontrivial automorphisms we cannot in general expect to find a scheme structure on the set M of isomorphism classes of objects we want to classify in such a way that it reflects faithfully the functorial properties of families. For example, in the case of projective nonsingular curves of genus 0 one should have $M = \mathrm{Spec}(\mathbf{k})$ and the only candidate for being the universal family is $\mathbb{P}^1 \to \mathrm{Spec}(\mathbf{k})$ because there is only one isomorphism class of such curves; but the families $F_m \to \mathbb{P}^1$ (Example B.11(iii)) cannot be pulled back from it, thus a universal family cannot exist in this case.

That's why it would be more natural, instead of taking isomorphism classes of deformations, to consider all families together and analyse them and their isomorphisms. This will result in a more general structure, called a *stack*, which contains all the information about families and deformations of the objects of the set M we want to classify. We refer to [43], [115] and [185] for foundational material about stacks, and to [18], [46], [53], [65] for expository treatments.

3

Examples of deformation functors

This chapter is devoted to the study of the most important functors of Artin rings which arise when one considers deformations of various algebro-geometric objects. We will verify Schlessinger's conditions and we will describe first-order deformations, i.e. the tangent spaces of these functors, and obstruction spaces. We will focus on several examples and applications. It will emerge from the treatment that there is a pattern common to almost all examples: the tangent space and the obstruction space of a given functor will be respectively isomorphic to H^i and to H^{i+1} of a certain sheaf which depends on the functor. It will be $i = 0$ if the deformation problem has no automorphisms, while $i = 1$ if there are automorphisms; in this case the H^0 will classify the infinitesimal automorphisms.

3.1 Affine schemes

In this section we study the deformation functor of an affine scheme. We already know that such a functor verifies Schlessinger's conditions H_0, $\bar{H}$, H_ϵ and we computed its tangent space (Proposition 2.4.1). In particular, we proved that it has a semiuniversal element if the scheme has isolated singularities (Corollary 2.4.2). Here we will analyse this case in more detail. We will start by recalling the description of the tangent space.

3.1.1 First-order deformations

Let B_0 be a $\mathbf{k}$-algebra, and let $X_0 = \mathrm{Spec}(B_0)$. We continue the study of infinitesimal deformations of X_0, equivalently of B_0, started in Section 1.2 in the nonsingular case. Let's recall the following fact (see 2.4.1(iii)).

Proposition 3.1.1. *There is a natural isomorphism*

$$\mathrm{Def}_{B_0}(\mathbf{k}[\epsilon]) \cong T^1_{B_0}$$

where the class of trivial deformations corresponds to $0 \in T^1_{B_0}$. *If* B_0 *is reduced then*

$$\mathrm{Def}_{B_0}(\mathbf{k}[\epsilon]) \cong \mathrm{Ext}^1_{\mathbf{k}}(\Omega_{B_0/\mathbf{k}}, B_0)$$

Proof. A first-order deformation of B_0 consists of a flat $\mathbf{k}[\epsilon]$-algebra B, plus a $\mathbf{k}$-isomorphism $B \otimes_{\mathbf{k}[\epsilon]} \mathbf{k} \cong B_0$. This set of data determines a $\mathbf{k}$-extension:

$$\begin{array}{ccccccccc} 0 & \to & B_0 & \xrightarrow{j} & B & \to & B_0 & \to & 0 \\ & & \| & & & & & & \\ & & \epsilon B & & & & & & \end{array} \tag{3.1}$$

obtained after tensoring the exact sequence

$$0 \to (\epsilon) \to \mathbf{k}[\epsilon] \to \mathbf{k} \to 0$$

by $\otimes_{\mathbf{k}[\epsilon]} B$. Isomorphic deformations give rise to isomorphic extensions. Conversely, given a $\mathbf{k}$-extension (3.1), B has a structure of $\mathbf{k}[\epsilon]$-algebra given by

$$\epsilon \mapsto j(1)$$

B is $\mathbf{k}[\epsilon]$-flat by Lemma A.9. If B_0 is reduced then

$$T^1_{B_0} = \mathrm{Ext}^1_{\mathbf{k}}(\Omega_{B_0/\mathbf{k}}, B_0)$$

by Corollary 1.1.11. □

The following is an immediate consequence of 3.1.1, 2.2.8 and Schlessinger's theorem (2.3.2).

Corollary 3.1.2. *If* $\dim_{\mathbf{k}}(T^1_{B_0}) < \infty$ *then* Def_{B_0} *has a semiuniversal formal element. This happens in particular if* $X_0 = \mathrm{Spec}(B_0)$ *has isolated singularities. If* $T^1_{B_0} = 0$ *then* B_0 *is rigid.*

We will give some indications for the practical computation of $T^1_{B_0}$ when B_0 is e.f.t..

Let $B_0 = P/J$, where P is a smooth $\mathbf{k}$-algebra of the form

$$P = \Delta^{-1}\mathbf{k}[X_1, \ldots, X_d]$$

for some multiplicative system $\Delta \subset \mathbf{k}[X_1, \ldots, X_d]$, and $J \subset P$ is an ideal.

Consider the exact sequence

$$0 \to \mathrm{Hom}(\Omega_{B_0/\mathbf{k}}, B_0) \to \mathrm{Hom}(\Omega_{P/\mathbf{k}} \otimes B_0, B_0) \xrightarrow{\delta^\vee} \mathrm{Hom}(J/J^2, B_0) \longrightarrow T^1_{B_0} \to 0$$

The module

$$\mathrm{Hom}(\Omega_{P/\mathbf{k}} \otimes B_0, B_0) = \mathrm{Der}_{\mathbf{k}}(P, B_0)$$

consists of all derivations D of the form

$$D(p) = \sum_{j=1}^{d} b_j \frac{\partial p}{\partial X_j}$$

for given $b_j \in B_0$, and

$$\delta^\vee(D)(\bar{f}) = D(f) = \sum_{j=1}^{d} b_j \frac{\partial f}{\partial X_j} \qquad f \in J$$

Assume that $J = (f_1, \ldots, f_n)$ and let

$$0 \to \mathbf{R} \xrightarrow{\imath} P^n \xrightarrow{j} J \to 0$$

be the corresponding presentation. We have the exact sequence:

$$0 \to \mathrm{Hom}(J/J^2, B_0) \xrightarrow{j^\vee} \mathrm{Hom}(B_0^n, B_0) \xrightarrow{\imath^\vee} \mathrm{Hom}(\mathbf{R}, B_0)$$

where $j^\vee$ identifies $\mathrm{Hom}(J/J^2, B_0)$ with the submodule of $\mathrm{Hom}(B_0^n, B_0)$ consisting of those homomorphisms which are 0 on $\mathbf{R}$. Identifying $\mathrm{Hom}(B_0^n, B_0) = B_0^n$, thereby viewing its elements as column vectors, we see that the condition for

$$\underline{q} = \begin{pmatrix} q_1 \\ \vdots \\ q_n \end{pmatrix} \in B_0^n$$

to be in $\mathrm{Hom}(J/J^2, B_0)$ is that ${}^t\underline{q} \cdot \underline{r} = 0$ for each $\underline{r} \in \mathbf{R}$ (where we are viewing $\mathbf{R}$ as consisting of column vectors as well). $j^\vee$ associates to a homomorphism

$$\begin{array}{rccc} \varphi : & J/J^2 & \to & B_0 \\ & \sum b_j \bar{f}_j & \mapsto & \sum b_j \varphi(\bar{f}_j) \end{array}$$

the column vector

$$\begin{pmatrix} \varphi(\bar{f}_1) \\ \vdots \\ \varphi(\bar{f}_n) \end{pmatrix}$$

Therefore $\mathrm{Im}(\delta^\vee) \subset B_0^n$ is generated by the column vectors corresponding to $\delta^\vee(\frac{\partial}{\partial X_1}), \ldots, \delta^\vee(\frac{\partial}{\partial X_d})$, i.e. by the classes mod J of:

$$\begin{pmatrix} \frac{\partial f_1}{\partial X_1} \\ \vdots \\ \frac{\partial f_n}{\partial X_1} \end{pmatrix}, \ldots, \begin{pmatrix} \frac{\partial f_1}{\partial X_d} \\ \vdots \\ \frac{\partial f_n}{\partial X_d} \end{pmatrix}$$

If $J = (f_1, \ldots, f_n)$ is generated by a regular sequence then $\iota^\vee = 0$, equivalently $j^\vee$ is an isomorphism, and it follows that

$$T^1_{B_0} \cong \frac{P^n}{\left(\begin{pmatrix} \frac{\partial f_1}{\partial X_1} \\ \vdots \\ \frac{\partial f_n}{\partial X_1} \end{pmatrix}, \ldots, \begin{pmatrix} \frac{\partial f_1}{\partial X_d} \\ \vdots \\ \frac{\partial f_n}{\partial X_d} \end{pmatrix}\right)} \otimes_P B_0$$

In particular, if $B_0 = P/(f)$ then

$$T^1_{B_0} \cong \frac{P}{\left(f, \frac{\partial f}{\partial X_1}, \ldots, \frac{\partial f}{\partial X_d}\right)}$$

It follows from this description that the hypersurface $V(f)$ is rigid if and only if it is nonsingular. A similar remark holds for complete intersections. In particular, recalling Definition D.1, we can state the following, for future reference:

Proposition 3.1.3. *An e.f.t. complete intersection ring B_0 such that* Spec(B_0) *is singular is not rigid.*

Example 3.1.4. Let P be the local ring of a nonsingular algebraic surface X at a $\mathbf{k}$-rational point p, $m = (x, y)$ its maximal ideal, and $B_0 = P/(f)$ the local ring of a curve $C \subset X$ at p. Let's compute $T^1_{B_0}$ in some cases.

(a) *Node (ordinary double point).* By definition $\hat{B}_0 \cong \mathbf{k}[[X, Y]]/(X^2 + Y^2)$. Then $f = x^2 + y^2 +$ higher-order terms, and

$$T^1_{B_0} \cong \frac{B_0}{(x, y)} \cong \mathbf{k}$$

if char$(\mathbf{k}) \neq 2$.

(b) *Ordinary cusp.* In this case $\hat{B}_0 \cong \mathbf{k}[[X, Y]]/(X^2 + Y^3)$. Then $f = x^2 + y^3 +$ higher-order terms, and

$$T^1_{B_0} \cong \frac{B_0}{(y^2, x)} \cong \mathbf{k}^2$$

if char$(\mathbf{k}) \neq 2, 3$.

(c) *Tacnode.* We have in this case $\hat{B}_0 \cong \mathbf{k}[[X, Y]]/(Y(Y + X^2))$ and

$$T^1_{B_0} \cong \frac{B_0}{(x^2 + 2y, xy)} \cong \mathbf{k}^3$$

if char$(\mathbf{k}) \neq 2$.

Conversely, we have the following result:

Proposition 3.1.5. *Assume char*$(\mathbf{k}) = 0$. *Let P be the local ring of a nonsingular algebraic surface X at a $\mathbf{k}$-rational point p, $m = (x, y)$ its maximal ideal, and $B_0 = P/(f)$ the local ring of a curve $C \subset X$ at p; let $t = \dim_{\mathbf{k}} T^1_{B_0}$. Then*

(a) $t = 0$ *if and only if* B_0 *is regular (a DVR).*
(b) $t = 1$ *if and only if* B_0 *is the local ring of a node.*
(c) $t = 2$ *if and only if* B_0 *is the local ring of an ordinary cusp.*
(d) $t = 3$ *if and only if* B_0 *is the local ring of a tacnode.*

Proof. The "if" implication follows from the above computations. We have

$$t = \dim_{\mathbf{k}} P/(f, f_x, f_y)$$

$f \in m_P^3$ immediately implies $t \geq 4$; then $f \in m_P^2$ and, after suitable choice of generators of m_P we may suppose $f = y^2 + x^n +$ higher-order terms, $n \geq 2$ or $f = y(y + x^n) +$ higher-order terms, $n \geq 2$. Now the conclusion follows easily. □

Example 3.1.6. *(The affine cone over* $\mathbb{P}^1 \times \mathbb{P}^2$*).* Let

$$P = \mathbf{k}[X_0, X_1, X_2, Y_0, Y_1, Y_2]$$

$$J = (X_1Y_2 - X_2Y_1, X_2Y_0 - X_0Y_2, X_0Y_1 - X_1Y_0)$$

Then $B_0 = P/J$ is the coordinate ring of the affine cone over the Segre embedding $\mathbb{P}^1 \times \mathbb{P}^2 \subset \mathbb{P}^5$. We have the following presentation:

$$0 \to P^2 \xrightarrow{A} P^3 \to J \to 0$$

where

$$A = \begin{pmatrix} X_0 & Y_0 \\ X_1 & Y_1 \\ X_2 & Y_2 \end{pmatrix}$$

A direct computation shows that $\mathrm{Hom}(J/J^2, B_0)$ is generated by the following column vectors:

$$\begin{matrix} Y_1 & Y_2 & 0 & X_1 & X_2 & 0 \\ -Y_0 & 0 & Y_2 & -X_0 & 0 & X_2 \\ 0 & -Y_0 & -Y_1 & 0 & -X_0 & -X_1 \end{matrix}$$

Since these vectors are, up to permutation,

$$\delta^\vee\left(\frac{\partial}{\partial X_0}\right), \delta^\vee\left(\frac{\partial}{\partial X_1}\right), \delta^\vee\left(\frac{\partial}{\partial X_2}\right), \delta^\vee\left(\frac{\partial}{\partial Y_0}\right), \delta^\vee\left(\frac{\partial}{\partial Y_1}\right), \delta^\vee\left(\frac{\partial}{\partial Y_2}\right)$$

we see that $T^1_{B_0} = 0$. This implies that B_0 is rigid (see Corollary 3.1.2).

More generally, one can prove that *the coordinate ring of the affine cone over the Segre embedding* $\mathbb{P}^n \times \mathbb{P}^m \subset \mathbb{P}^{(n+1)(m+1)-1}$ *is rigid whenever* $n + m \geq 3$. This has been computed for the first time in [67] in the case $n = m \geq 2$; the general case is in [157] (see Corollary 3.1.20 below).

Example 3.1.7. Let $P = \mathbf{k}[X_1, X_2, X_3]_{(\underline{X})}$, $J = (X_2X_3, X_1X_3, X_1X_2)$. Then $B_0 = P/J$ is the local ring at the origin of the union of the coordinate axes in $\mathbf{A}^3$. We have the presentation

$$P^3 \xrightarrow{A} P^3 \to J \to 0$$

where

$$A = \begin{pmatrix} X_1 & X_1 & 0 \\ -X_2 & 0 & X_2 \\ 0 & -X_3 & -X_3 \end{pmatrix}$$

and the columns of A generate $\mathbf{R}$.

$\mathrm{Hom}(J/J^2, B_0)$ is generated by the following column vectors mod J:

$$\begin{matrix} X_2 & X_3 & 0 & 0 & 0 & 0 \\ 0 & 0 & X_1 & X_3 & 0 & 0 \\ 0 & 0 & 0 & 0 & X_1 & X_2 \end{matrix}$$

and $\mathrm{Im}(\delta^\vee)$ is generated by the column vectors mod J:

$$\begin{matrix} 0 & X_3 & X_2 \\ X_3 & 0 & X_1 \\ X_2 & X_1 & 0 \end{matrix}$$

It follows at once that

$$T^1_{B_0} = \mathrm{Hom}(J/J^2, B_0)/\mathrm{Im}(\delta^\vee) \cong \mathbf{k}^3$$

because there are three generators of $\mathrm{Hom}(J/J^2, B_0)$ which are linearly independent modulo the generators of $\mathrm{Im}(\delta^\vee)$, and all other elements of $\mathrm{Hom}(J/J^2, B_0)$ are in $\mathrm{Im}(\delta^\vee)$.

In a similar vein one can consider, for any $d \geq 3$

$$B_0 = \mathbf{k}[X_1, \ldots, X_d]_{(\underline{X})}/J$$

where

$$J = (\ldots, X_1 X_2 \cdot \hat{X}_i \cdot X_d, \ldots)_{i=1,\ldots,d}$$

Then B_0 is the local ring at the origin of the union of the coordinate axes in $\mathbf{A}^d$. One computes easily, along the same lines of the case $d = 3$, that

$$T^1_{B_0} \cong \mathbf{k}^{d(d-2)}$$

Example 3.1.8. Let P be the local ring of a nonsingular algebraic variety V of dimension $n \geq 2$ at a $\mathbf{k}$-rational point p and let $B_0 = P/(f)$ be the local ring of a hypersurface $X \subset V$ at p. We call p a *node* for X if

$$\hat{B}_0 \cong \mathbf{k}[[X_1, \ldots, X_n]]/(X_1^2 + \cdots + X_n^2)$$

equivalently, if we can choose generators $x_1, \ldots, x_n$ of the maximal ideal m_P so that $f = \sum x_i^2 +$ higher-order terms. It can be proved immediately that if p is a node for X and $\mathrm{char}(\mathbf{k}) \neq 2$ then $T^1_{B_0} \cong \mathbf{k}$. In particular, $T^1_{\mathbf{k}[\epsilon]} \cong \mathbf{k}$.

3.1.2 The second cotangent module and obstructions

We will now describe an obstruction space for the functor Def_{B_0}.

Assume that $B_0 = P/J$ for a smooth $\mathbf{k}$-algebra P, and an ideal $J \subset P$. Consider a presentation:

$$\eta: \quad 0 \to \mathbf{R} \xrightarrow{\imath} F \xrightarrow{j} J \to 0 \tag{3.2}$$

where F is a finitely generated free P-module. Let $\lambda : \bigwedge^2 F \to F$ be defined by:

$$\lambda(x \wedge y) = (jx)y - (jy)x$$

and $\mathbf{R}^{tr} = \mathrm{Im}(\lambda)$. Obviously $\mathbf{R}^{tr} \subset \mathbf{R}$ and $\mathbf{R}^{tr} \subset JF$.

If $J = (f_1, \ldots, f_n)$ then $F = P^n$ and $\mathbf{R}$ is the module of relations among $f_1, \ldots, f_n$. $\mathbf{R}^{tr}$ is called the module of *trivial (or Koszul) relations*; it is generated by the relations of the form

$$(0, \quad \ldots, \quad \underset{i}{-f_j}, \quad \ldots, \quad \underset{j}{f_i}, \quad \ldots, \quad 0)$$

It follows that

$$\mathbf{R}/\mathbf{R}^{tr} = H_1(K_\bullet(f_1, \ldots, f_n))$$

the first homology module of the Koszul complex associated to $f_1, \ldots, f_n$.

Lemma 3.1.9. *The P-module $\mathbf{R}/\mathbf{R}^{tr}$ is annihilated by J and therefore it is a B_0-module in a natural way.*

Proof. Let $x \in \mathbf{R}$, $a \in J$. Let $y \in F$ be such that $j(y) = a$. Then

$$ax = j(y)x = j(y)x - j(x)y = \lambda(y \wedge x) \in \mathbf{R}^{tr}$$ □

Since $\mathbf{R}^{tr} \subset JF$ the presentation (3.2) induces an exact sequence of B_0-modules:

$$\mathbf{R}/\mathbf{R}^{tr} \xrightarrow{\bar{\imath}} F \otimes_P B_0 \xrightarrow{\bar{j}} J/J^2 \to 0 \tag{3.3}$$

Definition 3.1.10. The second cotangent module of B_0 *is the B_0-module $T^2_{B_0}$ defined by the exact sequence:*

$$\mathrm{Hom}_{B_0}(J/J^2, B_0) \to \mathrm{Hom}_{B_0}(F \otimes_P B_0, B_0) \to \mathrm{Hom}_{B_0}(\mathbf{R}/\mathbf{R}^{tr}, B_0) \to T^2_{B_0} \to 0$$

induced by (3.3). Obviously $T^2_{B_0}$ is a B_0-module of finite type.

Lemma 3.1.11. *For every e.f.t. $\mathbf{k}$-algebra B_0 the B_0-module $T^2_{B_0}$ is independent of the presentation (3.2).*

Proof. Assume that $F \cong P^n$ and that $j : P^n \to J$ is defined by the system of generators $f_1, \ldots, f_n$ of J. Let

$$0 \to \mathbf{S} \to P^m \to J \to 0$$

be another presentation of J, defined by the system of generators $g_1, \ldots, g_m$. We may assume that $m \geq n$ and that $g_k = f_k$, $k = 1, \ldots, n$. Let

$$g_s = \sum_k b_{ks} f_k \qquad s = n+1, \ldots, m$$

for some $b_{ks} \in P$. Denote by $\alpha : P^n \to P^m$ the map

$$\alpha(a_1, \ldots, a_n) = (a_1, \ldots, a_n, 0, \ldots, 0)$$

and by $\beta : P^m \to P^n$ the map

$$\beta(a_1, \ldots, a_m) = \left(a_1 + \sum_{s=n+1}^{m} b_{1s} a_s, \ldots, a_n + \sum_{s=n+1}^{m} b_{ns} a_s \right)$$

Evidently, $\alpha(\mathbf{R}) \subset \mathbf{S}$ and $\alpha(\mathbf{R}^{tr}) \subset \mathbf{S}^{tr}$. It is easy to verify that $\beta(\mathbf{S}) \subset \mathbf{R}$ and $\beta(\mathbf{S}^{tr}) \subset \mathbf{R}^{tr}$. It follows that α and β induce homomorphisms

$$\beta^\star : \mathrm{Hom}(\mathbf{R}/\mathbf{R}^{tr}, B_0) \to \mathrm{Hom}(\mathbf{S}/\mathbf{S}^{tr}, B_0)$$
$$\alpha^\star : \mathrm{Hom}(\mathbf{S}/\mathbf{S}^{tr}, B_0) \to \mathrm{Hom}(\mathbf{R}/\mathbf{R}^{tr}, B_0)$$

whence homomorphisms

$$\tilde{\beta} : \mathrm{Hom}(\mathbf{R}/\mathbf{R}^{tr}, B_0)/\mathrm{Hom}(P^n, B_0) \to \mathrm{Hom}(\mathbf{S}/\mathbf{S}^{tr}, B_0)/\mathrm{Hom}(P^m, B_0)$$
$$\tilde{\alpha} : \mathrm{Hom}(\mathbf{S}/\mathbf{S}^{tr}, B_0)/\mathrm{Hom}(P^m, B_0) \to \mathrm{Hom}(\mathbf{R}/\mathbf{R}^{tr}, B_0)/\mathrm{Hom}(P^n, B_0)$$

Since

$$\alpha^\star \beta^\star = \text{identity of } \mathrm{Hom}(\mathbf{R}/\mathbf{R}^{tr}, B_0)$$

it follows that

$$\tilde{\alpha}\tilde{\beta} = \text{identity of } \mathrm{Hom}(\mathbf{R}/\mathbf{R}^{tr}, B_0)/\mathrm{Hom}(P^n, B_0)$$

We now prove that

$$\tilde{\beta}\tilde{\alpha} = \text{identity of } \mathrm{Hom}(\mathbf{S}/\mathbf{S}^{tr}, B_0)/\mathrm{Hom}(P^m, B_0) \tag{3.4}$$

Let $g \in \mathrm{Hom}(\mathbf{S}/\mathbf{S}^{tr}, B_0)$ be induced by the homomorphism $G : \mathbf{S} \to B_0$. Then, if $(a_1, \ldots, a_m) \in \mathbf{S}$ and $(\bar{a}_1, \ldots, \bar{a}_m) \in \mathbf{S}/\mathbf{S}^{tr}$ is its class, we have:

$$(\beta^\star \alpha^\star)(g)(\bar{a}_1, \ldots, \bar{a}_m) = G(\alpha(\beta(a_1, \ldots, a_m))) =$$

$$G(a_1 + \textstyle\sum_{s=n+1}^{m} b_{1s} a_s, \ldots, a_n + \sum_{s=n+1}^{m} b_{ns} a_s, 0, \ldots, 0) =$$

$$G(a_1, \ldots, a_m) + G(\textstyle\sum_{s=n+1}^{m} b_{1s} a_s, \ldots, \sum_{s=n+1}^{m} b_{ns} a_s, -a_{n+1}, \ldots, -a_m) = (*)$$

Now note that

$$\left(\sum_{s=n+1}^{m} b_{1s}p_s, \ldots, \sum_{s=n+1}^{m} b_{ns}p_s, -p_{n+1}, \ldots, -p_m\right) \in \mathbf{S}$$

for every $(p_1, \ldots, p_m) \in P^m$. Therefore letting

$$\tau(p_1, \ldots, p_m) = G\left(\sum_{s=n+1}^{m} b_{1s}p_s, \ldots, \sum_{s=n+1}^{m} b_{ns}p_s, -p_{n+1}, \ldots, -p_m\right)$$

we define a homomorphism $\tau : P^m \to B_0$. It follows that

$$(*) = g(\bar{a}_1, \ldots, \bar{a}_m) + \tau(a_1, \ldots, a_m)$$

Hence

$$(\beta^\star\alpha^\star)(g) - g \in \mathrm{Im}[\mathrm{Hom}(P^m, B_0) \to \mathrm{Hom}(\mathbf{S}/\mathbf{S}^{tr}, B_0)]$$

or equivalently (3.4) holds. □

From the definition it easily follows that $T^2_{B_0}$ localizes. Namely, for every multiplicative subset $\Delta \subset P$ we have:

$$\Delta^{-1}T^2_{B_0} = T^2_{\Delta^{-1}B_0}$$

It follows that for any scheme X we can define in an obvious way the *second cotangent sheaf* which we will denote by T^2_X. It satisfies

$$T^2_{X,x} = T^2_{\mathcal{O}_{X,x}}$$

Proposition 3.1.12. *Assume that* $B_0 = P/J$ *for a smooth* **k**-*algebra* P. *Then* $T^2_{B_0}$ *is an obstruction space for the functor* Def_{B_0}.

Proof. Let A be an object of $\mathcal{A}$ and let

$$\xi: \begin{array}{ccc} B & \to & B_0 \\ \uparrow & & \uparrow \\ A & \to & \mathbf{k} \end{array}$$

be an infinitesimal deformation of B_0 over A.

We must associate to ξ a **k**-linear map

$$\xi_v : \mathrm{Ex}_{\mathbf{k}}(A, \mathbf{k}) \to T^2_{B_0}$$

satisfying the conditions of Definition 2.2.9. Let $B_0 = P/J$ for a smooth **k**-algebra P and an ideal $J = (f_1, \ldots, f_n) \subset P$. We have an exact sequence:

$$0 \to \mathbf{R} \to P^n \xrightarrow{\mathbf{f}} J \to 0$$

Then, by the smoothness, in particular flatness, of P we have

$$B = (P \otimes_{\mathbf{k}} A)/(F_1, \ldots, F_n)$$

where $f_j = F_j \pmod{m_A}$, $j = 1, \ldots, n$, by Theorem A.10. The flatness of B over A implies that for every $\underline{r} = (r_1, \ldots, r_n) \in \mathbf{R}$ there exist $R_1, \ldots, R_n \in P \otimes_{\mathbf{k}} A$ such that $r_j = R_j$ (mod. m_A), $j = 1, \ldots, n$, and $\sum_j R_j F_j = 0$, again by Theorem A.10.

Let $[\gamma] \in \mathrm{Ex}_{\mathbf{k}}(A, \mathbf{k})$ be represented by an extension

$$\gamma : \quad 0 \to (t) \to \tilde{A} \xrightarrow{\phi} A \to 0$$

Choose $\tilde{F}_1, \ldots, \tilde{F}_n, \tilde{R}_1, \ldots, \tilde{R}_n \in P \otimes_{\mathbf{k}} \tilde{A}$ liftings of $F_1, \ldots, F_n, R_1, \ldots, R_n$ respectively; then

$$\sum_j \tilde{R}_j \tilde{F}_j \in \ker[P \otimes_{\mathbf{k}} \tilde{A} \to P \otimes_{\mathbf{k}} A] \cong P$$

It is easy to check that a different choice of $\tilde{R}_1, \ldots, \tilde{R}_n$ or of $\tilde{F}_1, \ldots, \tilde{F}_n$ modifies $\sum_j \tilde{R}_j \tilde{F}_j$ by an element of J or by one of the form $\sum_j q_j r_j$, where $q_j \in P$, respectively. Therefore sending

$$\underline{r} \mapsto \sum_j \tilde{R}_j \tilde{F}_j \tag{3.5}$$

defines an element of

$$\mathrm{coker}[\mathrm{Hom}(B_0^n, B_0) \to \mathrm{Hom}(\mathbf{R}, B_0)]$$

Moreover, since if $\underline{r} = \underline{r}_{ij} = (0, \ldots, f_j, \ldots, -f_i, \ldots, 0)$ we can take

$$(\tilde{R}_1, \ldots, \tilde{R}_n) = (0, \ldots, \tilde{F}_j, \ldots, -\tilde{F}_i, \ldots, 0)$$

and we get $\sum_j \tilde{R}_j \tilde{F}_j = 0$, it follows that (3.5) is zero on $\mathrm{Hom}(\mathbf{R}^{tr}, B_0)$. Therefore the n-tuple of liftings $(\underline{\tilde{F}}) = (\tilde{F}_1, \ldots, \tilde{F}_n)$ defines an element $\xi_\upsilon(\gamma)$ of

$$\mathrm{coker}[\mathrm{Hom}(B_0^n, B_0) \to \mathrm{Hom}(\mathbf{R}/\mathbf{R}^{tr}, B_0)] = T^2_{B_0}$$

Let's prove that the map $\gamma \mapsto \xi_\upsilon(\gamma)$ is $\mathbf{k}$-linear.

Let $[\zeta] \in \mathrm{Ex}_{\mathbf{k}}(A, \mathbf{k})$ be another element defined by the extension:

$$\zeta : \quad 0 \to (t) \to A' \to A \to 0$$

and let $(\underline{F}') = (F'_1, \ldots, F'_n)$, $F'_j \in P \otimes_{\mathbf{k}} A'$ be the corresponding lifting, which defines $\xi_\upsilon(\zeta)$. Then $\xi_\upsilon(\gamma) + \xi_\upsilon(\zeta)$ is defined by

$$\underline{r} \mapsto \sum_j \tilde{R}_j \tilde{F}_j + \sum_j R'_j F'_j$$

where $R'_1, \ldots, R'_n \in P \otimes_{\mathbf{k}} A'$ are liftings of $R_1, \ldots, R_n$. Consider the diagram:

$$\begin{array}{rccccccc} & 0 \to & \mathbf{k}\oplus\mathbf{k} & \to & \tilde{A}\times_A A' & \to & A & \to 0 \\ & & \downarrow\delta & & \downarrow\sigma & & \| & \\ \gamma+\zeta: & 0\to & \mathbf{k} & \to & C & \to & A & \to 0 \end{array}$$

Then $\xi_\upsilon(\gamma+\zeta)$ is defined by

$$\underline{r} \mapsto \sum_j \Psi_j \Phi_j$$

where $\Psi_1, \ldots, \Psi_n, \Phi_1, \ldots, \Phi_n \in P \otimes_{\mathbf{k}} C$ are liftings of $R_1, \ldots, R_n, F_1, \ldots, F_n$.

Since

$$P \otimes_{\mathbf{k}} (\tilde{A} \times_A A') \cong (P \otimes_{\mathbf{k}} \tilde{A}) \times_A (P \otimes_{\mathbf{k}} A')$$

letting $\rho : P \otimes_{\mathbf{k}} (\tilde{A} \times_A A') \to P \otimes_{\mathbf{k}} C$ be the homomorphism induced by σ, we may assume that

$$\Phi_j = \rho(\tilde{F}_j, F'_j); \quad \Psi_j = \rho(\tilde{R}_j, R'_j)$$

Then:

$$\begin{aligned} \textstyle\sum_j \Psi_j\Phi_j &= \textstyle\sum_j \rho(\tilde{R}_j, R'_j)\rho(\tilde{F}_j, F'_j) = \rho\left[\sum_j (\tilde{R}_j, R'_j)(\tilde{F}_j, F'_j)\right] \\ &= \rho\left(\textstyle\sum_j \tilde{R}_j\tilde{F}_j, \sum_j R'_jF'_j\right) = \delta\left(\sum_j \tilde{R}_j\tilde{F}_j, \sum_j R'_jF'_j\right) \\ &= \textstyle\sum_j \tilde{R}_j\tilde{F}_j + \sum_j R'_jF'_j \end{aligned}$$

This proves that $\xi_\upsilon(\gamma+\zeta) = \xi_\upsilon(\gamma)+\xi_\upsilon(\zeta)$. A similar argument shows that $\xi_\upsilon(\lambda\gamma) = \lambda\xi_\upsilon(\gamma)$, $\lambda \in \mathbf{k}$.

Now assume that $[\gamma] \in \mathrm{Ex}_{\mathbf{k}}(A, \mathbf{k})$ is such that there exists an infinitesimal deformation

$$\tilde{\xi}: \begin{array}{ccc} \tilde{B} & \to & B_0 \\ \uparrow & & \uparrow \\ \tilde{A} & \to & \mathbf{k} \end{array}$$

such that

$$\begin{array}{ccc} \mathrm{Def}_{B_0}(\tilde{A}) & \to & \mathrm{Def}_{B_0}(A) \\ \\ \tilde{\xi} & \mapsto & \xi \end{array}$$

It follows that there exist liftings $\tilde{F}_j \in P \otimes_{\mathbf{k}} \tilde{A}$ of the F_j's such that every $\underline{r} \in \mathbf{R}$ has a lifting $\underline{\tilde{R}} \in (P \otimes_{\mathbf{k}} \tilde{A})^N$ such that $\sum_j \tilde{R}_j\tilde{F}_j = 0$. This means that $\xi_\upsilon(\gamma) = 0$.

Conversely, assume that $\xi_\upsilon(\gamma) = 0$, and let $\tilde{F}_1, \ldots, \tilde{F}_n \in P \otimes_{\mathbf{k}} \tilde{A}$ be arbitrary liftings of $F_1, \ldots, F_n$. Then there exists $(h_1, \ldots, h_n) \in P^n$ such that for every choice of a lifting $\underline{\tilde{R}} \in (P \otimes_{\mathbf{k}} \tilde{A})^n$ of a relation $\underline{r} \in \mathbf{R}$ we have:

$$\sum_j \tilde{R}_j\tilde{F}_j = -t\sum_j r_jh_j = -\sum_j \tilde{R}_jh_j$$

This means that the ideal $(\tilde{F}_1 + th_1, \ldots, \tilde{F}_n + th_n) \subset P \otimes_{\mathbf{k}} \tilde{A}$ defines a flat deformation of B_0 over $\tilde{A}$ lifting the deformation $B = (P \otimes_{\mathbf{k}} A)/(F_1, \ldots, F_n)$.

Any other choice of a lifting of the deformation ξ over $\tilde{A}$ is of the form

$$(\tilde{F}_1 + t(h_1 + k_1), \dots, \tilde{F}_n + t(h_n + k_n))$$

where $\underline{k} = (k_1, \dots, k_n) \in B_0^n$ satisfy $\sum_j r_j k_j = 0$ for every relation $\underline{r} \in \mathbf{R}$. Therefore $\underline{k} \in \operatorname{Hom}(J/J^2, B_0)$. It is straightforward to verify that if $\underline{k} \in \operatorname{Im}(\delta^\vee)$ then $\underline{\tilde{F}} + t\underline{h}$ and $\underline{\tilde{F}} + t(\underline{h} + \underline{k})$ define isomorphic liftings of ξ over $\tilde{A}$. This means that we have an action of $T^1_{B_0}$ on the set of liftings of ξ over $\tilde{A}$. By construction it follows that this action is transitive. □

Corollary 3.1.13. *Let* $X_0 = \operatorname{Spec}(B_0)$ *be an affine algebraic scheme such that* $\dim_{\mathbf{k}}(T^1_{B_0}) < \infty$, *and let* $(R, \{\eta_n\})$ *be a semiuniversal formal deformation of* X_0. *Then*

(i)

$$\dim_{\mathbf{k}}(T^1_{B_0}) \geq \dim(R) \geq \dim_{\mathbf{k}}(T^1_{B_0}) - \dim_{\mathbf{k}}(T^2_{B_0})$$

The first equality holds if and only if X_0 *is unobstructed.*

(ii) If B_0 *is an e.f.t. local complete intersection* **k**-*algebra then it is unobstructed. In particular, hypersurface singularities are unobstructed.*

Proof. (i) follows from Corollary 2.2.11.

(ii) If J is generated by a regular sequence then $\mathbf{R} = \mathbf{R}^{tr}$ and therefore $T^2_{B_0} = (0)$. □

Proposition 3.1.14. *Assume that* $B_0 = P/J$ *for a smooth* **k**-*algebra* P. *Then:*

(i) If $\operatorname{Spec}(B_0)$ *is reduced then*

$$T^2_{B_0} \cong \operatorname{Ext}^1_{B_0}(J/J^2, B_0)$$

(ii) If $\operatorname{Spec}(B_0)$ *is reduced and has depth at least 2 along the locus where it is not an l.c.i. (e.g.* $\operatorname{Spec}(B_0)$ *is normal of dimension* ≥ 2*) there is an isomorphism*

$$T^2_{B_0} \cong \operatorname{Ext}^2(\Omega_{B_0/\mathbf{k}}, B_0)$$

Proof. We keep the notations introduced on page 109.

(i) From the exact sequence (3.3) we deduce the following commutative diagram with exact rows and columns:

$$\begin{array}{ccccccc}
 & & & & \operatorname{Hom}(\ker(\bar{\imath}), B_0) & & \\
 & & & & \uparrow & & \\
\operatorname{Hom}(J/J^2, B_0) & \to & \operatorname{Hom}(F \otimes B_0, B_0) & \xrightarrow{\bar{j}^\vee} & \operatorname{Hom}(\mathbf{R}/\mathbf{R}^{tr}, B_0) & \to T^2 \to 0 \\
\| & & \| & & \cup & \cup \\
\operatorname{Hom}(J/J^2, B_0) & \to & \operatorname{Hom}(F \otimes B_0, B_0) & \to & \operatorname{Hom}(\operatorname{Im}(\bar{\imath}), B_0) & \to E^1 \to 0
\end{array}$$

where $E^1 = \operatorname{Ext}^1(J/J^2, B_0)$ and $T^2 = T^2_{B_0}$. Since the exact sequence

$$\eta : \quad 0 \to \mathbf{R} \xrightarrow{\imath} F \xrightarrow{j} J \to 0$$

from which (3.3) is obtained localizes, we see that $\ker(\bar{\imath})$ is supported on the locus where B_0 is not an l.c.i.; in particular $\ker(\bar{\imath})$ is torsion. Therefore we have $\mathrm{Hom}(\ker(\bar{\imath}), B_0) = (0)$ and the conclusion follows.

(ii) Consider the conormal sequence

$$J/J^2 \xrightarrow{\delta} \Omega_{P/\mathbf{k}} \otimes B_0 \to \Omega_{B_0/\mathbf{k}} \to 0$$

Since $\mathrm{Spec}(B_0)$ is reduced, $\ker(\delta)$ is supported in the locus where $\mathrm{Spec}(B_0) \subset \mathrm{Spec}(P)$ is not a regular embedding (Proposition D.1.4), and this locus coincides with the locus where $\mathrm{Spec}(B_0)$ is not an l.c.i. (Proposition D.2.5). From the assumption about the depth of $\mathrm{Spec}(B_0)$ it follows that

$$\mathrm{Hom}(\ker(\delta), B_0) = \mathrm{Ext}^1(\ker(\delta), B_0) = (0)$$

Using this fact and recalling that $\mathrm{Ext}^i(\Omega_{P/\mathbf{k}} \otimes B_0, B_0) = (0)$, $i > 0$, we obtain:

$$\mathrm{Ext}^2(\Omega_{B_0/\mathbf{k}}, B_0) \cong \mathrm{Ext}^1(\mathrm{Im}(\delta), B_0) \cong \mathrm{Ext}^1(J/J^2, B_0)$$ □

Example 3.1.15 (Schlessinger [154]). *(An obstructed affine curve.)* Let s be an indeterminate and let $B_0 = \mathbf{k}[s^7, s^8, s^9, s^{10}] \subset \mathbf{k}[s]$ be the coordinate ring of the affine rational curve $C \subset \mathbf{A}^4 = \mathrm{Spec}(\mathbf{k}[x, y, z, w])$ having parametric equations:

$$x = s^7, \quad y = s^8, \quad z = s^9, \quad w = s^{10}$$

Write $P = \mathbf{k}[x, y, z, w]$ and

$$B_0 = P/I$$

for an ideal $I \subset P$. One can check, using, for example, a computer algebra package, that I is generated by the six 2×2 minors of the following matrix:

$$\begin{pmatrix} x & y & z & w^2 \\ y & z & w & x^3 \end{pmatrix}$$

i.e. by the following polynomials:

$$f_1 = y^2 - xz; \qquad f_2 = xw - yz; \qquad f_3 = z^2 - yw;$$

$$f_4 = x^4 - w^2 y; \quad f_5 = x^3 y - zw^2; \quad f_6 = w^3 - x^3 z$$

This ideal is prime of height 3. Consider a presentation:

$$0 \to \mathbf{R} \to F \to I \to 0$$

where $F \cong P^6$ with generators, say $e_1, \ldots, e_6$, so that $e_i \mapsto f_i$. To describe $\mathbf{R}$ one can use the beginning of the free resolution of I given by the Eagon–Northcott complex (see [48]). One obtains a set of generators for $\mathbf{R}$ given by the rows of the following matrix:

$$\begin{array}{lcccccc}
R_1: & z & y & x & 0 & 0 & 0 \\
R_2: & w^2 & 0 & 0 & y & -x & 0 \\
R_3: & 0 & 0 & w^2 & 0 & z & y \\
R_4: & 0 & -w^2 & 0 & z & 0 & x \\
R_5: & w & z & y & 0 & 0 & 0 \\
R_6: & x^3 & 0 & 0 & z & -y & 0 \\
R_7: & 0 & 0 & x^3 & 0 & w & z \\
R_8: & 0 & -x^3 & 0 & w & 0 & y
\end{array}$$

Here each row gives the coefficients a_i of the linear combination $\sum_i a_i e_i \in \mathbf{R}$. We then have an exact sequence

$$0 \to \bar{\mathbf{R}} \to \bar{F} \to I/I^2 \to 0$$

where $\bar{F} = F/IF$ and $\bar{\mathbf{R}} = \mathbf{R}/(IF \cap \mathbf{R})$. Reducing mod I the above relations one gets the following set of generators of $\bar{\mathbf{R}}$ as elements of $\bar{F}$:

$$\begin{array}{lcccccc}
r_1: & s^9 & s^8 & s^7 & 0 & 0 & 0 \\
r_2: & s^{20} & 0 & 0 & s^8 & -s^7 & 0 \\
r_3: & 0 & 0 & s^{20} & 0 & s^9 & s^8 \\
r_4: & 0 & -s^{20} & 0 & s^9 & 0 & s^7 \\
r_5: & s^{10} & s^9 & s^8 & 0 & 0 & 0 \\
r_6: & s^{21} & 0 & 0 & s^9 & -s^8 & 0 \\
r_7: & 0 & 0 & s^{21} & 0 & s^{10} & s^9 \\
r_8: & 0 & -s^{21} & 0 & s^{10} & 0 & s^8
\end{array} \tag{3.6}$$

Since B_0 is reduced we have

$$T^2_{B_0} = \mathrm{Ext}^1_{B_0}(I/I^2, B_0) = \mathrm{Hom}(\bar{\mathbf{R}}, B_0)/\mathrm{Hom}(\bar{F}, B_0)$$

Representing an element $h \in \mathrm{Hom}(\bar{\mathbf{R}}, B_0)$ as the 8-tuple $(h(r_1), \ldots, h(r_8)) \in B_0^8$ we see that the submodule $\mathrm{Hom}(\bar{F}, B_0)$ is generated by the columns of (3.6). Therefore to prove that B_0 is obstructed it will suffice to produce a first-order deformation $\xi \in \mathrm{Def}_{B_0}(\mathbf{k}[\epsilon])$ whose obstruction to lifting to $\mathbf{k}[t]/(t^3)$ is represented by an $h : \bar{\mathbf{R}} \to B_0$ not in the submodule generated by the columns of (3.6). We define ξ by the ideal

$$(f_1 + \Delta f_1, \ldots, f_6 + \Delta f_6) \subset \mathbf{k}[\epsilon, x, y, z, w]$$

where

$$\Delta f := (\Delta f_1, \ldots, \Delta f_6) = (0, 0, 0, zw, w^2, -x^3)$$

this defines a deformation because $R_j \cdot \Delta f \in I$ for all $j = 1, \ldots, 8$. More precisely:

$$(R_1 \cdot \Delta f, \ldots, R_8 \cdot \Delta f) = (0, -wf_2, -f_5, f_4 - wf_3, 0, wf_3, f_6, -f_5)$$

Therefore we see that the obstruction to lifting ξ to second order is defined by the homomorphism $h : \bar{\mathbf{R}} \to B_0$ represented by

$$(0, 0, -t^{20}, t^{19}, 0, 0, -t^{21}, -t^{20})$$

Now we can immediately check that this vector is not in $\mathrm{Hom}(\bar{F}, B_0)$ and therefore the deformation ξ cannot be lifted: thus B_0 is obstructed.

3.1.3 Comparison with deformations of the nonsingular locus

Under certain conditions it is possible to compare the deformations of an affine scheme with the deformations of the open subscheme of its nonsingular points. In this and the following subsections we will describe the analysis made in [156] and [157], with applications to the study of certain quotient singularities. We will need a preliminary lemma.

Lemma 3.1.16. *Let X be an affine scheme, $Z \subset X$ a closed subscheme and G a coherent sheaf on X. Let $G^\vee = Hom(G, \mathcal{O}_X)$. If $depth_Z(\mathcal{O}_X) \geq 2$ then $depth_Z(G^\vee) \geq 2$ and therefore*

$$H^0(X, G^\vee) \cong H^0(X \backslash Z, G^\vee)$$

Proof. Consider a presentation

$$0 \to \mathbf{R} \to F \to G \to 0$$

where F is a free $\mathcal{O}_X$-module. Then we obtain an exact sequence

$$0 \to G^\vee \to F^\vee \to Q \to 0 \tag{3.7}$$

where $Q \subset \mathbf{R}^\vee$. Since $F^\vee$ is free we have $\mathrm{depth}_Z(F^\vee) = \mathrm{depth}_Z(\mathcal{O}_X) \geq 2$ and therefore $H^0_Z(F^\vee) = 0 = H^1_Z(F^\vee)$ ([75], Theorem 3.8, p. 44); it follows that $H^0_Z(G^\vee) = 0$. Similarly, one proves that $H^0_Z(\mathbf{R}^\vee) = 0$, and therefore $H^0_Z(Q) = 0$. From the sequence of local cohomology associated to (3.7) we obtain $H^1_Z(G^\vee) = 0$ and therefore $\mathrm{depth}_Z(G^\vee) \geq 2$ by [75], Theorem 3.8, p. 44. The last assertion follows from the exact sequence

$$0 \to H^0(X, G^\vee) \to H^0(X \backslash Z, G^\vee) \to H^1_X(G^\vee)$$

(see [84], p. 212). □

Consider an affine scheme $X = \mathrm{Spec}(B)$ where $B = P/J$ for a smooth $\mathbf{k}$-algebra P. Let $Z = \mathrm{Sing}(X)$ be the singular locus of X and $U = X \backslash Z$.

Let $Y = \mathrm{Spec}(P)$ and consider the exact sequence

$$0 \to T_X \to T_{Y|X} \to N_X \to T^1_X \to 0 \tag{3.8}$$

where $N_X = N_{X/Y}$. Since T^1_X is supported on Z, by restricting to U we get the exact sequence:

$$0 \to T_U \to T_{Y|U} \to N_{X|U} \to 0 \tag{3.9}$$

Proposition 3.1.17. *(i) If* $\mathrm{depth}_Z(\mathcal{O}_X) \geq 2$ *(e.g. X is normal of dimension* ≥ 2*) we have an exact sequence*

$$0 \to T^1_B \to H^1(U, T_U) \to H^1(U, T_{Y|U})$$

(ii) If $\mathrm{depth}_Z(\mathcal{O}_X) \geq 3$ *then*

$$T^1_B \cong H^1(U, T_U)$$

Proof. (i) We have the local cohomology exact sequences (see [84], p. 212):

$$0 \to H^0(X, N_X) \to H^0(U, N_{X|U}) \to H^1_Z(N_X)$$

$$0 \to H^0(X, T_{Y|X}) \to H^0(U, T_{Y|U}) \to H^1_Z(T_{Y|X})$$

If $\mathrm{depth}_Z(\mathcal{O}_X) \geq 2$ then from Lemma 3.1.16 we deduce that $\mathrm{depth}_Z(N_X) \geq 2$ and $\mathrm{depth}_Z(T_{Y|X}) \geq 2$. Therefore we have $H^1_Z(N_X) = 0 = H^1_Z(T_{Y|X})$ ([75], Theorem 3.8, p. 44) and

$$H^0(X, N_X) \cong H^0(U, N_{X\ U}), \quad H^0(X, T_{Y|X}) \cong H^0(U, T_{Y|U})$$

Comparing the exact cohomology sequences of (3.8) and (3.9) we get an exact and commutative diagram:

$$\begin{array}{ccccccc} H^0(X, T_{Y|X}) & \to & H^0(X, N_X) & \to & T^1_B & \to & 0 \\ \| & & \| & & \cap & & \\ H^0(U, T_{Y|U}) & \to & H^0(U, N_{X|U}) & \to & H^1(U, T_U) & \to & H^1(U, T_{Y|U}) \end{array}$$

which proves (i).

If $\mathrm{depth}_Z(\mathcal{O}_X) \geq 3$ then X is normal and $H^1_Z(T_{Y|X}) = 0 = H^2_Z(T_{Y|X})$ because $T_{Y|X}$ is locally free; from the local cohomology exact sequence we get

$$H^1(U, T_{Y|U}) \cong H^1(X, T_{Y|X}) = 0$$

because X is affine. Using (i) we deduce (ii). □

The above proposition can be applied to prove the rigidity of a large class of cones over projective varieties. We will need the following well-known lemmas, which we include for the reader's convenience.

Lemma 3.1.18. *Let* $W \subset I\!P^r$ *be a projective nonsingular variety,* CW *the affine cone over* W*,* $v \in CW$ *the vertex,* $U = CW \backslash \{v\}$ *and* $p : U \to W$ *the projection. If* G *is a coherent sheaf on* CW *such that* $G_{|U} = p^*F$ *for some coherent* $F \neq (0)$ *on* W*, then the following conditions are equivalent:*

(i) $\mathrm{depth}_v(G) \geq d$ *for some* $d \geq 2$*;*
(ii) $H^0(CW, G) = \oplus_{\nu \in \mathbb{Z}} H^0(W, F(\nu))$ *and* $H^k(W, F(\nu)) = 0$ *for all* $1 \leq k \leq d-2$ *and* $\nu \in \mathbb{Z}$*.*

Proof. We will use the equivalence

$$\mathrm{depth}_v(G) \geq d \quad \Leftrightarrow \quad H_v^k(G) = 0, \quad k < d$$

([75], Theorem 3.8, p. 44). We have an exact local cohomology sequence:

$$0 \to H_v^0(G) \to H^0(CW, G) \to H^0(U, G_{|U}) \to H_v^1(G) \to 0$$

and isomorphisms:

$$H^{k-1}(U, G_{|U}) \cong H_v^k(G), \qquad k \geq 2$$

Since $G_{|U} = p^*F$ with $F \neq (0)$ we have $\mathrm{depth}_v(G) \geq 1$, thus $H_v^0(G) = 0$. On the other hand, since $p_*G_{|U} = p_*p^*F = \bigoplus_{\nu\in\mathbb{Z}} F(\nu)$, we have $H^0(U, G_{|U}) = \oplus_{\nu\in\mathbb{Z}} H^0(W, F(\nu))$.
Now the conclusion follows. □

Lemma 3.1.19. *Let* $\underline{0} = [0, \ldots, 0, 1] \in \mathbb{P}^{r+1}$, $V = \mathbb{P}^{r+1}\backslash\{\underline{0}\}$ *and let* $\pi : V \to \mathbb{P}^r$ *be the projection. Then*

$$T_{V/\mathbb{P}^r} = \pi^*\mathcal{O}(1)$$

Proof. It is an immediate consequence of the commutative exact diagram

$$\begin{array}{ccccccc}
 & & 0 & & 0 & & \\
 & & \uparrow & & \uparrow & & \\
0 \to & T_{V/\mathbb{P}^r} & \to & T_V & \to & \pi^*T_{\mathbb{P}^r} & \to 0 \\
 & \| & & \uparrow & & \uparrow & \\
0 \to & \mathcal{O}_V(1) & \to & \mathcal{O}_V(1)^{r+2} & \to & \mathcal{O}_V(1)^{r+1} & \to 0 \\
 & & & \uparrow & & \uparrow & \\
 & & & \mathcal{O} & & \mathcal{O} & \\
 & & & \uparrow & & \uparrow & \\
 & & & 0 & & 0 &
\end{array}$$

where the vertical sequences are restrictions of Euler sequences. □

Corollary 3.1.20. *Let* $W \subset \mathbb{P}^r$ *be a projective nonsingular variety of dimension* ≥ 2. *Assume that*

(i) $H^0(\mathbb{P}^r, \mathcal{O}(\nu)) \to H^0(W, \mathcal{O}_W(\nu))$ *surjective for all* $\nu \in \mathbb{Z}$ *(W is projectively normal).*
(ii) $H^1(W, \mathcal{O}_W(\nu)) = 0$ *for all* $\nu \in \mathbb{Z}$.
(iii) $H^1(W, T_W(\nu)) = 0$ *for all* $\nu \in \mathbb{Z}$.

Then the affine cone CW over W is rigid.

Proof. CW has dimension ≥ 3 and hypothesis (i) implies that it is normal ([84], p. 126). Hypotheses (i) and (ii) imply that $\mathrm{depth}_v(\mathcal{O}_{CW}) \geq 3$, by Lemma 3.1.18.

Therefore by 3.1.17(ii) it suffices to show that $H^1(U, T_U) = 0$ where $U = CW\backslash\{v\}$. Let $p : U \to W$ be the projection. We have

$$\begin{aligned} H^1(U, \mathcal{O}_U) &= \oplus_{\nu\in\mathbb{Z}} H^1(W, \mathcal{O}_W(\nu)) = 0 \\ H^1(U, p^*T_W) &= \oplus_{\nu\in\mathbb{Z}} H^1(W, T_W(\nu)) = 0 \end{aligned} \tag{3.10}$$

by conditions (ii) and (iii). The relative tangent sequence of p takes the following form:

$$0 \to \mathcal{O}_U \to T_U \to p^*T_W \to 0 \tag{3.11}$$

In fact it follows from Lemma 3.1.19 and from Proposition B.1(i) that we have $T_{\pi^{-1}(W)/W} = \pi^*\mathcal{O}_W(1)$ and therefore, since $U = \pi^{-1}(W)\backslash W$, we have $T_{U/W} = p^*\mathcal{O}_W(1)$ and this is clearly equal to $\mathcal{O}_U$. The conclusion follows from (3.10) and from the cohomology sequence of (3.11). □

Corollary 3.1.21. *(i) The affine cone over $\mathbb{P}^n \times \mathbb{P}^m$ in its Segre embedding is rigid for every positive n, m such that $n + m \geq 3$.*
(ii) The affine cone over any Veronese embedding of $\mathbb{P}^n$, $n \geq 2$, is rigid.
(iii) If $W \subset \mathbb{P}^r$ is a projective nonsingular variety of dimension ≥ 2, such that $h^1(W, \mathcal{O}_W) = 0 = h^1(W, T_W)$ then the affine cone over the m-th Veronese embedding $W^{(m)}$ of W is rigid for every $m \gg 0$.

Proof. Using 3.1.20, (i) and (ii) are easy computations and (iii) follows from Serre's vanishing theorem ([84], Th. III.5.2). □

The affine cone over the quadric $\mathbb{P}^1 \times \mathbb{P}^1 \subset \mathbb{P}^3$ is not rigid because it is a hypersurface (see Proposition 3.1.3, page 106); it does not satisfy both conditions (ii) and (iii) of Corollary 3.1.20. The affine cone over a rational normal curve $\Gamma_r \subset \mathbb{P}^r$, $r \geq 2$, is not rigid either (for $r = 2$ it is a singular quadric surface in $\mathbf{A}^3$; for $r \geq 3$ see [131], [143], [144]). For a generalization of Corollary 3.1.20 and applications see [177].

3.1.4 Quotient singularities

The analysis of the previous subsection can be applied to the study of deformations of a class of affine singular schemes obtained as quotients of nonsingular ones by the action of a finite group.

Let $Y = \mathrm{Spec}(P)$ be an affine nonsingular algebraic variety on which a finite group G acts. Let $X = Y/G$ be the quotient variety, $q : Y \to X$ the projection. Assume that the action is free outside a G-invariant closed subscheme $W \subset Y$. Set $V = Y\backslash W$.

Proposition 3.1.22. *Assume that $\mathrm{depth}_W(\mathcal{O}_Y) \geq 2$. Then*

$$T_X \cong (q_*T_Y)^G$$

where G acts on $T_Y = \mathrm{Der}_{\mathbf{k}}(P, P)^\sim$ by $D \mapsto D^g := gDg^{-1}$ for all $D \in \mathrm{Der}_{\mathbf{k}}(P, P)$ and $g \in G$.

Proof. Consider the exact sequence of coherent sheaves on Y:

$$0 \to T_{Y/X} \to T_Y \to q^*T_X \to T^1_{Y/X}$$

$\Omega^1_{Y/X}$ is supported on W since q is etale outside W; then we have $T_{Y/X} = 0$. Similarly, $T^1_{Y/X}$ is supported on W so that from the above exact sequence restricted to V we deduce an isomorphism

$$H^0(V, T_Y) \cong H^0(V, q^*T_X)$$

Then by Lemma 3.1.16 we deduce that

$$H^0(Y, T_Y) \cong H^0(Y, q^*T_X)$$

Note that, letting $A = P^G$ the ring of invariant elements, we have $X = \mathrm{Spec}(A)$ and the above isomorphism is equivalent to an isomorphism

$$\mathrm{Der}_{\mathbf{k}}(P, P) \cong \mathrm{Der}_{\mathbf{k}}(A, P)$$

Therefore it will suffice to show that

$$\mathrm{Der}_{\mathbf{k}}(A, A) \cong \mathrm{Der}_{\mathbf{k}}(A, P)^G$$

So let $D \in \mathrm{Der}_{\mathbf{k}}(A, P)$ be such that $D = gDg^{-1}$ for all $g \in G$. Then for every $a \in A$ we have

$$D(a) = g(D(g^{-1}a)) = g(D(a))$$

so $D(a) \in A$ and therefore $\mathrm{Der}_{\mathbf{k}}(A, P)^G \subset \mathrm{Der}_{\mathbf{k}}(A, A)$. Conversely, if $D \in \mathrm{Der}_{\mathbf{k}}(A, A)$ then it defines a $\mathbf{k}$-derivation of A in P which is clearly G-invariant and we also have $\mathrm{Der}_{\mathbf{k}}(A, A) \subset \mathrm{Der}_{\mathbf{k}}(A, P)^G$. □

Corollary 3.1.23. *Let n be the order of G. Under the assumptions of 3.1.22, if char($\mathbf{k}$) does not divide n then T_X is a direct summand of q_*T_Y.*

Proof. Define a homomorphism $q_*T_Y \to T_X = (q_*T_Y)^G$ by

$$D \mapsto \frac{1}{n}\sum_{g \in G} D^g$$

This defines the splitting. □

Theorem 3.1.24. *In the above situation, if the action is free outside a G-invariant closed subscheme W of codimension ≥ 3, and char($\mathbf{k}$) does not divide the order of G then $X = Y/G$ is rigid.*

Proof. Let $Z = q(W)$ where $q : Y \to X$ is the projection, $V = Y \backslash W$, $U = X \backslash Z = V/G$. We have $\mathrm{depth}_Z(\mathcal{O}_X) \geq 2$ because X is normal, being the quotient of a nonsingular variety by a finite group (for this elementary fact see e.g. [166], p. 58). Therefore

$$T^1_X \subset H^1(U, T_U) \cong H^1(U, (q_* T_Y)^G) \tag{3.12}$$

where the inclusion follows from 3.1.17(i) and the isomorphism is Proposition 3.1.22. We also have an exact sequence

$$\begin{array}{ccccc} H^1(Y, T_Y) & \to & H^1(V, T_Y) & \to & H^2_W(T_Y) \\ & & \| & & \\ & & H^1(U, q_* T_Y) & & \end{array}$$

where the left vector space is 0 because Y is affine and the right one is 0 because of the depth assumption on W. It follows that $H^1(U, q_* T_Y) = 0$ and therefore $H^1(U, (q_* T_Y)^G) = 0$ as well because it is a direct summand of it by Corollary 3.1.23. The conclusion now follows from (3.12). □

In the theorem the hypothesis on the codimension of W cannot be removed. In fact all rational two-dimensional double points are quotient singularities and they are hypersurfaces, therefore they are not rigid (see [15]).

Example 3.1.25 (Kollar [110]). Let $n \geq 3$. Consider the weighted projective space

$$X = I\!P(1, 1, 1, a_3, \ldots, a_n) = \mathrm{Proj}(\mathbf{k}[X_0, \ldots, X_n])$$

where the X_i's are indeterminates with weights $a_i = 1$ if $i = 0, 1, 2$ and $a_i \geq 1$ if $i = 3, \ldots, n$ (see [45] for details on weighted projective spaces). Then X is locally the quotient of an affine space by a finite cyclic group and by the choice of the weights all its singularities have codimension ≥ 3. It follows from Theorem 3.1.24 that $T^1_X = 0$. Therefore the first-order deformations of X are classified by $H^1(X, T_X)$ by the exact sequence (2.17). But the exact sequence on X:

$$0 \to \mathcal{O}_X \to \bigoplus_i \mathcal{O}(a_i) \to T_X \to 0$$

implies $H^1(X, T_X) = 0$. Therefore X is rigid.

NOTES

1. The main references for this section are [154], [120], [14].

2. ([174]) Let B be an e.f.t. local **k**-algebra, $I \subset B$ an ideal generated by a regular sequence. Prove that if B/I is rigid then B is rigid as well.

3.2 Closed subschemes

3.2.1 The local Hilbert functor

Let $X \subset Y$ be a closed embedding of algebraic schemes. A cartesian diagram of morphisms of schemes

$$\eta : \begin{array}{ccccc} X & \to & \mathcal{X} & \subset & Y \times S \\ \downarrow & & \downarrow \pi & & \\ \mathrm{Spec}(\mathbf{k}) & \to & S & & \end{array}$$

where π is flat, and it is induced by the projection from $Y \times S$, is called a *(flat) family of deformations of X in Y* parametrized by S, or over S. We call S and $\mathcal{X}$ respectively the *parameter scheme* and the *total scheme* of the family. When $S = \mathrm{Spec}(A)$ with A in $\mathrm{ob}(\mathcal{A}^*)$ (resp. in $\mathrm{ob}(\mathcal{A})$, resp. $A = \mathbf{k}[\epsilon]$) we say that η is a *local*, resp. *infinitesimal*, resp. *first-order* family of deformations of X in Y. We will also say that η is a deformation of X in Y over A. The family is called *trivial* if $\mathcal{X} = X \times S$. X is *rigid in Y* if every infinitesimal deformation of X in Y is trivial. When considering a family η we speak generally of *a family of closed subschemes of Y*.

Let $X \subset Y$ be a closed embedding of algebraic schemes. For each A in $\mathrm{ob}(\mathcal{A})$ we let

$$H_X^Y(A) = \{\text{deformations of } X \text{ in } Y \text{ over } A\}$$

We can immediately verify that this defines a functor of Artin rings

$$H_X^Y : \mathcal{A} \to (\text{sets})$$

called the *local Hilbert functor* of X in Y.

Proposition 3.2.1. *Given a closed embedding of algebraic schemes $X \subset Y$ then:*

(i) The local Hilbert functor H_X^Y satisfies conditions H_0, H_ϵ, $\bar{H}$, H of Theorem 2.3.2.
(ii) There is a natural identification

$$H_X^Y(\mathbf{k}[\epsilon]) = H^0(X, N_{X/Y})$$

where $N_{X/Y}$ is the normal sheaf of X in Y.

Proof. (i) Obviously H_X^Y satisfies condition H_0. Let

$$\begin{array}{ccc} A' & & A'' \\ & \searrow\swarrow & \\ & A & \end{array}$$

be homomorphisms in $\mathcal{A}$, with $A'' \to A$ a small extension. Letting $\bar{A} = A' \times_A A''$ we have a commutative diagram with exact rows:

$$\begin{array}{ccccccccc} 0 & \to & (\epsilon) & \to & \bar{A} & \to & A' & \to & 0 \\ & & \| & & \downarrow & & \downarrow & & \\ 0 & \to & (\epsilon) & \to & A'' & \to & A & \to & 0 \end{array}$$

Take an element of

$$H_X^Y(A') \times_{H_X^Y(A)} H_X^Y(A'')$$

which is represented by a pair of deformations $\mathcal{X}' \subset Y \times \mathrm{Spec}(A')$ and $\mathcal{X}'' \subset Y \times \mathrm{Spec}(A'')$ such that

$$\mathcal{X}' \times_{\mathrm{Spec}(A')} \mathrm{Spec}(A) = \mathcal{X}'' \times_{\mathrm{Spec}(A'')} \mathrm{Spec}(A) \subset Y \times \mathrm{Spec}(A)$$

Consider the sheaf of $\bar{A}$-algebras $\mathcal{O}_{\mathcal{X}'} \times_{\mathcal{O}_{\mathcal{X}}} \mathcal{O}_{\mathcal{X}''}$ on X. Then $\bar{\mathcal{X}} := (|X|, \mathcal{O}_{\mathcal{X}'} \times_{\mathcal{O}_{\mathcal{X}}} \mathcal{O}_{\mathcal{X}''})$ is a scheme over $\mathrm{Spec}(\bar{A})$, flat over $\mathrm{Spec}(\bar{A})$ (see the proof of 2.4.1). Therefore $\bar{\mathcal{X}}$ is a deformation of X over $\mathrm{Spec}(\bar{A})$ inducing $\mathcal{X}'$ and $\mathcal{X}''$. We have a commutative diagram:

$$\begin{array}{ccc} \mathcal{X} & \to & \mathcal{X}' \\ \downarrow & & \downarrow \\ \mathcal{X}'' & \to & Y \times \mathrm{Spec}(\bar{A}) \end{array}$$

and the universal property of the fibred sum implies that there is a morphism

$$\Phi : \bar{\mathcal{X}} \to Y \times \mathrm{Spec}(\bar{A})$$

Pulling back Φ over $\mathrm{Spec}(A'')$ (resp. $\mathrm{Spec}(A')$) we obtain the closed embedding $\mathcal{X}'' \subset Y \times \mathrm{Spec}(A'')$ (resp. $\mathcal{X}' \subset Y \times \mathrm{Spec}(A')$). Since $\mathrm{Spec}(A'') \subset \mathrm{Spec}(\bar{A})$ is a closed embedding defined by a square zero ideal, it follows that Φ is a closed embedding as well (details are left to the reader). Therefore $\Phi : \bar{\mathcal{X}} \subset Y \times \mathrm{Spec}(\bar{A})$ defines an element of $H_X^Y(\bar{A})$ which is mapped to $(\mathcal{X}', \mathcal{X}'')$ by the map:

$$\alpha : H_X^Y(A' \times_A A'') \to H_X^Y(A') \times_{H_X^Y(A)} H_X^Y(A'')$$

It follows that α is surjective. Now let $\tilde{\mathcal{X}} \subset Y \times \mathrm{Spec}(\bar{A})$ be another element of $H_X^Y(\bar{A})$ which is mapped to $(\mathcal{X}', \mathcal{X}'')$. Then by the universal property of the fibred sum there is a morphism $\tau : \bar{\mathcal{X}} \to \tilde{\mathcal{X}}$ which, as in the proof of 2.4.1, is easily seen to be an isomorphism. Moreover, the following diagram commutes:

$$\begin{array}{ccccc} & & Y \times \mathrm{Spec}(\bar{A}) & & \\ & \nearrow & & \nwarrow & \\ \bar{\mathcal{X}} & & \stackrel{\tau}{\longrightarrow} & & \tilde{\mathcal{X}} \end{array}$$

Since the diagonal arrows are closed embeddings it follows that $\bar{\mathcal{X}} = \tilde{\mathcal{X}}$ as closed subschemes of $Y \times \mathrm{Spec}(\bar{A})$. This proves that α is actually a bijection and (i) follows.

(ii) Let $\mathcal{I} \subset \mathcal{O}_Y$ be the ideal sheaf of X. A first-order deformation of X in Y, i.e. a flat family:

$$\begin{array}{ccccc} X & \to & \mathcal{X} & \subset & Y \times \mathrm{Spec}(\mathbf{k}[\epsilon]) \\ \downarrow & & \downarrow & & \\ \mathrm{Spec}(\mathbf{k}) & \to & \mathrm{Spec}(\mathbf{k}[\epsilon]) & & \end{array} \tag{3.13}$$

is defined by a sheaf $\mathcal{O}_{\mathcal{X}}$ of flat $\mathbf{k}[\epsilon]$-algebras, with an isomorphism $\mathcal{O}_{\mathcal{X}} \otimes_{\mathbf{k}[\epsilon]} \mathbf{k} \cong \mathcal{O}_X$. The closed embedding $\mathcal{X} \subset Y \times \mathrm{Spec}(\mathbf{k}[\epsilon])$ is determined by a sheaf of ideals $\mathcal{I}_\epsilon \subset \mathcal{O}_Y[\epsilon] := \mathcal{O}_Y \otimes_{\mathbf{k}} \mathbf{k}[\epsilon]$ such that $\mathcal{O}_{\mathcal{X}} = \mathcal{O}_Y[\epsilon]/\mathcal{I}_\epsilon$. The above data are obtained by gluing together their restrictions to an affine open cover.

On an affine open set $U = \mathrm{Spec}(P) \subset Y$, let $X \cap U = \mathrm{Spec}(B)$ where $B = P/J$ for an ideal $J = (f_1, \ldots, f_N) \subset P$.

Consider the exact sequence

$$0 \to \mathbf{R} \xrightarrow{\upsilon} P^N \xrightarrow{\mathbf{f}} J \to 0$$

where $\mathbf{R}$ is the module of relations among $f_1, \ldots, f_N$. Taking $\mathrm{Hom}_P(-, B)$ we obtain the exact sequence:

$$0 \to \mathrm{Hom}_B(J/J^2, B) \to \mathrm{Hom}_P(P^N, B) \xrightarrow{\upsilon^\vee} \mathrm{Hom}_P(\mathbf{R}, B)$$

which identifies $\mathrm{Hom}_B(J/J^2, B)$ with $\ker(\upsilon^\vee)$. An element of $\ker(\upsilon^\vee)$ can be represented as an N-tuple $\underline{h} = (h_1, \ldots, h_N)$ of elements of P which, interpreted as an element of $\mathrm{Hom}_P(P^N, B)$ by scalar product (i.e. $\underline{h}(p_1, \ldots, p_N) = \sum_j h_j p_j \mod J$) must be zero on $\mathbf{R}$. Hence

$$\sum_j h_j r_j \in J \text{ for every } (r_1, \ldots, r_N) \in \mathbf{R}$$

This means that there exist $\Delta r_1, \ldots, \Delta r_N \in P$ such that

$$\sum_j h_j r_j = -\sum_j f_j \Delta r_j$$

or, equivalently, such that

$$(\underline{f} + \epsilon \underline{h})^t (\underline{r} + \epsilon \Delta \underline{r}) = 0$$

in $P \otimes_{\mathbf{k}} \mathbf{k}[\epsilon]$. Therefore from Corollary A.11 it follows that $\underline{f} + \epsilon \underline{h}$ generates an ideal in $P \otimes_{\mathbf{k}} \mathbf{k}[\epsilon]$ which defines a first-order deformation of $\mathrm{Spec}(B)$ in $\mathrm{Spec}(P)$ because every relation among $f_1, \ldots, f_N$ extends to a relation among $f_1 + \epsilon h_1, \ldots, f_N + \epsilon h_N$. Using the same argument backwards one sees that every first-order deformation of $\mathrm{Spec}(B)$ in $\mathrm{Spec}(P)$ defines an element of $\mathrm{Hom}_B(J/J^2, B)$.

It follows that at the global level we have a canonical 1–1 correspondence between first-order deformations of X in Y and $H^0(X, N_{X/Y})$. □

Corollary 3.2.2. *Let $X \subset Y$ be a closed embedding of algebraic schemes. If $h^0(X, N_{X/Y}) < \infty$, for example if X is projective, then H_X^Y is prorepresentable.*

Proof. Follows from Proposition 3.2.1 and from Theorem 2.3.2. □

If $X \subset Y$ is a closed embedding of projective schemes then the prorepresentability of H_X^Y follows directly from the existence of the Hilbert scheme Hilb^Y because H_X^Y is prorepresented by the complete local ring $\hat{\mathcal{O}}_{\mathrm{Hilb}^Y,[X]}$ (see § 4.3).

Let $\bar{A}$ be in $\mathrm{ob}(\hat{A})$. *A formal deformation of X in Y* is a sequence

$$\xi_n : \begin{array}{ccc} \mathcal{X}_n & \subset & Y \times \mathrm{Spec}(\bar{A}_n) \\ \downarrow & & \\ \mathrm{Spec}(\bar{A}_n) & & \end{array}$$

of infinitesimal deformations of X in Y over $\bar{A}_n = \bar{A}/m_{\bar{A}}^{n+1}$ such that ξ_n induces ξ_{n-1} by pullback under the natural inclusion $\mathrm{Spec}(\bar{A}_{n-1}) \to \mathrm{Spec}(\bar{A}_n)$ for all $n \geq 1$. We can describe the formal deformation $(\bar{A}, \{\xi_n\})$ of X in Y as a diagram of formal schemes

$$\begin{array}{ll} \mathcal{X} & \subset Y \times \mathrm{Specf}(\bar{A}) \\ \downarrow & \\ \mathrm{Specf}(\bar{A}) & \end{array}$$

As in the case of the functor Def_X, a formal deformation $(\bar{A}, \{\xi_n\})$ of X in Y defines an element $\hat{\xi} \in \widehat{H_X^Y}(\bar{A})$, i.e. a formal couple $(\bar{A}, \hat{\xi})$ for H_X^Y, and conversely, every such element is defined by a formal deformation of X in Y.

Considering not necessarily infinitesimal deformations of a closed subscheme one has the notion of *characteristic map*, as follows.

Definition 3.2.3. *Let $X \subset Y$ be a closed embedding of algebraic schemes, (S, s) a pointed scheme and*

$$\begin{array}{cccccc} & X & \subset & \mathcal{X} & \subset & Y \times S \\ \xi : & \downarrow & & \downarrow \pi & & \\ & \mathrm{Spec}(\mathbf{k}) & \xrightarrow{s} & S & & \end{array}$$

be a family of deformations of X in Y. By pulling back this family by morphisms $\mathrm{Spec}(\mathbf{k}[\epsilon]) \to S$ *with image s and applying Proposition 3.2.1 we obtain a linear map*

$$\chi_\xi : T_s S \to H^0(X, N_{X/Y})$$

called the characteristic map *of the family ξ at s.*

The characteristic map is the analogous for embedded deformations of the Kodaira–Spencer map for abstract ones.

Examples 3.2.4. (i) If $X \subset I\!P^r$, $r \geq 1$, is the complete intersection of $r-n$ hypersurfaces $f_1, \ldots, f_{r-n}$ of degrees $d_1 \leq d_2 \leq \ldots \leq d_{r-n}$ respectively, we have a presentation

$$\bigwedge^2 [\oplus_j \mathcal{O}(-d_j)] \xrightarrow{\mathbf{f}} \oplus_j \mathcal{O}(-d_j) \to \mathcal{I}_X \to 0 \tag{3.14}$$

Taking $Hom(-, \mathcal{O}_X)$ we obtain an exact sequence:

$$0 \to N_X \to \oplus_j \mathcal{O}_X(d_j) \xrightarrow{\check{\mathbf{f}}} \bigwedge^2 [\oplus_j \mathcal{O}_X(d_j)]$$

where $N_X = N_{X/I\!P^r}$. Since each nonzero entry of the matrix defining $\mathbf{f}$ is one of the f_j's, the map $\check{\mathbf{f}}$ is zero. Therefore

$$N_X \cong \oplus_j \mathcal{O}_X(d_j)$$

Equivalently, one can remark that (3.14) induces a surjective homomorphism

$$\oplus_j \mathcal{O}_X(-d_j) \to \mathcal{I}_X/\mathcal{I}_X^2$$

of locally free sheaves of the same rank, which must therefore be an isomorphism. In particular, if X is a hypersurface of degree d we have $N_X \cong \mathcal{O}_X(d)$ and therefore

$$h^0(X, N_X) = \binom{d+r}{r} - 1$$

confirming the fact that X can be deformed in $\mathbb{P}^r$ only inside the linear system of hypersurfaces of degree d, which is a projective space of dimension $\binom{d+r}{r} - 1$.

If X is a linear subspace, i.e. $d_1 = \cdots = d_{r-n} = 1$, then $N_X \cong \mathcal{O}_X(1)^{\oplus r-n}$ and therefore

$$h^0(X, N_X) = (r-n)(n+1)$$

as expected, since such linear subspaces X are parametrized by the grassmannian $G(n+1, r+1)$, which is nonsingular of dimension $(r-n)(n+1)$.

(ii) Let X be a Cartier divisor on a connected projective scheme Y. Consider the exact sequence

$$0 \to \mathcal{O}_Y \to \mathcal{O}_Y(X) \to N_{X/Y} \to 0$$

and the cohomology sequence:

$$0 \to H^0(Y, \mathcal{O}_Y) \to H^0(Y, \mathcal{O}_Y(X)) \xrightarrow{\chi} H^0(X, N_{X/Y}) \to H^1(Y, \mathcal{O}_Y) \quad (3.15)$$

Classically, the map χ was called the *characteristic map* of the linear system $|X|$. By definition, $\mathrm{Im}(\chi) \cong H^0(Y, \mathcal{O}_Y(X))/H^0(Y, \mathcal{O}_Y)$ is naturally identified with the tangent space to the linear system $|X|$. We can verify that this is so by identifying $\mathrm{Im}(\chi)$ as a subvector space of first-order deformations of X in Y, as follows.

Assume that X is defined by a system of local equations $\{f_i\}$, $f_i \in \Gamma(U_i, \mathcal{O}_Y)$ not 0-divisor, with respect to an affine cover $\{U_i\}$ of Y. We have $f_{ij} := f_i f_j^{-1} \in \Gamma(U_i \cap U_j, \mathcal{O}_Y^*)$ for all i, j, and $\{f_{ij}\}$ is a Cech 1-cocycle which defines the line bundle $\mathcal{O}_Y(X)$. A first-order deformation of X in Y is a Cartier divisor $\mathcal{X} \subset Y \times \mathrm{Spec}(\mathbf{k}[\epsilon])$ which is determined by a system $\{F_i = f_i + \epsilon g_i\}$, $g_i \in \Gamma(U_i, \mathcal{O}_Y)$, such that there exist $F_{ij} = f_{ij} + \epsilon g_{ij} \in \Gamma(U_i \cap U_j, \mathcal{O}^*_{Y\times \mathrm{Spec}(\mathbf{k}[\epsilon])})$ (hence $g_{ij} \in \Gamma(U_i \cap U_j, \mathcal{O}_Y))$ satisfying $F_i = F_{ij}F_j$ for all i, j. Therefore on $U_i \cap U_j$ we have:

$$f_i + \epsilon g_i = (f_{ij} + \epsilon g_{ij})(f_j + \epsilon g_j)$$

which is equivalent to the identity:

$$g_i = f_{ij} g_j + g_{ij} f_j$$

This identity shows that the system $\{\bar{g}_i = g_i \bmod \mathcal{I}_X\}$ defines a section of $N_{X/Y}$, as expected. We also see that $\{\bar{g}_i\} \in \mathrm{Im}(\chi)$ if and only if $g_i = f_{ij} g_j$, in which case $F_{ij} = f_{ij}$, i.e. X deforms inside the linear system $|X|$, as asserted.

(iii) In $\mathbb{P}^2 \times \mathbb{P}^1$ with bihomogeneous coordinates $(u, v, w; x, y)$ consider the hypersurface Σ_m, $m \geq 0$, defined by the equation:

$$x^m v - y^m u = 0$$

One can show that $\Sigma_m \cong F_m$, and that the structure morphism $\pi : F_m \to \mathbb{P}^1$ is induced by the projection $\mathbb{P}^2 \times \mathbb{P}^1 \to \mathbb{P}^1$. Therefore the surface Σ_m is a realization of the rational ruled surface F_m as the total scheme of a family of lines of $\mathbb{P}^2$ parametrized by $\mathbb{P}^1$. When $m > 0$ this family is nontrivial because $F_m \ncong \mathbb{P}^1 \times \mathbb{P}^1$ (for details see [5]).

Examples 3.2.5. The following set of examples deals with properties of projective curves. Proofs are straightforward and are left to the reader.

(i) If Y is a projective scheme and $C \subset Y$ is a projective integral l.c.i. curve, then the normal sheaf $N_{C/Y}$ is torsion free. If C is nonsingular then $N_{C/Y}$ is locally free.

(ii) Consider a nonsingular curve $C \subset \mathbb{P}^3$ and a (possibly singular) surface $S \subset \mathbb{P}^3$ of degree n containing C. Prove that there is an exact sequence of locally free sheaves on C:

$$0 \to \mathbf{a}^{-1} \otimes K_C(-n+4) \to N_{C/\mathbb{P}^3} \xrightarrow{\psi} \mathcal{O}_C(n) \to [\mathcal{O}_C/\mathbf{a}](n) \to 0 \qquad (3.16)$$

where $\mathbf{a} \subset \mathcal{O}_C$ is the ideal sheaf generated by the restriction to C of the partial derivatives

$$\frac{\partial F}{\partial X_0}, \ldots, \frac{\partial F}{\partial X_3}$$

where $F = 0$ is an equation of S.
(*Hint:* $\mathrm{Im}(\psi) = \mathcal{O}_C \otimes \mathrm{Im}(T_{\mathbb{P}^3|S} \to N_{S/\mathbb{P}^3})$).

In the case where S is nonsingular we obtain the sequence:

$$0 \to K_C(-n+4) \to N_C \to \mathcal{O}_C(n) \to 0 \qquad (3.17)$$

Deduce from (3.16) that if $\mathbf{a} \neq \mathcal{O}_C$ (i.e. if $C \cap \mathrm{Sing}(S) \neq \emptyset$) then C is not regularly embedded in S (yet the normal sheaf $N_{C/S}$ is locally free.)

(iii) Consider a nonsingular curve $C \subset \mathbb{P}^3$ and a point $p \in \mathbb{P}^3 \backslash C$. Then there is an exact sequence

$$0 \to \mathcal{O}_C(1) \to N_{C/\mathbb{P}^3} \to \omega_C(3) \to 0$$

which is obtained as a special case of (3.16) by taking as S the cone projecting C from p. Deduce that

$$h^1(C, N_{C/\mathbb{P}^3}) \leq h^1(C, \mathcal{O}_C(1))$$

(iv) Let $C \subset \mathbb{P}^r$, $r \geq 4$, be a nonsingular irreducible projective curve. Let $p_1, \ldots, p_k \in \mathbb{P}^r$, $1 \leq k \leq r-3$, be general points and let $\pi : C \to \mathbb{P}^{r-k}$ be the projection with centre the linear span $\langle p_1, \ldots, p_k \rangle$. Prove that there is an exact sequence

$$0 \to \mathcal{O}_C(1)^k \to N_{C/\mathbb{P}^r} \to \pi^* N_{\pi(C)/\mathbb{P}^{r-k}} \to 0 \tag{3.18}$$

which splits if and only if C is contained in a hyperplane. Deduce that

$$h^1(C, N_{C/\mathbb{P}^r}) \leq (r-2)h^1(C, \mathcal{O}_C(1)) \tag{3.19}$$

In particular, $H^1(C, N_{C/\mathbb{P}^r}) = (0)$ if $h^1(C, \mathcal{O}_C(1)) = 0$. (Solution: see [163], Prop. 11.2). Inequality (3.19) was already known classically as a bound on the dimension of the Hilbert scheme ([169], s. 8). For a sharpening of (3.19) see [47].

(v) Let $C \subset \mathbb{P}^r$ be a nonsingular irreducible projective curve and let $L = \mathcal{O}_C(1)$. Show that if the Petri map $\mu_0(L)$ (see Example 3.3.9) is injective then $H^1(C, N_{C/\mathbb{P}^r}) = (0)$.

(vi) Let $C \subset \mathbb{P}^r$ be a projective irreducible nonsingular curve of degree d and genus g. Prove that

$$\chi(N_{C/\mathbb{P}^r}) := h^0(C, N_{C/\mathbb{P}^r}) - h^1(C, N_{C/\mathbb{P}^r}) = (r+1)d + (r-3)(1-g)$$

In particular, for a canonical curve of genus $r+1 \geq 3$ and degree $2r$ in $\mathbb{P}^r$ we have

$$\chi(N_{C/\mathbb{P}^r}) = h^0(C, N_{C/\mathbb{P}^r}) = r(r+5)$$

because $h^1(C, N_{C/\mathbb{P}^r}) = 0$ since the Petri map $\mu_0(\omega_C)$ is injective (see (v) above).

3.2.2 Obstructions

Let $X \subset Y$ be a closed embedding, A in ob($\mathcal{A}$) and let

$$\xi: \quad \begin{array}{ccccc} X & \to & \mathcal{X} & \subset & Y \times \mathrm{Spec}(A) \\ \downarrow & & \downarrow f & & \\ \mathrm{Spec}(\mathbf{k}) & \to & \mathrm{Spec}(A) & & \end{array}$$

be a deformation of X in Y over A. Let

$$e: 0 \to \mathbf{k} \to \tilde{A} \to A \to 0$$

be a small extension. A *lifting* of ξ to $\tilde{A}$ is a deformation of X in Y over $\tilde{A}$:

$$\tilde{\xi}: \quad \begin{array}{ccccc} X & \to & \tilde{\mathcal{X}} & \subset & Y \times \mathrm{Spec}(\tilde{A}) \\ \downarrow & & \downarrow f & & \\ \mathrm{Spec}(\mathbf{k}) & \to & \mathrm{Spec}(\tilde{A}) & & \end{array}$$

whose pullback to Spec(A) is ξ.

Proposition 3.2.6. *Let $X \subset Y$ be a regular closed embedding (Definition D.1.1) of algebraic schemes with X projective. Then $H^1(X, N_{X/Y})$ is an obstruction space for the local Hilbert functor H_X^Y.*

Proof. Given

$$\xi : \begin{array}{ccccc} X & \to & \mathcal{X} & \subset & Y \times \mathrm{Spec}(A) \\ \downarrow & & \downarrow f & & \\ \mathrm{Spec}(\mathbf{k}) & \to & \mathrm{Spec}(A) & & \end{array}$$

where A is in ob($\mathcal{A}$), an infinitesimal deformation of X in Y, we will show that there is a natural linear map

$$o_{\xi/Y} : \mathrm{Ex}_{\mathbf{k}}(A, \mathbf{k}) \to H^1(X, N_{X/Y})$$

such that, for every extension

$$e : 0 \to \epsilon\mathbf{k} \to \tilde{A} \to A \to 0$$

we have $o_{\xi/Y}(e) = 0$ if and only if ξ has a lifting to $\tilde{A}$; this will prove the proposition. Since X is regularly embedded in Y we can find an affine open cover $\mathcal{U} = \{U_i\}_{i \in I}$ of Y such that $X_i := X \cap U_i$ is a complete intersection in U_i for each i. Let $U_i = \mathrm{Spec}(P_i)$, $X_i = \mathrm{Spec}(P_i/\mathcal{I}_i)$ where $\mathcal{I}_i = (f_{i1}, \ldots, f_{iN})$ with $\{f_{i1}, \ldots, f_{iN}\}$ a regular sequence in P_i. We then have $\mathcal{X}_{|U_i} = \mathrm{Spec}(P_{iA}/\mathcal{I}_{iA})$ where

$$\mathcal{I}_{iA} = (F_{i1}, \ldots, F_{iN}) \subset P_{iA} := P_i \otimes A$$

and $f_{i\alpha} = F_{i\alpha} \bmod m_A$, $\alpha = 1, \ldots, N$. Choose arbitrarily $\tilde{F}_{i1}, \ldots, \tilde{F}_{iN} \in P_{i\tilde{A}}$ such that $F_{i\alpha} = \tilde{F}_{i\alpha} \bmod \epsilon$. By Example A.12 $\{F_{i1}, \ldots, F_{iN}\}$ and $\{\tilde{F}_{i1}, \ldots, \tilde{F}_{iN}\}$ are regular sequences in P_{iA} and in $P_{i\tilde{A}}$ respectively; in particular, letting $\mathcal{I}_{i\tilde{A}} = (\tilde{F}_{i1}, \ldots, \tilde{F}_{iN})$,

$$\tilde{\mathcal{X}}_i := \mathrm{Spec}(P_{i\tilde{A}}/\mathcal{I}_{i\tilde{A}}) \subset U_i \times \mathrm{Spec}(\tilde{A})$$

is a lifting of $\mathcal{X}_{|U_i} \subset U_i \times \mathrm{Spec}(A)$ to $\tilde{A}$. In order to find a lifting of $\mathcal{X}$ to $\tilde{A}$ we must be able to choose the $\tilde{F}_{i\alpha}$'s in such a way that

$$\tilde{\mathcal{X}}_{i|U_{ij}} = \tilde{\mathcal{X}}_{j|U_{ij}} \subset U_{ij} \times \mathrm{Spec}(\tilde{A}) \tag{3.20}$$

for each $i, j \in I$. Letting $U_{ij} = \mathrm{Spec}(P_{ij})$ and viewing the $f_{i\alpha}$'s and $f_{j\alpha}$'s as elements of P_{ij} via the natural maps

$$\begin{array}{ccc} P_i & & P_j \\ & \searrow \quad \swarrow & \\ & P_{ij} & \end{array}$$

we have

$$\tilde{F}_{j\alpha} - \tilde{F}_{i\alpha} =: \epsilon h_{ij\alpha}$$

and $\underline{h}_{ij} := (h_{ij1}, \ldots, h_{ijN}) \in \Gamma(U_{ij}, N_{X/Y})$, because $N_{X/Y}$ is locally free of rank N and is trivial on each U_i. By construction, $\{\underline{h}_{ij}\} \in Z^1(\mathcal{U}, N_{X/Y})$. The condition (3.20) means that we can choose the $\tilde{F}_{i\alpha}$'s so that $\underline{h}_{ij} = \underline{0}$ all i, j. A different choice of the $\tilde{F}_{i\alpha}$'s is of the form $\tilde{F}_{i\alpha} + \epsilon h_{i\alpha}$ and $\underline{h}_i := (h_{i1}, \ldots, h_{iN}) \in \Gamma(U_i, N_{X/Y})$. Since we have

$$(\tilde{F}_{j\alpha} + \epsilon h_{j\alpha}) - (\tilde{F}_{i\alpha} + \epsilon h_{i\alpha}) = \epsilon(h_{ij\alpha} + h_{j\alpha} - h_{i\alpha}) \tag{3.21}$$

we see that $\{\underline{h}_{ij}\}$ defines an element $o_{\xi/Y}(e) \in H^1(X, N_{X/Y})$ which is zero if and only if the $\tilde{\mathcal{X}}_i$'s satisfy condition (3.20) and define a lifting $\tilde{\mathcal{X}}$ of $\mathcal{X}$ to $\tilde{A}$. It is easy to show that the map o_ξ is **k**-linear. □

The element $o_{\xi/Y}(e) \in H^1(X, N_{X/Y})$ is called *the obstruction to lifting* ξ *to* $\tilde{A}$; we call ξ *obstructed* if $o_{\xi/Y}(e) \neq 0$ for some $e \in \text{Ex}_{\mathbf{k}}(A, \mathbf{k})$; otherwise it is *unobstructed*. X is said to be *unobstructed in* Y if all its infinitesimal deformations in Y are unobstructed; otherwise X is said to be *obstructed in* Y. Examples of obstructed closed subschemes are usually quite subtle, especially if one is interested in nonsingular obstructed subvarieties. In order to be able to describe them in a natural way it is necessary to know the existence of the Hilbert scheme of a projective scheme. We will give examples in § 4.6.

Corollary 3.2.7. *Let $j : X \subset Y$ be a regular closed embedding of algebraic schemes with X projective and let $(R, \{\xi_n\})$ be the formal universal deformation of X in Y. Then*

(i)

$$h^0(X, N_{X/Y}) \geq \dim(R) \geq h^0(X, N_{X/Y}) - h^1(X, N_{X/Y})$$

The first equality holds if and only if X is unobstructed in Y.
(ii) X is rigid in Y if and only if $H^0(X, N_{X/Y}) = 0$.
(iii) If $H^1(X, N_{X/Y}) = (0)$ then X is unobstructed in Y.

The proof is left to the reader. In the case of a closed embedding $X \subset Y$ which is not regular, Proposition 3.2.6 says nothing about the obstruction space of H_X^Y. We refer the reader to § 4.4 and § 4.6 for some information about the general case.

Examples 3.2.8. (i) Let C be a projective nonsingular curve contained in a nonsingular surface S, and assume that C is *negatively embedded* in S, i.e. $\deg(\mathcal{O}_C(C)) < 0$. Then $H^0(C, \mathcal{O}_C(C)) = 0$ and therefore C is rigid in S.

This happens in particular when $C \cong \mathbb{P}^1$ is an exceptional curve of the first kind. Another example is when C has genus $g \geq 2$, $S = C \times C$ and C is identified to the diagonal $\Delta \subset S$. In this case $N_{C/S} = T_C$, which has degree $2 - 2g < 0$. Note that $H^1(C, T_C) \neq (0)$ but C is unobstructed in S, being rigid in S. This example shows that the sufficient condition of Corollary 3.2.7 is not necessary.

(ii) Hypersurfaces of $\mathbb{P}^r$ are unobstructed. In fact, if $X \subset \mathbb{P}^r$ has degree d then

$$h^1(X, N_{X/\mathbb{P}^r}) = h^1(X, \mathcal{O}_X(d)) = 0$$

More generally, complete intersections in $\mathbb{P}^r$ are unobstructed (see Subsection 4.6.1).

(iii) Let $Q \subset \mathbb{P}^3$ be a quadric cone with vertex v, and $L \subset Q$ a line. Then we have an inclusion

$$N_{L/Q} \subset N_{L/\mathbb{P}^3} = \mathcal{O}_L(1) \oplus \mathcal{O}_L(1)$$

whose cokernel is $\mathcal{O}_L(2)(-v)$ (see Example 3.2.5(ii)). It follows that $N_{L/Q} = \mathcal{O}_L(1)$; in particular, it is locally trivial and $H^1(L, N_{L/Q}) = 0$, despite the fact that $L \subset Q$ is not a regular embedding (Example 3.2.5(ii)) and L is obstructed in Q (see Note 1 of § 4.6).

(iv) Let $C \subset \mathbb{P}^r$ be a nonsingular irreducible projective curve and let $L = \mathcal{O}_C(1)$. If the Petri map $\mu_0(L)$ (see Example 3.3.9) is injective then C is unobstructed in $\mathbb{P}^r$ because $h^1(C, N_{C/\mathbb{P}^r}) = 0$ (Example 3.2.5(v)). In particular canonical curves of genus $g \geq 3$ are unobstructed in $\mathbb{P}^{g-1}$ (Example 3.2.5(vi)).

3.2.3 The forgetful morphism

Let $X \subset Y$ be a closed embedding of algebraic schemes. The *forgetful morphism*

$$\Phi : H_X^Y \to \mathrm{Def}_X$$

is the morphism which associates to an infinitesimal deformation of X in Y:

$$\xi : \begin{array}{ccccc} X & \to & \mathcal{X} & \subset & Y \times \mathrm{Spec}(A) \\ \downarrow & & \downarrow & & \\ \mathrm{Spec}(\mathbf{k}) & \to & \mathrm{Spec}(A) & & \end{array}$$

the isomorphism class of the deformation of X:

$$\begin{array}{ccc} X & \to & \mathcal{X} \\ \downarrow & & \downarrow \\ \mathrm{Spec}(\mathbf{k}) & \to & \mathrm{Spec}(A) \end{array}$$

Proposition 3.2.9. *Assume that X and Y are nonsingular and that X is projective. Consider the exact sequence*

$$0 \to T_X \to T_{Y|X} \to N_{X/Y} \to 0 \tag{3.22}$$

Then:

(i) $d\Phi = \delta : H^0(X, N_{X/Y}) \to H^1(X, T_X)$ is the coboundary map coming from (3.22).

(ii) The coboundary map

$$\delta_1 : H^1(X, N_{X/Y}) \to H^2(X, T_X)$$

arising from the exact sequence (3.22) is an obstruction map for Φ (see Definition 2.3.5).

(iii) If $H^1(X, T_{Y|X}) = 0$ then Φ is smooth.

(iv) If X is unobstructed in Y and δ is surjective then Φ is smooth. In particular, X is unobstructed as an abstract variety and has $rk(\delta)$ number of moduli.

Proof. (i) Let

$$\begin{array}{ccccc} & X & \subset & \mathcal{X} & \subset Y \times \mathrm{Spec}(\mathbf{k}[\epsilon]) \\ \xi : & \downarrow & & \downarrow \pi & \\ & \mathrm{Spec}(\mathbf{k}) & \xrightarrow{s} & \mathrm{Spec}(\mathbf{k}[\epsilon]) & \end{array}$$

be a first-order deformation of X in Y. We must show that $\delta\chi(\xi) = \kappa(\xi)$. Let $\chi(\xi) = \underline{h} \in H^0(X, N_{X/Y})$. Consider an affine open cover $\mathcal{U} = \{U_i = \mathrm{Spec}(P_i)\}$ of Y, and let $X_i = X \cap U_i = \mathrm{Spec}(P_i/(f_{i1}, \dots, f_{iN}))$. We have

$$\mathcal{X}_i := \mathcal{X}_{|U_i} = \mathrm{Spec}(P[\epsilon]/(f_{i1} + \epsilon h_{i1}, \dots, f_{iN} + \epsilon h_{iN}))$$

Then $(h_{i1}, \dots, h_{iN}) =: \underline{h}_i = \underline{h}_{|U_i} \in \Gamma(U_i, N_{X/Y})$. Since X_i is affine and nonsingular the abstract deformation $\mathcal{X}_i$ of X_i is trivial: thus there exist isomorphisms $\theta_i : X_i \times \mathrm{Spec}(\mathbf{k}[\epsilon]) \to \mathcal{X}_i$ and $\kappa(\xi) \in H^1(X, T_X)$ is defined by the 1-cocycle $\{d_{ij}\} \in Z^1(\mathcal{U}, T_X)$ corresponding to the system of automorphisms

$$\theta_{ij} = \theta_i^{-1}\theta_j : X_{ij} \times \mathrm{Spec}(\mathbf{k}[\epsilon]) \to X_{ij} \times \mathrm{Spec}(\mathbf{k}[\epsilon])$$

where $X_{ij} = X \cap U_{ij}$. Let's compute $\delta(\underline{h})$. The isomorphism θ_i is given by an isomorphism of $\mathbf{k}[\epsilon]$-algebras

$$t_i : P_i[\epsilon]/(f_{i1} + \epsilon h_{i1}, \dots, f_{iN} + \epsilon h_{iN}) \to P_i[\epsilon]/(f_{i1}, \dots, f_{iN})$$

which, by the smoothness of P_i, is induced by a $\mathbf{k}[\epsilon]$-automorphism

$$T_i : P_i[\epsilon] \to P_i[\epsilon]$$

of the form $T_i(p + \epsilon q) = p + \epsilon(q + d_i(p))$ where $d_i \in \mathrm{Der}_{\mathbf{k}}(P_i, P_i) = \Gamma(U_i, T_Y)$ is such that $d_i(h_{i\alpha}) = -f_{i\alpha}$. We have $\{d_i\} \in C^0(\mathcal{U}, T_Y)$ and $\delta(\underline{h})$ is defined by $\{(d_j - d_i)_{|X_i}\}$. Since $(d_j - d_i)_{|X_i} = d_{ij}$ we conclude that $\delta(\underline{h}) = \kappa(\xi)$.

(ii) One must show that, given an infinitesimal deformation

$$\begin{array}{ccccc} & X & \subset & \mathcal{X} & \subset Y \times \mathrm{Spec}(A) \\ \xi : & \downarrow & & \downarrow \pi & \\ & \mathrm{Spec}(\mathbf{k}) & \xrightarrow{s} & \mathrm{Spec}(A) & \end{array}$$

where A is in $\mathcal{A}$, we have a commutative diagram

$$\begin{array}{ccc} & \mathrm{Ex}_{\mathbf{k}}(A, \mathbf{k}) & \\ \swarrow o_{\xi/Y} & & \searrow o_\xi \\ H^1(X, N_{X/Y}) & \xrightarrow{\delta_1} & H^2(X, T_X) \end{array}$$

The proof of this fact is similar to the proof of part (i) and will be omitted.

(iii) The exact sequence (3.22) shows that the hypothesis implies

$$d\Phi : H^0(X, N_{X/Y}) \to H^1(X, T_X) \qquad \text{surjective and}$$

$$\delta_1 : H^1(X, N_{X/Y}) \to H^2(X, T_X) \qquad \text{injective}$$

Therefore the assertion follows from Proposition 2.3.6.

(iv) is left to the reader. □

Remark 3.2.10. Let $X \subset Y$ be a regular embedding of algebraic schemes with X reduced and Y nonsingular, $\mathcal{I} \subset \mathcal{O}_Y$ the ideal sheaf of X, and let $\Phi : H_X^Y \to \mathrm{Def}_X$ be the forgetful morphism. The differential

$$d\Phi : H^0(X, N_{X/Y}) = \mathrm{Hom}_{\mathcal{O}_X}(\mathcal{I}/\mathcal{I}^2, \mathcal{O}_X) \to \mathrm{Ext}^1_{\mathcal{O}_X}(\Omega^1_X, \mathcal{O}_X)$$

is the **k**-linear map which associates to $\sigma : \mathcal{I}/\mathcal{I}^2 \to \mathcal{O}_X$ the pushout $\sigma_*(S)$ where

$$S : 0 \to \mathcal{I}/\mathcal{I}^2 \to \Omega^1_{Y|X} \to \Omega^1_X \to 0$$

is the conormal sequence of $X \subset Y$. This generalizes Proposition 3.2.9. The proof consists in considering, for a first-order deformation of X in Y

$$X \subset \mathcal{X} \subset Y \times \mathrm{Spec}(\mathbf{k}[\epsilon])$$

the induced diagram of conormal sequences

$$\begin{array}{ccccccc} 0 \to & \mathcal{I}/\mathcal{I}^2 \oplus \mathcal{O}_X & \to & \Omega^1_{Y|X} \oplus \mathcal{O}_X & \to & \Omega^1_X & \to 0 \\ & \downarrow & & \downarrow & & \| & \\ 0 \to & \mathcal{O}_X & \to & \Omega^1_{\mathcal{X}|X} & \to & \Omega^1_X & \to 0 \end{array}$$

and in recognizing the second row as the pushout of the first.

Part (iii) of the proposition often gives a very effective way of proving that a given $X \subset Y$ is unobstructed as an abstract variety. If X is a curve in $\mathbb{P}^r$ the vanishing of $H^1(X, T_{\mathbb{P}^r|X})$ is related to the Petri map (see Example 3.3.9).

The following examples are further applications of this principle.

Examples 3.2.11. (i) ([107]) Let $X \subset \mathbb{P}^r$, $r \geq 3$, be a nonsingular hypersurface of degree $d \geq 2$. Then $h^1(N_{X/\mathbb{P}^r}) = h^1(\mathcal{O}_X(d)) = 0$ and therefore X is unobstructed in $\mathbb{P}^r$. On the other hand, from the exact sequence:

$$0 \to T_{\mathbb{P}^r}(-d) \to T_{\mathbb{P}^r} \to T_{\mathbb{P}^r|X} \to 0$$

and the Euler sequence:

$$0 \to \mathcal{O}_{\mathbb{P}^r} \to \mathcal{O}_{\mathbb{P}^r}(1)^{r+1} \to T_{\mathbb{P}^r} \to 0$$

we deduce that:

$$h^1(T_{\mathbb{P}^r|X}) = h^2(T_{\mathbb{P}^r}(-d)) = 0 \ \text{ if } r \geq 4$$

while for $r = 3$ we have the exact sequence:

$$\begin{array}{ccccc} 0 \leftarrow & H^2(T_{\mathbb{P}^3}(-d))^\vee & \leftarrow & H^0(\mathcal{O}_{\mathbb{P}^3}(d-4)) & \leftarrow & H^0(\mathcal{O}_{\mathbb{P}^3}(d-5))^4 \\ & \| & & & & \\ & H^1(T_{\mathbb{P}^3|X})^\vee & & & & \end{array}$$

Therefore we see that

$$h^1(T_{\mathbb{P}^r|X}) = \begin{cases} 1 & \text{if } r = 3 \text{ and } d = 4; \\ 0 & \text{otherwise} \end{cases}$$

From 3.2.9 we therefore deduce that $H_X^{\mathbb{P}^r} \to \mathrm{Def}_X$ is smooth and X is unobstructed as an abstract variety unless $r = 3$ and $d = 4$ (this is precisely the case when X is a K3 surface). An analogous result holds more generally for complete intersections ([161]).

Using (3.22) one computes easily that $H^2(T_X) \neq 0$ if X is a nonsingular surface of degree $d \geq 5$ in $\mathbb{P}^3$: therefore the unobstructedness of X could not have been deduced from 2.4.6 in this case.

One can generalize to singular hypersurfaces as follows. Consider a reduced hypersurface $X \subset \mathbb{P}^r$, $r \geq 3$, of degree $d \geq 2$. Then the conormal sequence is

$$0 \to \mathcal{O}_X(-d) \to \Omega^1_{\mathbb{P}^r|X} \to \Omega^1_X \to 0$$

so that $H^1(X, N_{X/\mathbb{P}^r}) = 0$ and we have the exact sequence

$$\begin{array}{ccccc} H^0(X, N_{X/\mathbb{P}^r}) & \xrightarrow{d\Phi} & \mathrm{Ext}^1_{\mathcal{O}_X}(\Omega^1_X, \mathcal{O}_X) & \to & \mathrm{Ext}^1(\Omega^1_{\mathbb{P}^r|X}, \mathcal{O}_X) \\ \| & & & & \| \\ H^0(X, \mathcal{O}_X(d)) & & & & H^1(X, T_{\mathbb{P}^r|X}) \end{array}$$

where the equality on the right is because $\Omega^1_{\mathbb{P}^r|X}$ is locally free. Therefore as before, we see that $\Phi : H_X^Y \to \mathrm{Def}_X$ is smooth and X is unobstructed if $(r, d) \neq (3, 4)$.

(ii) The previous example can be easily generalized to nonsingular hypersurfaces of $\mathbb{P}^n \times \mathbb{P}^m$, $1 \leq n \leq m$, $n + m \geq 3$. Let

$$\begin{array}{ccc} \mathbb{P}^n \times \mathbb{P}^m & \xrightarrow{q} & \mathbb{P}^m \\ \downarrow p & & \\ \mathbb{P}^n & & \end{array}$$

be the projections. Consider a nonsingular hypersurface $X \subset \mathbb{P}^n \times \mathbb{P}^m$ of bidegree (a, b), i.e. defined by an equation $\sigma = 0$ for some $\sigma \in H^0(\mathcal{O}(a, b))$, where

$$\mathcal{O}(a, b) := p^*\mathcal{O}(a) \otimes q^*\mathcal{O}(b)$$

From the exact sequence

$$0 \to \mathcal{O} \to \mathcal{O}(a, b) \to N_{X/\mathbb{P}^n \times \mathbb{P}^m} \to 0$$

one deduces that

$$H^1(N_{X/\mathbb{P}^n \times \mathbb{P}^m}) = (0)$$

and therefore X is unobstructed in $\mathbb{P}^n \times \mathbb{P}^m$. For any coherent sheaf $\mathcal{F}$ on $\mathbb{P}^n \times \mathbb{P}^m$ we use the notation

$$\mathcal{F}(\alpha, \beta) = \mathcal{F} \otimes \mathcal{O}(\alpha, \beta)$$

Using the fact that

$$T_{\mathbb{P}^n \times \mathbb{P}^m} = p^*T_{\mathbb{P}^n} \oplus q^*T_{\mathbb{P}^m}$$

and the Leray spectral sequence with respect to any one of the projections, one easily computes that

$$h^i(T_{\mathbb{P}^n\times\mathbb{P}^m}(\alpha,\beta)) = 0$$

when $n+m\geq 4$, $i=1,2$ and (α,β) arbitrary. Moreover, when $(n,m)=(1,2)$ one finds:

$$\begin{array}{lll} h^i(T_{\mathbb{P}^1\times\mathbb{P}^2}) & = 0 & \text{all } i\geq 1 \\ h^2(T_{\mathbb{P}^1\times\mathbb{P}^2}(-a,-b)) & = 0 & \text{unless } (a,b)=(2,3) \end{array}$$

Putting all this information together and using the exact sequence

$$0 \to T_{\mathbb{P}^n\times\mathbb{P}^m}(-a,-b) \to T_{\mathbb{P}^n\times\mathbb{P}^m} \to T_{\mathbb{P}^n\times\mathbb{P}^m|X} \to 0$$

one deduces that

$$h^1(T_{\mathbb{P}^n\times\mathbb{P}^m|X}) = 0$$

unless $(n,m) = (1,2)$ and $(a,b) = (2,3)$ (this is precisely the case when X is a K3 surface). Now as before, we conclude that the forgetful morphism $H_X^{\mathbb{P}^n\times\mathbb{P}^m} \to \mathrm{Def}_X$ is smooth and X is unobstructed as an abstract variety.

3.2.4 The local relative Hilbert functor

Given a projective morphism $p:\mathcal{X}\to S$ of schemes and a $\mathbf{k}$-rational point $s\in S$, consider the fibre $\mathcal{X}(s)$ and a closed subscheme $Z\subset\mathcal{X}(s)$. For each A in ob$(\mathcal{A})$ an *infinitesimal deformation of Z in $\mathcal{X}$ relative to p parametrized by A* is a commutative diagram:

$$\begin{array}{ccccc} \mathcal{Z} & \subset & \mathcal{X}_A & \to & \mathcal{X} \\ & \searrow & \downarrow & & \downarrow p \\ & & \mathrm{Spec}(A) & \xrightarrow{s} & S \end{array}$$

where the right square is cartesian, the left diagonal morphism is flat and its closed fibre is Z; this means in particular that the morphism s has image $\{s\}$ and therefore that A is an $\mathcal{O}_{S,s}$-algebra. Then, letting $\Lambda = \mathcal{O}_{S,s}$, we can define the *local relative Hilbert functor*

$$H_Z^{\mathcal{X}/S} : \mathcal{A}_\Lambda \to (\text{sets})$$

by

$$H_Z^{\mathcal{X}/S}(A) = \left\{ \begin{array}{c} \text{infinitesimal deformations of } Z \text{ in } \mathcal{X} \\ \text{relative to } p \text{ parametrized by } A \end{array} \right\}$$

We have the following generalization of Corollaries 3.2.2 and 2.4.7:

Theorem 3.2.12. *Let $p:\mathcal{X}\to S$ be a projective morphism of schemes, $s\in S$ a $\mathbf{k}$-rational point, and $Z\subset\mathcal{X}(s)$ a closed subscheme of the fibre $\mathcal{X}(s)$. Denote $\mathcal{O}_{S,s}$ by Λ. Then:*

(i) The local relative Hilbert functor $H_Z^{\mathcal{X}/S}:\mathcal{A}_\Lambda\to$ (sets) is prorepresentable and has tangent space $H^0(Z,N_{Z/\mathcal{X}(s)})$.

(ii) If Z is regularly embedded in $\mathcal{X}(s)$ and p is flat then $H^1(Z, N_{Z/\mathcal{X}(s)})$ is an obstruction space for $H_Z^{\mathcal{X}/S}$, and we have an exact sequence:

$$0 \to H^0(Z, N_{Z/\mathcal{X}(s)}) \to t_R \to T_s S \to H^1(Z, N_{Z/\mathcal{X}(s)}) \tag{3.23}$$

where R is the local Λ-algebra prorepresenting $H_Z^{\mathcal{X}/S}$.

Proof. (i) The proof of 3.2.1 can be followed almost verbatim, showing that $H_Z^{\mathcal{X}/S}$ satisfies conditions H_0, H_ϵ, $\bar{H}$, H and that $H^0(Z, N_{Z/\mathcal{X}(s)})$ is its tangent space.

(ii) The proof of 3.2.6 can be easily adapted to this case. The exact sequence (3.23) follows from the above and from (2.2). □

Another generalization of the local Hilbert functors can be obtained with no extra effort. Consider a projective scheme X and a formal deformation of X

$$\bar{\pi} : \mathcal{X} \to \mathrm{Specf}(R)$$

where R is in $\mathrm{ob}(\hat{\mathcal{A}})$ and $\bar{\pi}$ is a flat projective morphism of formal schemes; let $Z \subset X$ be a closed subscheme. For each A in $\mathrm{ob}(\mathcal{A})$ define an *infinitesimal deformation of Z in $\mathcal{X}$ relative to $\bar{\pi}$ parametrized by A* as a commutative diagram:

$$\begin{array}{ccccc} \mathcal{Z} & \subset & \mathcal{X}_A & \to & \mathcal{X} \\ & \searrow & \downarrow & & \downarrow \bar{\pi} \\ & & \mathrm{Spec}(A) & \xrightarrow{s} & \mathrm{Specf}(R) \end{array}$$

where the right square is cartesian, the left diagonal morphism is flat and its closed fibre is Z. Note that $\mathrm{Spec}(A) = \mathrm{Specf}(A)$ and the morphism s is defined by a surjective homomorphism $R \to A$, so that $\mathcal{X}_A$ is just an ordinary scheme projective and flat over $\mathrm{Spec}(A)$. We can define the *local relative Hilbert functor*

$$H_Z^{\mathcal{X}/\mathrm{Specf}(R)} : \mathcal{A}_R \to (\mathrm{sets})$$

as above. A result analogous to 3.2.12 can be proved in this case as well with a similar proof. Details of this straightforward generalization are left to the reader.

3.3 Invertible sheaves

3.3.1 The local Picard functors

The local Picard functors are related to the Picard schemes in the same way as the local Hilbert functors are to the Hilbert schemes, in the sense that if the Picard scheme of a given scheme exists then the local Picard functors describe its infinitesimal and local properties. In this section we will prove the basic properties of the local Picard functors. The Picard schemes will not be treated here. For a full treatment the reader is referred to Kleiman [100].

Let X be a scheme. Recall that the *Picard group* of X is defined to be the group $\mathrm{Pic}(X)$ of isomorphism classes of invertible sheaves on X; we have the identification $\mathrm{Pic}(X) = H^1(X, \mathcal{O}_X^*)$ (see [84]). We denote by $[L] \in \mathrm{Pic}(X)$ the class of an invertible sheaf L. For every A in $\mathrm{ob}(\mathcal{A})$ we will write for brevity X_A instead of $X \times \mathrm{Spec}(A)$. For every morphism $A \to B$ in $\mathcal{A}$, and for every invertible sheaf L on X_A we denote by $L \otimes_A B$ the invertible sheaf it induces on X_B by pullback. Similarly, given $\lambda \in \mathrm{Pic}(X_A)$ we denote by $\lambda \otimes_A B \in \mathrm{Pic}(X_B)$ its image under the pullback operation. This defines a homomorphism of group

$$\mathrm{Pic}(X_A) = H^1(X_A, \mathcal{O}_{X_A}^*) \to H^1(X_B, \mathcal{O}_{X_B}^*) = \mathrm{Pic}(X_B)$$

which makes $A \mapsto \mathrm{Pic}(X_A)$ a covariant functor on $\mathcal{A}$ with values in the category of abelian groups.

We fix an element $\lambda_0 \in \mathrm{Pic}(X)$ once and for all and, for each A in $\mathrm{ob}(\mathcal{A})$, we let

$$P_{\lambda_0}(A) := \{\lambda \in \mathrm{Pic}(X_A) : \lambda \otimes_A \mathbf{k} = \lambda_0\}$$

Then $P_{\lambda_0}(A)$ is a *subset* of $\mathrm{Pic}(X_A)$ whose definition is functorial in A so that we have a functor of Artin rings:

$$P_{\lambda_0} : \mathcal{A} \to (\mathrm{sets})$$

which will be called the *local Picard functor* of (X, λ_0).

Let L be an invertible sheaf on X such that $\lambda_0 = [L]$. For a given $A \in \mathrm{ob}(\mathcal{A})$ the elements of $P_{\lambda_0}(A)$ are the isomorphism classes of infinitesimal deformations of L over A: an *infinitesimal deformation of* L over A is an invertible sheaf $\mathcal{L}$ on $X \times \mathrm{Spec}(A)$ such that $L = \mathcal{L}_{|X \times \mathrm{Spec}(\mathbf{k})}$. In the case $A = \mathbf{k}[\epsilon]$ we speak of a *first-order deformation* of L.

The main result about this functor is the following.

Theorem 3.3.1. *Let X be an algebraic scheme, $\lambda_0 \in \mathrm{Pic}(X)$. Assume that the following conditions are satisfied:*

(a) $H^0(X, \mathcal{O}_X) \cong \mathbf{k}$
(b) $\dim_{\mathbf{k}} H^1(X, \mathcal{O}_X) < \infty$

Then P_{λ_0} is prorepresentable and $P_{\lambda_0}(\mathbf{k}[\epsilon]) = H^1(X, \mathcal{O}_X)$. Moreover, $H^2(X, \mathcal{O}_X)$ is an obstruction space for P_{λ_0}.

Proof. Let's check the conditions of Theorem 2.3.2. It is clear that $P = P_{\lambda_0}$ satisfies condition H_0. Let

$$\begin{array}{ccc} A' & & A'' \\ & \searrow \quad \swarrow & \\ & A & \end{array}$$

be a diagram in $\mathcal{A}$ with $A'' \to A$ surjective. Consider an element

$$(\lambda', \lambda'') \in P(A') \times_{P(A)} P(A'')$$

and let $\lambda = \lambda' \otimes_{A'} A = \lambda'' \otimes_{A''} A$. Let L', L'', L be invertible sheaves on $X_{A'}, X_{A''}, X_A$ respectively such that $[L'] = \lambda', [L''] = \lambda'', [L] = \lambda$. Then we have homomorphisms $L' \to L$ and $L'' \to L$ of sheaves on $|X|$ inducing isomorphisms $L' \otimes_{A'} A \cong L, L'' \otimes_{A''} A \cong L$. Let $B = A' \times_A A''$.

Claim: $\mathcal{O}_{X_B} \cong \mathcal{O}_{X_{A'}} \times_{\mathcal{O}_{X_A}} \mathcal{O}_{X_{A''}}$.

For every open set $U \subset |X|$ we have by definition:

$$[\mathcal{O}_{X_{A'}} \times_{\mathcal{O}_{X_A}} \mathcal{O}_{X_{A''}}](U) = \mathcal{O}_{X_{A'}}(U) \times_{\mathcal{O}_{X_A}(U)} \mathcal{O}_{X_{A''}}(U)$$

and by the universal property of the fibred sum we have a homomorphism

$$\phi : \mathcal{O}_{X_B} \to \mathcal{O}_{X_{A'}} \times_{\mathcal{O}_{X_A}} \mathcal{O}_{X_{A''}}$$

which is induced by the homomorphisms $\mathcal{O}_{X_B} \to \mathcal{O}_{X_{A'}}$ and $\mathcal{O}_{X_B} \to \mathcal{O}_{X_{A''}}$ coming from $B \to A'$ and $B \to A''$ respectively. Since $A'' \to A$ is surjective

$$\phi \otimes_B A' : \mathcal{O}_{X_B} \otimes_B A' \to \mathcal{O}_{X_{A'}}$$

is an isomorphism. From Lemma A.4 it follows that ϕ is an isomorphism.

From the claim we deduce that $N := L' \times_L L''$ is an invertible sheaf on X_B and the projections induce isomorphisms $N \otimes_B A' \cong L', N \otimes_B A'' \cong L''$. Therefore

$$P(B) \ni [N] \mapsto (\lambda', \lambda'') \in P(A') \times_{P(A)} P(A'')$$

This shows that the map

$$\alpha : P(B) \to P(A') \times_{P(A)} P(A'')$$

is surjective.

Assume that M is an invertible sheaf such that $\alpha([M]) = \alpha([N])$. This means that there are homomorphisms $M \to L'$ and $M \to L''$ inducing isomorphisms $M \otimes_B A' \cong L', M \otimes_B A'' \cong L''$. It follows that we have an automorphism $\theta : L \to L$ given by the composition:

$$L \cong L' \otimes_{A'} A \cong M \otimes_B A \cong L'' \otimes_{A''} A \cong L$$

which makes the following diagram commutative:

$$\begin{array}{ccccccc} & & & M & & & \\ & \swarrow q' & & & & \searrow q'' & \\ L' & & & & & & L'' \\ & \searrow u' & & & & \swarrow u'' & \\ & & L & \xrightarrow{\theta} & L & & \end{array}$$

By hypothesis (a) the isomorphism θ is the multiplication by a unit $a \in A$. Since $A'' \to A$ is surjective we can lift a to $a'' \in A''$ and we can change q'' to $a''q''$ thus assuming that $u'q' = u''q''$. It follows that we have a commutative diagram

$$\begin{array}{ccccc} & & M & & \\ & \swarrow & & \searrow & \\ L' & & & & L'' \\ & \searrow & & \swarrow & \\ & & L & & \end{array}$$

and this implies that $M \cong N$. This shows that α is bijective. Therefore P satisfies also conditions H_ϵ and H.

We have an exact sequence

$$0 \to \mathcal{O}_X \xrightarrow{\exp} \mathcal{O}^*_{X_{\mathbf{k}[\epsilon]}} \to \mathcal{O}^*_X \to 0$$

where $\exp(f) = 1 + \epsilon f$. It follows that

$$P(\mathbf{k}[\epsilon]) = \ker[H^1(X_{\mathbf{k}[\epsilon]}, \mathcal{O}^*_{X_{\mathbf{k}[\epsilon]}}) \to H^1(X, \mathcal{O}^*_X)] = H^1(X, \mathcal{O}_X)$$

Finally, given A in $\mathrm{ob}(\mathcal{A})$ and $[e] \in \mathrm{Ex}_{\mathbf{k}}(A, \mathbf{k})$ represented by an extension

$$0 \to (t) \to \tilde{A} \to A \to 0$$

we have an exact sequence:

$$0 \to t\mathcal{O}_X \to \mathcal{O}^*_{X_{\tilde{A}}} \to \mathcal{O}^*_{X_A} \to 0$$

which induces an exact sequence

$$H^1(X_{\tilde{A}}, \mathcal{O}^*_{X_{\tilde{A}}}) \to H^1(X_A, \mathcal{O}^*_{X_A}) \xrightarrow{\delta} H^2(X, \mathcal{O}_X)$$

Given $\lambda \in P(A)$ it can be lifted to a $\tilde{\lambda} \in P(\tilde{A})$ if and only if $\delta(\lambda) = 0 \in H^2(X, \mathcal{O}_X)$. This shows that $H^2(X, \mathcal{O}_X)$ is an obstruction space for P. □

Corollary 3.3.2. *Let X be a projective integral scheme. Then P_{λ_0} is prorepresentable for every $\lambda_0 \in \mathrm{Pic}(X)$.*

Proof. A projective integral scheme satisfies both conditions (a) and (b) of the theorem. □

Remark 3.3.3. Let X be an algebraic scheme and $\lambda_0 \in \mathrm{Pic}(X)$. Then the tangent and obstruction spaces of the functor P_{λ_0}, as described by Theorem 3.3.1, depend only on X and not on λ_0. This is because, given any $\lambda_0, \mu_0 \in \mathrm{Pic}(X)$, there is a canonical isomorphism of functors $P_{\lambda_0} \cong P_{\mu_0}$. We leave to the reader the easy proof of this fact.

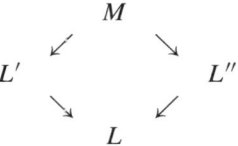

3.3.2 Deformations of sections, I

Let X be a projective integral scheme, and let L be an invertible sheaf on X. One can define a homomorphism of sheaves:

$$m : \mathcal{O}_X \to H^0(X, L)^\vee \otimes L \tag{3.24}$$

as follows. For every open set $U \subset X$

$$\begin{array}{rcl} m(U) : \mathcal{O}(U) & \to & H^0(X, L)^\vee \otimes \Gamma(U, L) \\ f & \mapsto & [s \mapsto f s_{|U}] \end{array}$$

for every $s \in H^0(X, L)$. The induced maps on global sections are just given by cup product:

$$\begin{array}{rcl} m_i : H^i(X, \mathcal{O}_X) & \to & \mathrm{Hom}(H^0(X, L), H^i(X, L)) \\ a & \mapsto & [s \mapsto a \cup s] \end{array}$$

If L is base point free and

$$\varphi_L : X \to \mathbb{P} := \mathbb{P}(H^0(X, L)^\vee)$$

is the morphism defined by the sections of L, then it is easy to check that the homomorphism (3.24) is the same as the one appearing in the pulled back Euler sequence of $\mathbb{P}$:

$$0 \to \mathcal{O}_X \to H^0(X, L)^\vee \otimes L \to \varphi_L^* T_{\mathbb{P}} \to 0$$

We leave this to the reader. The linear maps between cohomology groups induced by these sheaf homomorphisms have deformation theoretic interpretations which we will now explain.

Consider a deformation $\mathcal{L}$ of L over $A \in \mathrm{ob}(\mathcal{A})$: we have an induced restriction map

$$\rho_{\mathcal{L}} : H^0(X_A, \mathcal{L}) \to H^0(X, L)$$

We say that *a section* $\sigma \in H^0(X, L)$ *extends to* $\mathcal{L}$ if $\sigma \in \mathrm{Im}(\rho_{\mathcal{L}})$.

Proposition 3.3.4. *Let $\mathcal{L}_a$ be a first-order deformation of L, corresponding to an element $a \in H^1(X, \mathcal{O}_X)$. A section $s \in H^0(X, L)$ extends to $\mathcal{L}_a$ if and only if*

$$a \cup s = 0 \in H^1(X, L)$$

equivalently, if and only if $s \in \ker[m_1(a)]$ where

$$m_1 : H^1(X, \mathcal{O}_X) \to \mathrm{Hom}(H^0(X, L), H^1(X, L))$$

is the map induced by (3.24).

Proof. Let $\mathcal{U} = \{U_\alpha\}$ be an affine open covering of X such that L is represented by a system of transition functions $\{f_{\alpha\beta}\}$, $f_{\alpha\beta} \in \Gamma(U_{\alpha\beta}, \mathcal{O}_X^*)$. Then the first-order deformation $\mathcal{L}_a$ of L can be represented, in the same covering $\{U_\alpha\}$ of $X \times \mathrm{Spec}(\mathbf{k}[\epsilon])$, by transition functions:

$$\tilde{f}_{\alpha\beta} \in \Gamma(U_{\alpha\beta}, O^*_{X\times\mathrm{Spec}(\mathbf{k}[\epsilon])})$$

such that

$$\tilde{f}_{\alpha\beta}\tilde{f}_{\beta\gamma} = \tilde{f}_{\alpha\gamma} \tag{3.25}$$

and which restrict to the $f_{\alpha\beta}$'s modulo ϵ.

Since $\mathcal{O}^*_{X\times\mathrm{Spec}(\mathbf{k}[\epsilon])} = \mathcal{O}_X^* + \epsilon\mathcal{O}_X$ we can write

$$\tilde{f}_{\alpha\beta} = f_{\alpha\beta} + \epsilon g_{\alpha\beta} \tag{3.26}$$

for suitable $g_{\alpha\beta} \in \Gamma(U_{\alpha\beta}, \mathcal{O}_X)$. Identity (3.25) gives

$$\frac{g_{\alpha\beta}}{f_{\alpha\beta}} + \frac{g_{\beta\gamma}}{f_{\beta\gamma}} = \frac{g_{\alpha\gamma}}{f_{\alpha\gamma}}$$

and the system $(\frac{g_{\alpha\beta}}{f_{\alpha\beta}})$ is a Čech 1-cocycle which defines the element $a \in H^1(X, \mathcal{O}_X)$.

Let's assume that $s \in H^0(X, L)$ is represented by the cocycle (s_α), $s_\alpha \in \Gamma(U_\alpha, \mathcal{O}_X)$, such that $s_\alpha = f_{\alpha\beta}s_\beta$ on $U_{\alpha\beta}$. For s to extend to a section

$$\tilde{s} \in H^0(X \times \mathrm{Spec}(\mathbf{k}[\epsilon]), \mathcal{L}_a)$$

it is necessary and sufficient that there exist $t_\alpha \in \Gamma(U_\alpha, \mathcal{O}_X)$ such that

$$s_\alpha + \epsilon t_\alpha = \tilde{f}_{\alpha\beta}(s_\beta + \epsilon t_\beta)$$

on $U_{\alpha\beta}$. After replacing the $\tilde{f}_{\alpha\beta}$'s by the expressions (3.26) we obtain the identities

$$s_\alpha + \epsilon t_\alpha = (f_{\alpha\beta} + \epsilon g_{\alpha\beta})(s_\beta + \epsilon t_\beta)$$

which are equivalent to:

$$g_{\alpha\beta}s_\beta = t_\alpha - f_{\alpha\beta}t_\beta$$

These identities can be also written as:

$$\frac{g_{\alpha\beta}}{f_{\alpha\beta}}s_\alpha = t_\alpha - f_{\alpha\beta}t_\beta$$

and they mean exactly that the 1-cocycle

$$\left(\frac{g_{\alpha\beta}}{f_{\alpha\beta}}s_\alpha\right) \in Z^1(\mathcal{U}, L)$$

is a coboundary, i.e. that $a \cup s = 0$. □

Corollary 3.3.5. *In the above situation:*

(i) If $a \in \ker(m_1)$ *then all the sections of* L *extend to* $\mathcal{L}_a$.
(ii) For every $a \in H^1(X, \mathcal{O}_X)$, *at least* $\max\{0, h^0(X, L) - h^1(X, L)\}$ *linearly independent sections of* L *extend to* $\mathcal{L}_a$.

Proof. Immediate. □

Let's fix a section $s \in H^0(X, L)$ and let $D = \mathrm{div}(s) \subset X$ be the divisor of s. Consider the local Hilbert functor H_D^X. We have a morphism of functors:

$$\mathbf{a}_D : H_D^X \to P_{[L]}$$

associating to a deformation of D in X over A in ob($\mathcal{A}$), given by an effective Cartier divisor $\mathcal{D} \subset X_A$, the element $[\mathcal{O}_{X_A}(\mathcal{D})] \in P_{[L]}(A)$. $\mathbf{a}_D$ is called the *Abel–Jacobi morphism* of $D \subset X$. Consider the exact sequence

$$0 \to \mathcal{O}_X \xrightarrow{s} L \to L_D \to 0 \tag{3.27}$$

It induces linear maps:

$$\delta_0 : H^0(D, L_D) \to H^1(X, \mathcal{O}_X)$$

$$\delta_1 : H^1(D, L_D) \to H^2(X, \mathcal{O}_X)$$

Noting that $L_D = N_{D/X}$ we have:

Proposition 3.3.6. *In the above situation* δ_0 *is the differential of* $\mathbf{a}_D$ *and* δ_1 *is an obstruction map for* $\mathbf{a}_D$.

Proof. We keep the notations of the proof of Proposition 3.3.4 and therefore we assume D defined by local equations $s_\alpha = 0$ where (s_α) is a 0-cocycle defining s with respect to the covering $\mathcal{U}$; in particular, $s_\alpha = f_{\alpha\beta}s_\beta$. A first-order deformation $\mathcal{D} \subset X \times \mathrm{Spec}(\mathbf{k}[\epsilon])$ of D is defined by local equations

$$s_\alpha + \epsilon t_\alpha = 0$$

which satisfy the cocycle conditions:

$$s_\alpha + \epsilon t_\alpha = (f_{\alpha\beta} + \epsilon g_{\alpha\beta})(s_\beta + \epsilon t_\beta)$$

These conditions can be also written as:

$$t_\alpha = f_{\alpha\beta}t_\beta + \frac{g_{\alpha\beta}}{f_{\alpha\beta}}s_\alpha \tag{3.28}$$

They mean that

$$(\bar{t}_\alpha = t_\alpha \bmod s_\alpha)$$

define a section $\bar{t} \in H^0(D, L_D)$ which is the one corresponding to the first-order deformation $\mathcal{D}$ of D. From the identities (3.28) we also see that $\delta_0(\bar{t})$ is represented by the 1-cocycle $(\frac{g_{\alpha\beta}}{f_{\alpha\beta}})$. Since this cocycle represents $\mathbf{a}_D(\mathcal{D}) \in H^1(X, \mathcal{O}_X)$ this proves that $\delta_0(\bar{t}) = d_{\mathbf{a}_D}(\bar{t})$. The proof that δ_1 is an obstruction map is similar and will be left to the reader. □

Corollary 3.3.7. *Under the assumptions of Proposition 3.3.6:*

(i) If $H^1(X,L)=0$ then $\mathbf{a}_D$ is smooth.
(ii) If the natural map

$$H^1(X,L)\to H^1(D,L_D)$$

is zero then H^X_D is less obstructed than $P_{[L]}$. If, moreover, $P_{[L]}$ is smooth then H^X_D is smooth.

Proof. (i) If $H^1(X,L)=0$ then (3.27) implies that δ_0 is surjective and δ_1 is injective. Therefore $\mathbf{a}_D$ is smooth by Proposition 2.3.6. The proof of (ii) is similar using Proposition 2.3.6 again. □

Remarks 3.3.8. (i) Assume X to be projective and nonsingular. A Cartier divisor D on X is called *semiregular* if the natural map

$$H^1(X,\mathcal{O}_X(D))\to H^1(D,\mathcal{O}_D(D))$$

is zero, i.e. if condition (ii) of the corollary is satisfied. Part (ii) of the corollary can thus be rephrased by saying that D is unobstructed in X if it is semiregular and $P_{[L]}$ is smooth. The smoothness of $P_{[L]}$ is known to be true if $\mathrm{char}(\mathbf{k})=0$, by a general theorem of Cartier: we thus recover a celebrated theorem of Severi, Kodaira and Spencer claiming the unobstructedness of semiregular divisors on a projective nonsingular complex variety X.

These matters are covered in full detail in [130]. The original sources are [172] and [108]. For further developments and applications of the notion of semiregularity see [21], [149] and [23].

(ii) In view of Proposition 3.3.6 the cohomology sequence of (3.27) can be interpreted as a sequence of tangent spaces and differentials as follows:

$$\begin{array}{ccccccc} 0\to & \frac{H^0(X,L)}{\langle\sigma\rangle} & \to & H^0(D,L_D) & \xrightarrow{\delta_0} & H^1(X,\mathcal{O}_X) \\ & \| & & \| & & \| \\ 0\to & T_D|D| & \to & H^X_D(\mathbf{k}[\epsilon]) & \xrightarrow{d_{\mathbf{a}_D}} & P_{[L]}(\mathbf{k}[\epsilon]) \end{array}$$

where we used the identification of $\frac{H^0(X,L)}{\langle\sigma\rangle}$ with the tangent space to the linear system $|D|$ at D proved in Example 3.2.4(ii).

Example 3.3.9. Let X be a Gorenstein curve (e.g. a local complete intersection curve). Then the map m_1 is dual to the map:

$$\mu_0(L): H^0(X,L)\otimes H^0(X,\omega_X L^{-1})\to H^0(X,\omega_X) \tag{3.29}$$

given by multiplication of global sections (ω_X denotes the dualizing sheaf of X). $\mu_0(L)$ is called *the Petri map of L*. More generally, if $V\subset H^0(X,L)$ is a vector space of sections of L, one can consider the multiplication map:

$$\mu_0(V) : V \otimes H^0(X, \omega_X L^{-1}) \to H^0(X, \omega_X)$$

which is called *the Petri map of* V. From Proposition 3.3.4 it follows that $\mathrm{coker}(\mu_0(V))^\perp \subset H^1(X, \mathcal{O}_X)$ is the space of first-order deformations of L to which all sections of V extend. In particular, if $h^0(X, L) = r + 1$ and $\deg(L) = d$ then $\mathrm{coker}(\mu_0(L))^\perp$ is the tangent space at L to the scheme $W^r_d(X)$ of linear systems of degree d and dimension $\geq r$ on X (see [9] for details).

3.3.3 Deformations of pairs (X, L)

Let X be a nonsingular projective algebraic variety and let $d : \mathcal{O}_X \to \Omega^1_X$ be the canonical derivation. We can define a homomorphism of sheaves of abelian groups

$$\mathcal{O}^*_X \to \Omega^1_X$$

by the rule

$$u \mapsto \frac{du}{u}$$

for all open sets $U \subset X$ and $u \in \Gamma(U, \mathcal{O}^*_X)$. We have an induced group homomorphism:

$$c : H^1(X, \mathcal{O}^*_X) \to H^1(X, \Omega^1_X)$$

To simplify the notation we write $c(L)$ instead of $c([L])$ for a given invertible sheaf L on X. If $\mathbf{k} = \mathbf{C}$ then $c(L)$ is, up to a multiplicative constant, the *Chern class* of L (see [128], p. 127). Since Ω^1_X is locally free we have an identification

$$H^1(X, \Omega^1_X) = \mathrm{Ext}^1_{\mathcal{O}_X}(T_X, \mathcal{O}_X)$$

so that we can associate to $c(L)$ an extension

$$0 \to \mathcal{O}_X \to \mathcal{E}_L \to T_X \to 0 \tag{3.30}$$

defined up to isomorphism, called the *Atiyah extension* of L. The sheaf $\mathcal{E}_L$ is locally free of rank $\dim(X) + 1$ and

$$\mathcal{P}_L := \mathcal{E}^\vee_L \otimes_{\mathcal{O}_X} L$$

is called the *sheaf of (first-order) principal parts* of L.

Let $\mathcal{U} = \{U_\alpha\}$ be an affine open covering of X such that L is represented by a system of transition functions $\{f_{\alpha\beta}\}$, $f_{\alpha\beta} \in \Gamma(U_{\alpha\beta}, \mathcal{O}^*_X)$. Then $c(L)$ is represented by the Čech 1-cocycle

$$\left(\frac{df_{\alpha\beta}}{f_{\alpha\beta}}\right) \in Z^1(\mathcal{U}, \Omega^1_X)$$

The sheaf $\mathcal{E}_{L|U_\alpha}$ is isomorphic to $\mathcal{O}_{U_\alpha} \oplus T_{X|U_\alpha}$. A section (a_α, d_α) of $\mathcal{O}_{U_\alpha} \oplus T_{X|U_\alpha}$ and a section (a_β, d_β) of $\mathcal{O}_{U_\beta} \oplus T_{X|U_\beta}$ are identified on $U_{\alpha\beta}$ if and only if $d_\alpha = d_\beta$ and $a_\beta - a_\alpha = \frac{d_\alpha(f_{\alpha\beta})}{f_{\alpha\beta}}$.

Remark 3.3.10. We have $c(L \otimes M) = c(L) + c(M)$, in particular $c(L^n) = nc(L)$ for any $n \in \mathbb{Z}$. Therefore the Atiyah extension of L^n is a constant multiple of the extension (3.30). This means in particular that if $n \neq 0$ then $\mathcal{E}_{L^n} \cong \mathcal{E}_L$.

Consider, for example, $X = \mathbb{P} := \mathbb{P}(V)$ for some finite-dimensional **k**-vector space V. Then one can easily compute that the Euler sequence

$$0 \to \mathcal{O}_{\mathbb{P}} \to V \otimes \mathcal{O}_{\mathbb{P}}(1) \to T_{\mathbb{P}} \to 0$$

is the Atiyah extension of $\mathcal{O}_{\mathbb{P}}(1)$. Therefore $\mathcal{E}_L \cong \mathcal{E}_{\mathcal{O}(1)} = V \otimes \mathcal{O}_{\mathbb{P}}(1)$ for every nontrivial line bundle L on $\mathbb{P}$.

Let A be in $\mathrm{ob}(\mathcal{A})$. An *infinitesimal deformation of the pair* (X, L) over A consists of a pair $(\xi, \mathcal{L})$ (also denoted by $(\mathcal{X}, \mathcal{L})$), where

$$\xi: \begin{array}{ccc} X & \to & \mathcal{X} \\ \downarrow & & \downarrow \\ \mathrm{Spec}(\mathbf{k}) & \to & \mathrm{Spec}(A) \end{array}$$

is an infinitesimal deformation of X over A and $\mathcal{L}$ is an invertible sheaf on $\mathcal{X}$ such that $L = \mathcal{L}_{|X}$. One can also say that $\mathcal{L}$ is a *deformation of L along* ξ. In the case $A = \mathbf{k}[\epsilon]$ we speak of a *first order-deformation* of (X, L). Two deformations $(\mathcal{X}, \mathcal{L})$ and $(\mathcal{X}', \mathcal{L}')$ of (X, L) over A will be called *isomorphic* if there is an isomorphism of deformations $f : \mathcal{X} \to \mathcal{X}'$ and an isomorphism $\mathcal{L} \to f^*\mathcal{L}'$.

By letting

$$\mathrm{Def}_{(X,L)}(A) = \Big\{\text{deformations of } (X, L) \text{ over } A\Big\}/\text{isomorphism}$$

we define a functor of Artin rings

$$\mathrm{Def}_{(X,L)} : \mathcal{A} \to (\text{sets})$$

called *the functor of infinitesimal deformations of the pair* (X, L).

We will denote by $[\mathcal{X}, \mathcal{L}] \in \mathrm{Def}_{(X,L)}(A)$ the isomorphism class of a deformation $(\mathcal{X}, \mathcal{L})$ of (X, L) over $A \in \mathrm{ob}(\mathcal{A})$.

Theorem 3.3.11. *Let* (X, L) *be a pair consisting of a nonsingular projective algebraic variety X and an invertible sheaf L on X. Then:*

(i) The functor $\mathrm{Def}_{(X,L)}$ *has a semiuniversal formal element.*
(ii) There is a canonical isomorphism

$$\mathrm{Def}_{(X,L)}(\mathbf{k}[\epsilon]) = \frac{\{\textit{1st-order deformations of } (X, L)\}}{\textit{isomorphism}} \cong H^1(X, \mathcal{E}_L)$$

and $H^2(X, \mathcal{E}_L)$ *is an obstruction space for* $\mathrm{Def}_{(X,L)}$.

(iii) Given a first-order deformation ξ of X, there is a first-order deformation of L along ξ if and only if

$$\kappa(\xi) \cdot c(L) = 0$$

where "$\cdot$" denotes the composition:

$$H^1(X, T_X) \times H^1(X, \Omega^1_X) \xrightarrow{\cup} H^2(X, T_X \otimes \Omega^1_X) \to H^2(X, \mathcal{O}_X)$$

of the cup product of cohomology classes $\cup$ with the map induced by the duality pairing $T_X \otimes \Omega^1_X \to \mathcal{O}_X$ (therefore the left-hand side is an element of $H^2(X, \mathcal{O}_X)$).

Proof. Let's check the conditions of Theorem 2.3.2. It is clear that $\mathrm{Def}_{(X,L)}$ satisfies condition H_0. Let

$$\begin{array}{ccc} A' & & A'' \\ & \searrow \quad \swarrow & \\ & A & \end{array}$$

be a diagram in $\mathcal{A}$ with $A'' \to A$ surjective. Consider an element

$$([\mathcal{X}', \mathcal{L}'], [\mathcal{X}'', \mathcal{L}'']) \in \mathrm{Def}_{(X,L)}(A') \times_{\mathrm{Def}_{(X,L)}(A)} \mathrm{Def}_{(X,L)}(A'')$$

Then there is an $[\mathcal{X}, \mathcal{L}] \in \mathrm{Def}_{(X,L)}(A)$ and a diagram of deformations:

$$\begin{array}{ccccc} \mathcal{X}' & & & & \mathcal{X}'' \\ & \nwarrow f' & & \nearrow f'' & \\ \downarrow & & \mathcal{X} & & \downarrow \\ \mathrm{Spec}(A') & & \downarrow & & \mathrm{Spec}(A'') \\ & \nwarrow & & \nearrow & \\ & & \mathrm{Spec}(A) & & \end{array}$$

where the morphisms f' and f'' induce isomorphisms of deformations:

$$\mathcal{X}' \times_{\mathrm{Spec}(A')} \mathrm{Spec}(A) \cong \mathcal{X} \cong \mathcal{X}'' \times_{\mathrm{Spec}(A'')} \mathrm{Spec}(A)$$

Moreover, we have isomorphisms

$$f'^* \mathcal{L}' \cong \mathcal{L} \cong f''^* \mathcal{L}''$$

Let $B = A' \times_A A''$. Then, as in the proof of Theorem 2.4.1, one sees that

$$\bar{\mathcal{X}} := (|X|, \mathcal{O}_{\mathcal{X}'} \times_{\mathcal{O}_{\mathcal{X}}} \mathcal{O}_{\mathcal{X}''})$$

is a deformation of X over B inducing the pair $([\mathcal{X}'], [\mathcal{X}''])$. Define

$$\bar{\mathcal{L}} := \mathcal{L}' \times_{\mathcal{L}} \mathcal{L}''$$

Then $\bar{\mathcal{L}}$ is an invertible sheaf over $\bar{\mathcal{X}}$ which restricts on $\mathcal{X}'$ to $\mathcal{L}'$ and on $\mathcal{X}''$ to $\mathcal{L}''$. Therefore the pair $(\bar{\mathcal{X}}, \bar{\mathcal{L}})$ defines an element of $\mathrm{Def}_{(X,L)}(B)$ such that

$$[\bar{\mathcal{X}}, \bar{\mathcal{L}}] \mapsto ([\mathcal{X}', \mathcal{L}'], [\mathcal{X}'', \mathcal{L}'']) \tag{3.31}$$

under the map

$$\mathrm{Def}_{(X,L)}(B) \to Def_{(X,L)}(A') \times_{\mathrm{Def}_{(X,L)}(A)} \mathrm{Def}_{(X,L)}(A'')$$

This proves that $\mathrm{Def}_{(X,L)}$ satisfies condition $\bar{H}$ of Theorem 2.3.2.

In order to show that $\mathrm{Def}_{(X,L)}$ satisfies condition H_ϵ we must prove that the element $[\bar{\mathcal{X}}, \bar{\mathcal{L}}]$ constructed above is the unique one satisfying (3.31) if we assume that $A'' = \mathbf{k}[\epsilon]$. From the proof of Theorem 2.4.1 we know that $[\bar{\mathcal{X}}] \in \mathrm{Def}_X(B)$ is the unique element inducing $([\mathcal{X}'], [\mathcal{X}'']) \in \mathrm{Def}_X(A') \times_{\mathrm{Def}_X(A)} \mathrm{Def}_X(A'')$. Using this fact the proof can now be completed along the lines of the proof of Theorem 3.3.1. We still need to verify that $\mathrm{Def}_{(X,L)}$ satisfies condition H_f. This will result as a consequence of part (ii) to be proved next, because X is projective.

(ii) Let $(\xi, \mathcal{L})$ be a first-order deformation of (X, L), where

$$\xi: \begin{array}{ccc} X & \to & \mathcal{X} \\ \downarrow & & \downarrow \\ \mathrm{Spec}(\mathbf{k}) & \to & \mathrm{Spec}(\mathbf{k}[\epsilon]) \end{array}$$

Let $\mathcal{U} = \{U_\alpha\}$ be an affine open covering such that L is given by a system of transition functions $f_{\alpha\beta} \in Z^1(\mathcal{U}, \mathcal{O}_X^*)$ and $\kappa(\xi) \in H^1(X, T_X)$ is given by a Čech 1-cocycle $(d_{\alpha\beta}) \in Z^1(\mathcal{U}, T_X)$. Let $\theta_{\alpha\beta} = 1 + \epsilon d_{\alpha\beta}$ be the automorphism of $U_{\alpha\beta} \times \mathrm{Spec}(\mathbf{k}[\epsilon])$ corresponding to $d_{\alpha\beta}$.

The invertible sheaf $\mathcal{L}$ can be given by a system of transition functions $(F_{\alpha\beta}) \in Z^1(\mathcal{U}, \mathcal{O}_{\mathcal{X}}^*)$ which reduces to $\{f_{\alpha\beta}\}$ mod ϵ. Therefore it can be represented on $U_{\alpha\beta} \times \mathrm{Spec}(\mathbf{k}[\epsilon])$ as

$$F_{\alpha\beta} = f_{\alpha\beta} + \epsilon g_{\alpha\beta}, \qquad g_{\alpha\beta} \in \Gamma(U_{\alpha\beta}, \mathcal{O}_X)$$

and the cocycle condition translates into

$$F_{\alpha\beta}\theta_{\alpha\beta}(F_{\beta\gamma}) = F_{\alpha\gamma}$$

equivalently:

$$F_{\alpha\beta}(F_{\beta\gamma} + \epsilon d_{\alpha\beta} F_{\beta\gamma}) = F_{\alpha\gamma}$$

which means:

$$(f_{\alpha\beta} + \epsilon g_{\alpha\beta})[f_{\beta\gamma} + \epsilon g_{\beta\gamma} + \epsilon d_{\alpha\beta}(f_{\beta\gamma} + \epsilon g_{\beta\gamma})] = f_{\alpha\gamma} + \epsilon g_{\alpha\gamma}$$

After dividing by $f_{\alpha\gamma}$ and equating the coefficients of ϵ we obtain:

$$\frac{g_{\alpha\beta}}{f_{\alpha\beta}} + \frac{g_{\beta\gamma}}{f_{\beta\gamma}} - \frac{g_{\alpha\gamma}}{f_{\alpha\gamma}} + \frac{d_{\alpha\beta} f_{\beta\gamma}}{f_{\beta\gamma}} = 0 \tag{3.32}$$

This identity means that the data $\{(\frac{g_{\alpha\beta}}{f_{\alpha\beta}}, d_{\alpha\beta})\}$ define an element of $\mathcal{Z}^1(\mathcal{U}, \mathcal{E}_L)$, and that conversely such an element defines a first-order deformation of (X, L). This proves that

$$\mathrm{Def}_{(X,L)}(\mathbf{k}[\epsilon]) \cong H^1(X, \mathcal{E}_L)$$

modulo verifying that this correspondence is independent from the choice of the covering $\mathcal{U}$ and of the cocycles representing (X, L); we leave this to the reader.

Consider a small extension

$$e : 0 \to (t) \to \tilde{A} \to A \to 0$$

in $\mathcal{A}$ and let $(\xi, \mathcal{L})$ be an infinitesimal deformation of (X, L) over A, where

$$\begin{array}{cccc} & X & \to & \mathcal{X} \\ \xi : & \downarrow & & \downarrow \\ & \mathrm{Spec}(\mathbf{k}) & \to & \mathrm{Spec}(A) \end{array}$$

Let $\mathcal{U} = \{U_\alpha\}$ be an affine open cover of X; let

$$\theta_\alpha : U_\alpha \times \mathrm{Spec}(A) \to \mathcal{X}_{|U_\alpha}$$

be isomorphisms so that $\theta_{\alpha\beta} := \theta_\alpha^{-1}\theta_\beta$ is an automorphism of the trivial deformation $U_{\alpha\beta} \times \mathrm{Spec}(A)$. We may assume that $\mathcal{L}$ is given by a system of transition functions $(F_{\alpha\beta})$, where each $F_{\alpha\beta}$ is a nowhere zero function on $U_{\alpha\beta} \times \mathrm{Spec}(A)$, such that

$$F_{\alpha\beta}\theta_{\alpha\beta}(F_{\beta\gamma}) = F_{\alpha\gamma}$$

In order to see if a lifting $(\tilde{\xi}, \tilde{\mathcal{L}})$ of $(\xi, \mathcal{L})$ to $\mathrm{Spec}(\tilde{A})$ exists we choose arbitrarily a collection

$$\{\tilde{\theta}_{\alpha\beta}, \tilde{F}_{\alpha\beta}\}$$

where, for each α, β, γ :

(a) $\tilde{\theta}_{\alpha\beta}$ is an automorphism of the product family $U_{\alpha\beta} \times \mathrm{Spec}(\tilde{A})$ which restricts to $\theta_{\alpha\beta}$ on $U_{\alpha\beta} \times \mathrm{Spec}(A)$.
(b) $\tilde{F}_{\alpha\beta}$ is a nowhere zero function on $U_{\alpha\beta} \times \mathrm{Spec}(\tilde{A})$ which restricts to $F_{\alpha\beta}$ on $U_{\alpha\beta} \times \mathrm{Spec}(A)$.

Such collection exists by Lemma 1.2.8. Because of (a) we have

$$\tilde{\theta}_{\alpha\beta}\tilde{\theta}_{\beta\gamma}\tilde{\theta}_{\alpha\gamma}^{-1} = \mathrm{id} + td_{\alpha\beta\gamma}$$

where "id" here means the identity of $\Gamma(U_{\alpha\beta\gamma}, \mathcal{O}_X)$, and $(d_{\alpha\beta\gamma}) \in \mathcal{Z}^2(\mathcal{U}, T_X)$ is a 2-cocycle which represents the obstruction to lifting ξ over $\mathrm{Spec}(\tilde{A})$ (see the proof of Proposition 1.2.12). Because of (b), for each α, β, γ there is $g_{\alpha\beta\gamma} \in \Gamma(U_{\alpha\beta\gamma}, \mathcal{O}_X)$ such that

$$\tilde{F}_{\alpha\beta}\tilde{\theta}_{\alpha\beta}(\tilde{F}_{\beta\gamma})\tilde{F}_{\alpha\gamma}^{-1} = 1 + tg_{\alpha\beta\gamma}$$

Therefore we have:

$$\left(\tilde{\theta}_{\alpha\beta}\left[\tilde{F}_{\beta\gamma}\tilde{\theta}_{\beta\gamma}(\tilde{F}_{\gamma\delta})\tilde{F}_{\beta\delta}^{-1}\right]\right)\cdot$$
$$[\tilde{F}_{\alpha\gamma}\tilde{\theta}_{\alpha\gamma}(\tilde{F}_{\gamma\delta})\tilde{F}_{\alpha\delta}^{-1}]^{-1}[\tilde{F}_{\alpha\beta}\tilde{\theta}_{\alpha\beta}(\tilde{F}_{\beta\delta})\tilde{F}_{\alpha\delta}^{-1}][\tilde{F}_{\alpha\beta}\tilde{\theta}_{\alpha\beta}(\tilde{F}_{\beta\gamma})\tilde{F}_{\alpha\gamma}^{-1}]^{-1}$$
$$= 1 + t(g_{\beta\gamma\delta} - g_{\alpha\gamma\delta} + g_{\alpha\beta\delta} - g_{\alpha\beta\gamma})$$

On the other hand, the left side can be written as

$$\tilde{\theta}_{\alpha\beta}\tilde{\theta}_{\beta\gamma}(\tilde{F}_{\gamma\delta})[\tilde{\theta}_{\alpha\gamma}(\tilde{F}_{\gamma\delta})]^{-1} = t\frac{d_{\alpha\beta\gamma}(f_{\gamma\delta})}{f_{\gamma\delta}}$$

Therefore we see that

$$(g_{\alpha\beta\gamma}, d_{\alpha\beta\gamma}) \in Z^2(\mathcal{U}, \mathcal{E}_L) \tag{3.33}$$

We are free to modify our choice by replacing the $\tilde{\theta}_{\alpha\beta}$'s and the $\tilde{F}_{\alpha\beta}$'s by

$$\Phi_{\alpha\beta} = \tilde{\theta}_{\alpha\beta} + td_{\alpha\beta}, \qquad G_{\alpha\beta} = \tilde{F}_{\alpha\beta} + tg_{\alpha\beta} \tag{3.34}$$

for some $d_{\alpha\beta} \in \Gamma(U_{\alpha\beta}, T_X)$ and $g_{\alpha\beta} \in \Gamma(U_{\alpha\beta}, \mathcal{O}_X)$. One checks easily that the cocycle (3.33) is replaced by the cohomologous one:

$$(g_{\alpha\beta\gamma} + g_{\alpha\beta} - g_{\alpha\gamma} + g_{\beta\gamma}, d_{\alpha\beta\gamma} + d_{\alpha\beta} - d_{\alpha\gamma} + d_{\beta\gamma}) \tag{3.35}$$

and it is clear that the lifting $(\tilde{\xi}, \tilde{\mathcal{L}})$ exists if and only if the data (3.34) can be determined so that (3.35) is zero. Therefore associating to the extension e the cohomology class $o_{(\xi,\mathcal{L})}(e) \in H^2(X, \mathcal{E}_L)$ defined by the cocycle (3.33) we have defined an obstruction map

$$\mathrm{Ex}_{\mathbf{k}}(A, \mathbf{k}) \to H^2(X, \mathcal{E}_L)$$

which we leave to the reader to verify to be linear. This makes $H^2(X, \mathcal{E}_L)$ an obstruction space for the functor $\mathrm{Def}_{(X,L)}$.

(iii) Observing that

$$\left(\frac{d_{\alpha\beta}f_{\beta\gamma}}{f_{\beta\gamma}}\right) \in Z^2(\mathcal{U}, \mathcal{O}_X)$$

represents $\kappa(\xi)\cdot c(L)$, the identity (3.32) expresses the condition that this 2-cocycle is a coboundary, and proves (iii). □

If X is a connected nonsingular projective curve then $H^2(X, \mathcal{O}_X) = 0$ so that every line bundle can be deformed along any first-order deformation of X. Moreover, $H^2(X, \mathcal{E}_L) = 0$ for any L, thus $\mathrm{Def}_{(X,L)}$ is smooth and from the exact sequence (3.30) it follows that

$$h^1(X, \mathcal{E}_L) = 4g - 3$$

if X has genus $g \geq 2$.

For higher-dimensional varieties the situation is more complicated in general. For example, if X is a K3-surface then the cup product

$$H^1(X, T_X) \times H^1(X, \Omega^1_X) \to H^2(X, \mathcal{O}_X) \cong \mathbf{k}$$

coincides with Serre duality. Therefore

$$H^1(X, T_X) \xrightarrow{\cdot c(L)} H^2(X, \mathcal{O}_X)$$

is surjective for every nontrivial line bundle L. This means that L deforms along a 19-dimensional subspace of $H^1(X, T_X)$, because $h^1(X, T_X) = 20$ (see Example 2.4.11(ii), page 73).

We have a natural *forgetful morphism*

$$\Phi : \mathrm{Def}_{(X,L)} \to \mathrm{Def}_X$$

defined in the obvious way. The differential and the obstruction map of this morphism are described as follows.

Proposition 3.3.12. *In the situation of Theorem 3.3.11:*

(i) The differential

$$d\Phi : \mathrm{Def}_{(X,L)}(\mathbf{k}[\epsilon]) \to \mathrm{Def}_X(\mathbf{k}[\epsilon])$$

coincides with the linear map

$$H^1(X, \mathcal{E}_L) \to H^1(X, T_X)$$

coming from the exact sequence (3.30).

(ii) The map

$$H^2(X, \mathcal{E}_L) \to H^2(X, T_X)$$

coming from the exact sequence (3.30) is an obstruction map for Φ.

(iii) If $H^2(X, \mathcal{O}_X) = 0$ the morphism Φ is smooth.

Proof. (i) and (ii) are left to the reader. For (iii) use Proposition 2.3.6. □

Example 3.3.13. Assume $\mathbf{k} = \mathbf{C}$. Let $A = V/\Lambda$ be an abelian variety of dimension g, represented as the quotient of a g-dimensional complex vector space V by a lattice $\Lambda \subset V$. Then, letting $\Omega = V^\vee$ we have

$$T_A = V \otimes \mathcal{O}_A, \qquad \Omega^1_A = \Omega \otimes \mathcal{O}_A$$

Moreover,

$$H^1(A, \mathcal{O}_A) = \bar{\Omega} := \mathrm{Hom}_{\bar{\mathbf{C}}}(V, \mathbf{C})$$

is the space of $\mathbf{C}$-antilinear forms on V and

$$H^2(A, \mathcal{O}_A) = \bigwedge^2 \bar{\Omega}$$

(see [20], Theorem 1.4.1). Therefore

$$H^1(A, T_A) = V \otimes H^1(A, \mathcal{O}_A) = V \otimes \bar{\Omega}$$

In particular, $h^1(A, T_A) = g^2$. We also have:

$$H^1(A, \Omega_A^1) = \Omega \otimes H^1(A, \mathcal{O}_A) = \Omega \otimes \bar{\Omega}$$

which can be identified with the space of hermitian forms on V. Let L be an ample invertible sheaf on X. Then

$$c(L) \in H^1(A, \Omega_A^1) = \Omega \otimes \bar{\Omega}$$

is identified with a positive definite hermitian form $h_L : V \times V \to \mathbf{C}$ ([20], § 4.1). It follows that the map

$$H^1(A, T_A) \xrightarrow{\cdot c(L)} H^2(A, \mathcal{O}_A) \tag{3.36}$$

of Theorem 3.3.11 is just the composition of the map

$$\begin{array}{ccc} V \otimes \bar{\Omega} & \to & \bar{\Omega} \otimes \bar{\Omega} \\ v \otimes \ell & \mapsto & h_L(v, -) \otimes \ell \end{array} \tag{3.37}$$

with the canonical surjection

$$\bar{\Omega} \otimes \bar{\Omega} \to \bigwedge^2 \bar{\Omega}$$

Since h_L is positive definite the map (3.37) is an isomorphism and it follows that the map (3.36) is surjective. The conclusion is that L deforms along a subspace of $H^1(A, T_A)$ of dimension

$$\frac{g(g+1)}{2} = g^2 - \binom{g}{2}$$

For the related case of jacobians see Example 3.4.24(iii), page 182.

3.3.4 Deformations of sections, II

Consider again a projective nonsingular variety X and an invertible sheaf L on X. Assume as above that L is given, in an affine open cover $\mathcal{U} = \{U_\alpha\}$, by transition functions $(f_{\alpha\beta}) \in Z^1(\mathcal{U}, \mathcal{O}_X^*)$. We can define a homomorphism of sheaves

$$M : \mathcal{E}_L \to H^0(X, L)^\vee \otimes L$$

in the following way. Consider a section $\eta \in \Gamma(U, \mathcal{E}_L)$, where $U \subset X$ is an open set; it is given by a system (a_α, d_α) where $a_\alpha \in \Gamma(U \cap U_\alpha, \mathcal{O}_X)$, $d_\alpha \in \Gamma(U \cap U_\alpha, T_X)$, subject to the conditions that $d_\beta = d_\alpha$ and $a_\beta - a_\alpha = d_\alpha(f_{\alpha\beta})/f_{\alpha\beta}$ on $U \cap U_\alpha \cap U_\beta$. Then, for every $s = (s_\alpha) \in H^0(X, L)$ we let

$$M(\eta)(s_\alpha) = a_\alpha s_\alpha + d_\alpha(s_\alpha)$$

On $U \cap U_\alpha \cap U_\beta$ we find:

$$\begin{aligned} f_{\alpha\beta} M(\eta)(s_\beta) &= f_{\alpha\beta}(a_\beta s_\beta + d_\beta(s_\beta)) = a_\beta s_\alpha + f_{\alpha\beta} d_\beta(s_\beta) \\ &= s_\alpha(a_\alpha + d_\alpha(f_{\alpha\beta})/f_{\alpha\beta}) + f_{\alpha\beta} d_\beta(s_\beta) = s_\alpha a_\alpha + d_\alpha(f_{\alpha\beta}) s_\beta + f_{\alpha\beta} d_\beta(s_\beta) \\ &= s_\alpha a_\alpha + d_\alpha(f_{\alpha\beta} s_\beta) = s_\alpha a_\alpha + d_\alpha(s_\alpha) = M(\eta)(s_\alpha) \end{aligned}$$

Therefore the functions $M(\eta)(s_\alpha) \in \Gamma(U \cap U_\alpha, \mathcal{O}_X)$ patch together to define a section $M(\eta)(s) \in \Gamma(U, L)$. This defines M. It is obvious from the definition that we have a commutative diagram:

$$\begin{array}{rcl} m : \mathcal{O}_X & \to & H^0(X, L)^\vee \otimes L \\ \downarrow & & \| \\ M : \mathcal{E}_L & \to & H^0(X, L)^\vee \otimes L \end{array} \tag{3.38}$$

namely M extends m (see (3.24), page 141). The linear map

$$M_1 : H^1(X, \mathcal{E}_L) \to \mathrm{Hom}(H^0(X, L), H^1(X, L))$$

induced by M can be explicitly described as follows. Let $\eta_1 \in H^1(X, \mathcal{E}_L)$ be represented by the Čech cocycle $(a_{\alpha\beta}, d_{\alpha\beta}) \in Z^1(\mathcal{U}, \mathcal{E}_L)$. Then

$$\begin{array}{rcl} M_1(\eta_1) : H^0(X, L) & \to & H^1(X, L) \\ (s_\alpha) & \mapsto & \overline{(a_{\alpha\beta} s_\alpha + d_{\alpha\beta} s_\alpha)} \end{array}$$

The map M_1 has the following deformation theoretic interpretation. Let $A \in \mathrm{ob}(\mathcal{A})$ and let $(\mathcal{X}, \mathcal{L})$ be an infinitesimal deformation of (X, L) over $\mathrm{Spec}(A)$. Then we say that *a section $s \in H^0(X, L)$ extends to $\mathcal{L}$* if

$$s \in \mathrm{Im}[H^0(\mathcal{X}, \mathcal{L}) \to H^0(X, L)]$$

Proposition 3.3.14. *Let X be a projective nonsingular variety, L a line bundle on X, and $(\mathcal{X}, \mathcal{L})$ a first-order deformation of the pair (X, L) defined by a cohomology class $\eta_1 \in H^1(X, \mathcal{E}_L)$ according to Theorem 3.3.11(ii). Then a section $s \in H^0(X, L)$ extends to $\mathcal{L}$ if and only if $s \in \ker(M_1(\eta_1))$.*

Proof. With respect to an affine cover $\mathcal{U} = \{U_\alpha\}$ of X we assume L represented by $(f_{\alpha\beta}) \in Z^1(\mathcal{U}, \mathcal{O}_X^*)$ and η_1 by $(a_{\alpha\beta}, d_{\alpha\beta}) \in Z^1(\mathcal{U}, \mathcal{E}_L)$. Then, according to the description given in the proof of Theorem 3.3.11, $\mathcal{X}$ is determined by gluing the products $U_\alpha \times \mathrm{Spec}(\mathbf{k}[\epsilon])$ along the open subsets $U_{\alpha\beta} \times \mathrm{Spec}(\mathbf{k}[\epsilon])$ by means of the automorphisms $\theta_{\alpha\beta} = 1 + \epsilon d_{\alpha\beta}$, and $\mathcal{L}$ is determined by transition functions of the form $F_{\alpha\beta} = f_{\alpha\beta}(1 + \epsilon a_{\alpha\beta})$ satisfying the patching identities on $U_{\alpha\beta\gamma} \times \mathrm{Spec}(\mathbf{k}[\epsilon])$:

$$F_{\alpha\beta} \theta_{\alpha\beta}(F_{\beta\gamma}) = F_{\alpha\gamma}$$

The condition that $s = (s_\alpha) \in H^0(X, L)$ extends to $\mathcal{L}$ is equivalent to the existence of functions $t_\alpha \in \Gamma(U_\alpha, \mathcal{O}_X)$ such that

$$\theta_{\alpha\beta}[F_{\alpha\beta}(s_\beta + \epsilon t_\beta)] = s_\alpha + \epsilon t_\alpha$$

for all α, β. After replacing the expressions for $F_{\alpha\beta}$ and $\theta_{\alpha\beta}$ we obtain the identities:

$$f_{\alpha\beta}(1 + \epsilon a_{\alpha\beta})(s_\beta + \epsilon t_\beta) + \epsilon d_{\alpha\beta}(f_{\alpha\beta} s_\beta) = s_\alpha + \epsilon t_\alpha$$

By equating the coefficients of ϵ on both sides we obtain:

$$a_{\alpha\beta} s_\alpha + d_{\alpha\beta}(s_\alpha) = t_\alpha - f_{\alpha\beta} t_\beta$$

and this means exactly that $M_1(\eta_1)(s) = 0$. □

Corollary 3.3.15. *In the above situation:*

(i) If $\eta_1 \in \ker(M_1)$ then all the sections of L extend to the first-order deformation of (X, L) corresponding to η_1.
(ii) For any $\eta_1 \in H^1(X, \mathcal{E}_L)$ at least $\max\{0, h^0(X, L) - h^1(X, L)\}$ linearly independent sections of L extend to the first-order deformation of (X, L) corresponding to η_1.

Proof. Immediate. □

This statement, as well as Proposition 3.3.14, show the deformation theoretic interest of the map M_1. It can be useful to have a closer picture of the relation between the maps m and M. Assuming that L is base point free and denoting by $\varphi_L : X \to \mathbb{P} := \mathbb{P}(H^0(X, L)^\vee)$ the morphism defined by the sections of L, we have a commutative and exact diagram, which extends (3.38):

$$\begin{array}{ccccccccc} & & & & & & 0 & & \\ & & & & & & \downarrow & & \\ & & 0 & & & & T_X & & \\ & & \downarrow & & & & \downarrow & & \\ 0 & \to & \mathcal{O}_X & \xrightarrow{m} & H^0(X, L)^\vee \otimes L & \to & \varphi_L^* T_{\mathbb{P}} & \to & 0 \\ & & \downarrow & & \| & & \downarrow & & \\ 0 & \to & \mathcal{E}_L & \xrightarrow{M} & H^0(X, L)^\vee \otimes L & \to & N_{\varphi_L} & \to & 0 \\ & & \downarrow & & & & \downarrow & & \\ & & T_X & & & & 0 & & \\ & & \downarrow & & & & & & \\ & & 0 & & & & & & \end{array} \tag{3.39}$$

The first column is the Atiyah extension. The sheaf N_{φ_L} is the normal sheaf of the morphism φ_L. It will be considered more systematically in § 3.4, page 162. In the case where φ_L is a closed embedding, $N_{\varphi_L} = N_{\varphi_L(X)/\mathbb{P}}$ and this diagram shows that

$$\operatorname{coker}(m_1) \subset H^1(X, \varphi^* T_{\mathbb{P}})$$

where we recall that

$$m_1 : H^1(X, \mathcal{O}_X) \to \operatorname{Hom}(H^0(X, L), H^1(X, L))$$

is the map induced by m (see Proposition 3.3.4, page 141). Therefore, according to Proposition 3.2.9(iii), $h^2(X, \mathcal{O}_X) = 0$ and the surjectivity of m_1 are sufficient conditions for the smoothness of the forgetful morphism $\Phi : H^{I\!P}_{\varphi_L(X)} \to \mathrm{Def}_X$. In particular, if X is a nonsingular curve and m_1 is surjective then Φ is smooth (see also Example 3.2.5(v)).

Example 3.3.16. Let X be a connected projective nonsingular curve of genus g. Then the map M_1 is dual to

$$P_L : H^0(X, L) \otimes H^0(X, \omega_X L^{-1}) \to H^0(X, \omega_X \otimes \mathcal{E}_L^\vee)$$

More generally, for any vector subspace $V \subset H^0(X, L)$ we can consider the map obtained by restricting P_L:

$$P_V : V \otimes H^0(X, \omega_X L^{-1}) \to H^0(X, \omega_X \otimes \mathcal{E}_L^\vee)$$

It follows directly from Proposition 3.3.14 that $\mathrm{coker}(P_V)^\perp \subset H^1(X, \mathcal{E}_L)$ is the space of first-order deformations of (X, L) to which all sections of V extend. P_L (resp. P_V) is called *the extended Petri map of* L (resp. *of* V). After dualizing the cohomology diagram of (3.39) we deduce the following commutative diagram which relates the Petri map and the extended Petri map in the case where L is globally generated:

$$\begin{array}{ccccccc} & & & & & & 0 \\ & & & & & & \downarrow \\ & & 0 & & & & H^0(\omega_X^2) \\ & & \downarrow & & & & \downarrow \\ 0 \to & & H^1(N_{\varphi_L})^\vee & \to & H^0(L) \otimes H^0(\omega_X L^{-1}) & \xrightarrow{P_L} & H^0(\omega_X \otimes \mathcal{E}_L^\vee) \\ & & \downarrow & & \| & & \downarrow \\ 0 \to & & H^1(\varphi_L^* T_{I\!P})^\vee & \to & H^0(L) \otimes H^0(\omega_X L^{-1}) & \xrightarrow{\mu_0(L)} & H^0(\omega_X) \end{array} \tag{3.40}$$

If $\varphi_L : X \subset I\!P$ is an embedding then this diagram contains some significant information on the relations between $H^{I\!P}_X$ and Def_X. From what has been remarked just before this example it follows that the injectivity of $\mu_0(L)$, being equivalent to the surjectivity of its dual m_1, is a sufficient condition for the smoothness of the forgetful morphism $\Phi : H^{I\!P}_X \to \mathrm{Def}_X$. Diagram (3.40) also shows that we have an exact sequence

$$\begin{array}{ccccc} 0 \to H^1(N_{X/I\!P})^\vee & \to & H^1(T_{I\!P|X})^\vee & \xrightarrow{\mu_1(L)} & H^0(\omega_X^2) \\ \| & & \| & & \\ \ker(P_L) & & \ker(\mu_0(L)) & & \end{array}$$

thus the injectivity of P_L is implied by the injectivity of $\mu_0(L)$ or more generally, by the injectivity of $\mu_1(L)$.

For more about this topic we refer the reader to [8].

NOTES

1. The coboundary maps δ_k in the cohomology sequence of the Atiyah extension (3.30) are induced by cup product with $c(L)$, since the extension is defined by $c(L)$. In particular,

3.3.11(iii) just says that L deforms along ζ if and only if

$$\kappa(\xi) \in \ker[\delta_1 : H^1(X, T_X) \to H^2(X, \mathcal{O}_X)],$$

which is obvious in view of (i) and (ii).

2. The content of Theorem 3.3.11(iii) is outlined in the Appendix by Mumford to Chapter V of [189]. See also [89]. This result is related to the notion of *deformation of a polarization* (see [132]).

3. For a discussion of Example 3.3.13 for abelian varieties defined over a field of positive characteristic see Oort [137], remark on page 226.

4. The Petri map has been considered classically in [141] and it reappeared in the modern literature for the first time in [10]. It is of great importance in the study of the Brill–Noether loci on a curve and of their relation to moduli (see [9] for more details).

3.4 Morphisms

In this section we study deformations of a morphism between algebraic schemes. In the analytic case the corresponding theory has been developed by Horikawa in [85], [86], [88] and [89]. For a treatment in the analytic case we refer to [134]. Related work in the algebraic case is in [114] and [148].

Definition 3.4.1. *Let $f : X \to Y$ be a morphism of algebraic schemes and let A be in* ob($\mathcal{A}$) *(resp. $A = \mathbf{k}[\epsilon]$, resp. A in* ob($\mathcal{A}^*$)*). An* infinitesimal *(resp.* first-order, *resp.* local) family of deformations of f parametrized by *A or by S (more briefly a* deformation of f over A or over S) *is a cartesian diagram*

$$\begin{array}{ccc} X & \to & \mathcal{X} \\ \downarrow f & & \downarrow F \\ Y & \to & \mathcal{Y} \\ \downarrow & & \downarrow \psi \\ \operatorname{Spec}(\mathbf{k}) & \xrightarrow{s} & S \end{array} \tag{3.41}$$

where S = Spec(A)*, and ψ and ψF are flat ("cartesian diagram" in this case means that the horizontal morphisms induce an isomorphism of the left column with the pullback of the right column by s). If we replace S by a pointed scheme (S, s) we will call (3.41) a* family of deformations of f.

Essentially, a deformation of f consists of a morphism F between deformations of X and of Y and an assigned identification with f of the restriction of F to the closed fibre. Note that if $Y = \operatorname{Spec}(\mathbf{k})$ then a family of deformations of $f : X \to \operatorname{Spec}(\mathbf{k})$ is just a family of deformations of X in the sense of the definition given at the beginning of § 1.2.

The notion of *trivial deformation* of f can be given in a obvious way, as well as the notion of *rigid morphism*.

Given an infinitesimal deformation (3.41) and a small extension

$$e: \qquad 0 \to (t) \to \tilde{A} \to A \to 0$$

a *lifting* of (3.41) to $\tilde{A}$ is a cartesian diagram

$$\begin{array}{ccccc} X & \to & \mathcal{X} & \to & \tilde{\mathcal{X}} \\ \downarrow f & & \downarrow F & & \downarrow \tilde{F} \\ Y & \to & \mathcal{Y} & \to & \tilde{\mathcal{Y}} \\ \downarrow & & \downarrow \psi & & \downarrow \tilde{\psi} \\ \mathrm{Spec}(\mathbf{k}) & \xrightarrow{s} & \mathrm{Spec}(A) & \to & \mathrm{Spec}(\tilde{A}) \end{array}$$

(i.e. the horizontal morphisms induce an isomorphism of each column with the pull-back of the column on the right by the lowest horizontal morphisms) where $\tilde{F}$ and $\tilde{\psi}\tilde{F}$ are flat. Of course such a lifting defines in particular a deformation of f over $\tilde{A}$.

Definition 3.4.1 gives the most general notion of deformation of a morphism. It can be modified in several ways so to obtain more restricted notions each having independent interest and applications. We will study some of them starting from the simplest ones. In each situation we will consider a corresponding functor of Artin rings which classifies deformations modulo an appropriate equivalence relation.

3.4.1 Deformations of a morphism leaving domain and target fixed

If in Definition 3.4.1 we impose that both $\mathcal{X}$ and $\mathcal{Y}$ are the product families, i.e. if we consider a cartesian diagram of the form:

$$\begin{array}{cc} X & \to X \times S \\ \downarrow f & \downarrow F \\ Y & \to Y \times S \\ \downarrow & \downarrow \psi \\ \mathrm{Spec}(\mathbf{k}) \to & S \end{array}$$

where $S = \mathrm{Spec}(A)$ and ψ is the projection, then we obtain the notion of *deformation of f with fixed domain and target* (*wfdat*).

Deformations of f wfdat can be interpreted as deformations of the graph of f in $X \times Y$ so that the methods introduced in § 3.2 apply. Precisely, let's define a functor of Artin rings by setting

$$\mathrm{Def}_{X/f/Y}(A) = \left\{\text{deformations of } f \text{ over } A \text{ wfdat}\right\}$$

for all A in $\mathrm{ob}(\mathcal{A})$. Then we have the following:

Proposition 3.4.2. *Let $f : X \to Y$ be a morphism of algebraic schemes, with X projective and reduced and Y nonsingular. Then:*

(i) There is a natural isomorphism of functors

$$\mathrm{Def}_{X/f/Y} \cong H^{X\times Y}_{\Gamma_f}$$

where $\Gamma_f \subset X \times Y$ is the graph of f. In particular, $\mathrm{Def}_{X/f/Y}$ is prorepresentable.
(ii) We have a natural isomorphism of vector spaces:

$$\mathrm{Def}_{X/f/Y}(\mathbf{k}[\epsilon]) \to H^0(X, f^*T_Y)$$

(iii) $H^1(X, f^*T_Y)$ is an obstruction space for the functor $\mathrm{Def}_{X/f/Y}$.

Proof. (i) Let A be in $\mathrm{ob}(\mathcal{A})$ and let $F : X \times \mathrm{Spec}(A) \to Y \times \mathrm{Spec}(A)$ be a deformation of f. Its graph $\Gamma_F \subset X \times Y \times \mathrm{Spec}(A)$ defines a deformation of $\Gamma_f \subset X \times Y$ over A because $\Gamma_F \cong X \times \mathrm{Spec}(A)$ and the projection $\Gamma_F \to \mathrm{Spec}(A)$ equals the composition

$$\Gamma_F \cong X \times \mathrm{Spec}(A) \to \mathrm{Spec}(A)$$

in particular, it is flat. Therefore we can define

$$\mathrm{Def}_{X/f/Y}(A) \to H^{X\times Y}_{\Gamma_f}(A)$$

by

$$F \mapsto (\Gamma_F \subset X \times Y \times \mathrm{Spec}(A))$$

This is an isomomorphism of functors. In fact, given a deformation

$$\Gamma_A \subset X \times Y \times \mathrm{Spec}(A)$$

of Γ_F in $X \times Y$ the projection

$$X \times Y \times \mathrm{Spec}(A) \to X \times \mathrm{Spec}(A)$$

induces a morphism $\Gamma_A \to X \times \mathrm{Spec}(A)$ which is an isomorphism of deformations. Therefore the composition

$$X \times \mathrm{Spec}(A) \cong \Gamma_A \subset X \times Y \times \mathrm{Spec}(A) \to Y \times \mathrm{Spec}(A)$$

can be identified with a deformation of f. This defines the inverse of the morphism of functors of the statement.

(ii) We have natural isomorphisms of vector spaces:

$$\mathrm{Def}_{X/f/Y}(\mathbf{k}[\epsilon]) \cong H^{X\times Y}_{\Gamma_f}(\mathbf{k}[\epsilon]) \cong H^0(\Gamma, N_{\Gamma/X\times Y})$$

Since the projection $p : X \times Y \to X$ is smooth and the composition $pj : \Gamma \to X$ is an isomorphism, from Proposition D.2.5 it follows that j is a regular embedding. Therefore applying Proposition D.1.4 we obtain the exact sequence:

$$0 \to \mathcal{I}_\Gamma/\mathcal{I}_\Gamma^2 \to j^*\Omega^1_{X\times Y} \to \Omega^1_\Gamma \to 0$$

On the other hand, we have the exact sequence:

$$0 \to f^*\Omega^1_Y \to j^*\Omega^1_{X\times Y} \to (pj)^*\Omega^1_X \to 0$$

obtained by restricting to Γ the sequence

$$0 \to q^*\Omega^1_Y \to \Omega^1_{X\times Y} \to p^*\Omega^1_X \to 0$$

(where $q : X \times Y \to Y$ is the second projection). Since $(pj)^*\Omega^1_X \cong \Omega^1_\Gamma$, comparing the two sequences we deduce that $f^*\Omega^1_Y \cong \mathcal{I}_\Gamma/\mathcal{I}_\Gamma^2$. Therefore

$$H^0(\Gamma, N_{\Gamma/X\times Y}) = \mathrm{Hom}(\mathcal{I}_\Gamma/\mathcal{I}_\Gamma^2, \mathcal{O}_\Gamma) \cong \mathrm{Hom}(f^*\Omega^1_Y, \mathcal{O}_X) = H^0(X, f^*T_Y)$$

and (ii) follows.

Similarly, $H^1(\Gamma, N_{\Gamma/X\times Y}) = H^1(X, f^*T_Y)$ and (iii) follows as well. □

The notions of *obstructed* (resp. *unobstructed*) *deformation*, and of *obstructed* (resp. *unobstructed*) *morphism* can be given in the usual way. One can give the notion of *rigid morphism wfdat* in an obvious way. We leave to the reader the task of proving that, under the hypothesis of Proposition 3.4.2, $H^0(X, f^*T_Y) = 0$ implies that f is rigid wfdat.

Let $f : X \to Y$ be a morphism of algebraic schemes with X reduced and Y nonsingular; let (3.41) be a family of deformations of f wfdat parametrized by a pointed scheme (S, s). To every tangent vector $t \in T_{S,s}$, viewed as a morphism $\mathrm{Spec}(\mathbf{k}[\epsilon]) \to S$ with image $\{s\}$ we can associate the pullback of the family F over $\mathrm{Spec}(\mathbf{k}[\epsilon])$ by t and, using the correspondence of 3.4.2(i), we obtain an element of $H^0(X, f^*T_Y)$ associated to s. This defines a natural linear map

$$T_{S,s} \to H^0(X, f^*T_Y)$$

which will be called the *characteristic map* of the family (3.41) wfdat.

The next corollary follows immediately from the proposition. It can also be deduced as a consequence of Lemma 1.2.6.

Corollary 3.4.3. *If X is a nonsingular projective scheme, then the space of first-order deformations of the identity $X \to X$ is $H^0(X, T_X)$.*

Examples 3.4.4. (i) Let $f : X \to Y$ be a nonconstant morphism of projective nonsingular connected curves, with $g(Y) \geq 2$. Then $\deg(T_Y) < 0$ and therefore $h^0(X, f^*T_Y) = 0$. Thus f is rigid as a morphism wfdat.

(ii) A morphism f from a scheme X to a projective space is given by a linear system on X: deformations of f wfdat can thus be interpreted as "deformations of linear systems" in an appropriate way. We briefly discuss the case of curves. Let X be a projective irreducible and nonsingular curve, $f : X \to \mathbb{P}^r$ a morphism and let $L = f^*\mathcal{O}_{\mathbb{P}^r}(1)$, $\deg(L) = n$. Then f is defined by a vector subspace $V \subset H^0(X, L)$

of dimension $r+1$ plus the choice of a basis of V. From the Euler sequence pulled back to X we have:

$$\chi(f^*T_{\mathbb{P}^r}) = (r+1)\chi(L) - \chi(\mathcal{O}_X) = \rho(g,r,n) + r(r+2)$$

where

$$\rho(g,r,n) := g - (r+1)(g-n+r)$$

is the *Brill–Noether number* and

$$r(r+2) = h^0(T_{\mathbb{P}^r}) = \dim[PGL(r+1)]$$

Assume that $r+1 = h^0(L)$, i.e. that f is defined by the complete linear series $|L|$, and consider the exact sequence

$$H^1(\mathcal{O}_X) \to H^1(L)^{r+1} \to H^1(f^*T_{\mathbb{P}^r}) \to 0$$

obtained from the Euler sequence. It dualizes as:

$$0 \to H^1(f^*T_{\mathbb{P}^r})^\vee \to H^0(L) \otimes H^0(\omega_X L^{-1}) \overset{\mu_0(L)}{\longrightarrow} H^0(\omega_X)$$

where $\mu_0(L)$ is the Petri map (see Example 3.3.9, page 144). Therefore we see that f is unobstructed if $\mu_0(L)$ is injective. A necessary condition for this to be true is that

$$(r+1)(g-n+r) = \dim[H^0(L) \otimes H^0(\omega_X L^{-1})] \leq h^0(\omega_X) = g$$

i.e. that $\rho(g,r,n) \geq 0$. This necessary condition is not sufficient. The simplest example is given by a nonsingular complete intersection $X = Q \cap S \subset \mathbb{P}^3$ of a quadric cone Q and of a cubic surface S. Then X is a canonical curve of genus 4 and the projection from the vertex of the cone defines a complete g^1_3 $|L|$ such that $\omega_X L^{-1} \cong L$. In this case $\rho(4,1,3) = 0$ but $\dim[\ker(\mu_0(L))] = 1$. The morphism $f : X \to \mathbb{P}^1$ is obstructed because $h^0(f^*T_{\mathbb{P}^1}) = 4$ but L is the unique g^1_3 on X. Therefore the unobstructed first-order deformations of f are only those in the three-dimensional space coming from the automorphisms of $\mathbb{P}^1$.

(iii) If $\mathbb{P}^1 \cong E \subset S$ is a nonsingular projective rational curve negatively embedded in a projective nonsingular surface S with $E^2 = -n < 0$, $n \geq 1$, then we have $h^0(E, N_{E/S}) = 0$ and $h^0(E, T_{S|E}) = 3$. More precisely, the exact sequence

$$0 \to T_E \to T_{S|E} \to N_{E/S} \to 0$$

splits because $\mathrm{Ext}^1_{\mathcal{O}_E}(N_{E/S}, T_E) = H^1(E, \mathcal{O}_E(n+2)) = 0$: therefore

$$T_{S|E} \cong \mathcal{O}_E(2) \oplus \mathcal{O}_E(-n)$$

This means that, despite the fact that E is rigid in S, the morphism $f : E \to S$ has a three-dimensional family of deformations obtained by composing it with the automorphisms of E.

More generally, whenever we have an embedding $f : \mathbb{P}^1 \to Y$ with Y nonsingular algebraic variety, we have an inclusion

$$H^0(\mathbb{P}^1, T_{\mathbb{P}^1}) \subset H^0(\mathbb{P}^1, f^*T_Y)$$

which implies $h^0(\mathbb{P}^1, f^*T_Y) \geq 3$. In general, for any nonconstant morphism $f : \mathbb{P}^1 \to Y$ the sheaf f^*T_Y is locally free and splits as a direct sum of $\dim(Y) - 1$ invertible sheaves, by the structure theorem (see [136] p. 22). The study of such morphisms is closely related to the notions of *uniruledness* and *rational connectedness*. We refer to Debarre [42] and to Kollar [109] for a detailed treatment of these matters.

(iv) Similarly, if $E \subset Y$ is an embedding of a projective nonsingular curve of genus 1 into a nonsingular algebraic variety Y, from the inclusion $\mathcal{O}_E = T_E \subset T_{Y|E}$ we deduce

$$H^0(E, \mathcal{O}_E) \subset H^0(E, T_{Y|E})$$

which implies $h^0(E, T_{Y|E}) \geq 1$.

3.4.2 Deformations of a morphism leaving the target fixed

In this subsection we will follow quite faithfully the treatment given in [86].

Given a morphism $f : X \to Y$ of algebraic schemes, a notion slightly more general than the previous one is that of a *deformation of f with target Y*, obtained by specializing Definition 3.4.1 to the case when $\mathcal{Y}$ is the product family, i.e. by considering a cartesian diagram of the form

$$\begin{array}{ccc} X & \to & \mathcal{X} \\ \downarrow f & & \downarrow F \\ Y & \to & Y \times S \\ \downarrow & & \downarrow \psi \\ \mathrm{Spec}(\mathbf{k}) & \to & S \end{array}$$

where $S = \mathrm{Spec}(A)$ with A in $\mathrm{ob}(\mathcal{A})$ (resp. in $\mathrm{ob}(\mathcal{A}^*)$) and ψ is the projection.

Such a deformation can be denoted concisely by the diagram

$$\begin{array}{ccc} \mathcal{X} & \xrightarrow{F} & Y \times S \\ & \searrow \quad \swarrow & \\ & S & \end{array} \tag{3.42}$$

The deformation (3.42) will be called *locally trivial* if its domain $\mathcal{X}$ defines a locally trivial deformation of X. Given a deformation (3.42) and another deformation of f with target Y parametrized by S:

$$\begin{array}{ccc} \mathcal{X}' & \xrightarrow{F'} & Y \times S \\ & \searrow \quad \swarrow & \\ & S & \end{array}$$

an *isomorphism* between them is an isomorphism of deformations of X:

$$\begin{array}{ccc} & X & \\ \swarrow & & \searrow \\ \mathcal{X} & \xrightarrow{\Phi} & \mathcal{X}' \\ \searrow & & \swarrow \\ & S & \end{array}$$

which makes the following diagram commutative:

$$\begin{array}{ccc} \mathcal{X} & & \\ & \searrow & \\ \downarrow \Phi & & Y \times S \\ & \nearrow & \\ \mathcal{X}' & & \end{array}$$

Definition 3.4.5. *To a morphism $f : X \to Y$ of algebraic schemes there is associated an exact sequence of coherent sheaves on X*

$$0 \to T_{X/Y} \xrightarrow{J} T_X \xrightarrow{df} Hom(f^*\Omega^1_Y, \mathcal{O}_X) \xrightarrow{P} N_f \to 0 \tag{3.43}$$

which defines the sheaf N_f called the normal sheaf of f. *The morphism f is called non-degenerate when $T_{X/Y} = 0$.*

The sequence (3.43) is of course related to the exact sequence (1.5) on page 16. Comparing (3.43) with (1.5) we see that $N_f = T^1_{X/Y}$ if X is nonsingular. If f is smooth then $N_f = 0$. The condition that f is non-degenerate is equivalent to f being unramified on a dense open subset of X.

We now introduce vector spaces $D_{X/Y}$ and $D^1_{X/Y}$ which will be used to describe infinitesimal deformations of the morphism f.

Definition 3.4.6. *Let $f : X \to Y$ be a morphism between algebraic schemes, with X projective. Let $\mathcal{U} = \{U_i\}_{i \in I}$ be an affine open cover of X and define*

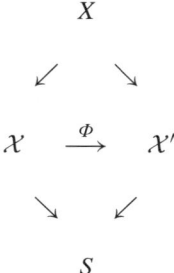

$$D_{X/Y} = \frac{\{(v,t) \in \mathcal{C}^0(\mathcal{U}, f^*T_Y) \times \mathcal{Z}^1(\mathcal{U}, T_X) : \delta v = df(t)\}}{\{(df(w), \delta w) : w \in \mathcal{C}^0(\mathcal{U}, T_X)\}}$$

and

$$D^1_{X/Y} = \frac{\{(\zeta,s) \in \mathcal{C}^1(\mathcal{U}, f^*T_Y) \times \mathcal{Z}^2(\mathcal{U}, T_X) : \delta \zeta = df(s)\}}{\{(df(u), \delta u) : u \in \mathcal{C}^1(\mathcal{U}, T_X)\}}$$

where δ is the coboundary map in Čech cohomology.

Lemma 3.4.7. *In the situation of Definition 3.4.6:*

(i) $D_{X/Y}$ and $D^1_{X/Y}$ don't depend on the choice of the affine cover $\mathcal{U}$ of X.
(ii) We have the following exact sequences:

$$\begin{aligned}
&(a)\ H^0(X,T_X) \to H^0(X, f^*T_Y) \to D_{X/Y} \to H^1(X,T_X) \to H^1(X, f^*T_Y)\\
&(b)\qquad 0 \to H^1(X, T_{X/Y}) \to D_{X/Y} \to H^0(X, N_f) \to H^2(X, T_{X/Y})\\
&(c)\ H^1(X,T_X) \to H^1(X, f^*T_Y) \to D^1_{X/Y} \to H^2(X,T_X) \to H^2(X, f^*T_Y)\\
&(d)\qquad 0 \to H^2(X, T_{X/Y}) \to D^1_{X/Y} \to H^1(X, N_f) \to H^3(X, T_{X/Y})
\end{aligned} \tag{3.44}$$

(iii) If f is non-degenerate then

$$D_{X/Y} \cong H^0(X, N_f)$$

$$D^1_{X/Y} \cong H^1(X, N_f)$$

(iv) If f is smooth then

$$D_{X/Y} \cong H^1(X, T_{X/Y})$$

$$D^1_{X/Y} \cong H^2(X, T_{X/Y})$$

Proof. It is clear that $(ii) \Rightarrow (i)$. Moreover, since f non-degenerate implies $T_{X/Y} = 0$, (iii) follows from the exact sequences (3.44)(b) and (3.44)(d). Similarly, f smooth implies $N_f = 0$ and (iv) follows again from (3.44)(b) and (3.44)(d).

Therefore it suffices to prove (ii). In the first sequence the homomorphism

$$D_{X/Y} \to H^1(X, T_X)$$

is given by sending $(v,t) \mapsto t$; similarly, we have

$$H^0(X, f^*T_Y) \to D_{X/Y}$$

which sends $u \mapsto (u, 0)$. The proof of exactness is left to the reader.

In the second exact sequence the map

$$H^1(X, T_{X/Y}) \to D_{X/Y}$$

is given by

$$\mathcal{Z}^1(\mathcal{U}, T_{X/Y}) \ni \psi \mapsto \overline{(0, J\psi)} \in D_{X/Y}$$

The map

$$D_{X/Y} \to H^0(X, N_f)$$

sends $(v, t) \mapsto Pv$. Every element of $H^0(X, N_f)$ is represented by some $v \in \mathcal{C}^0(\mathcal{U}, f^*T_Y)$ such that $\delta v = df(t)$ for some $t \in \mathcal{C}^1(\mathcal{U}, T_X)$. We have $df(\delta t) = \delta\delta t = 0$ so that δt can be viewed as an element of $\mathcal{Z}^2(\mathcal{U}, T_{X/Y})$. This defines the map

$$H^0(X, N_f) \to H^2(X, T_{X/Y})$$

The proof of exactness is left to the reader.

The other two sequences are defined and their exactness is checked similarly. □

Let $f : X \to Y$ be a morphism between algebraic schemes. Define functors of Artin rings:

$$\mathrm{Def}_{f/Y}, \mathrm{Def}'_{f/Y} : \mathcal{A} \to (\text{sets})$$

by

$$\mathrm{Def}_{f/Y}(A) = \left\{ \begin{array}{c} \text{isomorphism classes of} \\ \text{deformations of } f \text{ over } A \text{ with fixed target} \end{array} \right\}$$

$$\mathrm{Def}'_{f/Y}(A) = \left\{ \begin{array}{c} \text{isomorphism classes of locally trivial} \\ \text{deformations of } f \text{ over } A \text{ with fixed target} \end{array} \right\}$$

for all A in ob($\mathcal{A}$). Obviously, $\mathrm{Def}_{f/Y} = \mathrm{Def}'_{f/Y}$ when X is nonsingular. We will consider the locally trivial case. The main general result about $\mathrm{Def}'_{f/Y}$ is the following:

Theorem 3.4.8. *Let $f : X \to Y$ be a morphism of algebraic schemes with X projective. Then $\mathrm{Def}'_{f/Y}$ has a formal semiuniversal deformation. Its tangent space is $D_{X/Y}$ and $D^1_{X/Y}$ is an obstruction space (see Definition 3.4.6).*

Proof. Let's check the conditions of Theorem 2.3.2. $\mathrm{Def}'_{f/Y}$ trivially satisfies condition H_0. Consider a diagram in $\mathcal{A}$:

$$\begin{array}{ccc} A' & & A'' \\ & \searrow\swarrow & \\ & A & \end{array}$$

with $A'' \to A$ a small extension and let $\bar{A} = A' \times_A A''$. Let

$$(f_{A'}, f_{A''}) \in \mathrm{Def}'_{f/Y}(A') \times_{\mathrm{Def}'_{f/Y}(A)} \mathrm{Def}'_{f/Y}(A'')$$

where

$$f_{A'} : \mathcal{X}' \to Y \times \mathrm{Spec}(A'), \qquad f_{A''} : \mathcal{X}'' \to Y \times \mathrm{Spec}(A'')$$

Since Def'_X satisfies $\bar{H}$ there is a deformation $\bar{\mathcal{X}}$ of X over $\bar{A}$ such that

$$\mathcal{X}' \cong \bar{\mathcal{X}} \times_{\mathrm{Spec}(\bar{A})} \mathrm{Spec}(A'), \qquad \mathcal{X}'' \cong \bar{\mathcal{X}} \times_{\mathrm{Spec}(\bar{A})} \mathrm{Spec}(A'')$$

and we have a commutative diagram:

$$\begin{array}{ccccc} & & \mathcal{X}'' & \xrightarrow{f_{A''}} & Y \times \mathrm{Spec}(A'') \\ & & \downarrow & & \\ \mathcal{X}' & \to & \bar{\mathcal{X}} & & \swarrow \\ \downarrow f_{A'} & & & & \\ Y \times \mathrm{Spec}(A') & \to & Y \times \mathrm{Spec}(\bar{A}) & & \end{array}$$

By the universal property of the fibred sum we obtain a morphism

$$f_{\bar{A}} : \bar{\mathcal{X}} \to Y \times \mathrm{Spec}(\bar{A})$$

which pulls back over $\mathrm{Spec}(A')$ and $\mathrm{Spec}(A'')$ to $f_{A'}$ and $f_{A''}$ respectively. Therefore $f_{\bar{A}} \mapsto (f_{A'}, f_{A''})$ under the map

$$\alpha : \mathrm{Def}'_{f/Y}(\bar{A}) \to \mathrm{Def}'_{f/Y}(A') \times_{\mathrm{Def}'_{f/Y}(A)} \mathrm{Def}'_{f/Y}(A'')$$

and this proves $\bar{H}$.

Let $A = \mathbf{k}$ and $A'' = \mathbf{k}[\epsilon]$ and suppose that $f_{\bar{A}} : \bar{\mathcal{X}} \to Y \times \mathrm{Spec}(\bar{A})$ and $\tilde{f} : \tilde{\mathcal{X}} \to Y \times \mathrm{Spec}(\bar{A})$ are elements of $\mathrm{Def}'_{f/Y}(\bar{A})$ mapped to $(f_{A'}, f_{A''})$ by α. By the universal property of $\bar{\mathcal{X}}$ we have a morphism $\bar{\mathcal{X}} \to \tilde{\mathcal{X}}$ which, being an isomorphism mod ϵ, is an isomorphism. Moreover, we have a commutative diagram

$$\begin{array}{ccc} \bar{\mathcal{X}} \to & & \tilde{\mathcal{X}} \\ & \searrow & \downarrow \\ & & Y \times \mathrm{Spec}(\bar{A}) \end{array}$$

and therefore $f_{\bar{A}}$ and $\tilde{f}$ define the same element of $\mathrm{Def}'_{f/Y}(\bar{A})$; this implies that α is bijective in this case and H_ϵ holds.

In order to describe the tangent space to Def'_f, consider a first-order locally trivial deformation

$$\begin{array}{ccc} X & \to & \mathcal{X} \\ \downarrow f & & \downarrow F \\ Y & \to & Y \times \mathrm{Spec}(\mathbf{k}[\epsilon]) \\ \downarrow & & \downarrow \psi \\ \mathrm{Spec}(\mathbf{k}) & \to & \mathrm{Spec}(\mathbf{k}[\epsilon]) \end{array}$$

Then the deformation

$$\begin{array}{ccc} X & \to & \mathcal{X} \\ \downarrow & & \downarrow \psi F \\ \mathrm{Spec}(\mathbf{k}) & \to & \mathrm{Spec}(\mathbf{k}[\epsilon]) \end{array} \tag{3.45}$$

is locally trivial. Choose an affine open cover $\mathcal{U} = \{U_i\}_{i\in I}$ of X and, for each index i, let

$$\theta_i : U_i \times \mathrm{Spec}(\mathbf{k}[\epsilon]) \to \mathcal{X}_{|U_i}$$

be an isomorphism of deformations. The composition

$$F_i := F\theta_i : U_i \times \mathrm{Spec}(\mathbf{k}[\epsilon]) \to Y \times \mathrm{Spec}(\mathbf{k}[\epsilon])$$

is a deformation wfdat of $f_i := f_{|U_i}$ and therefore it corresponds to an element $v_i \in \Gamma(U_i, f_i^*T_Y) = \Gamma(U_i, f^*T_Y)$ by Proposition 3.4.2. Therefore we get an element $v = \{v_i\} \in C^0(\mathcal{U}, f^*T_Y)$. Restricting to U_{ij} we have:

$$F_{i|U_{ij}}\theta_i^{-1}\theta_j = F_{j|U_{ij}} \tag{3.46}$$

and, by Lemma 1.2.6, $\theta_i^{-1}\theta_j$ corresponds to a section $t_{ij} \in \Gamma(U_{ij}, T_X)$; the collection $\{t_{ij}\}$ is an element $t \in Z^1(\mathcal{U}, T_X)$ which defines the Kodaira–Spencer class of the deformation (3.45). The identity (3.46) means that $v_j - v_i = d(t_{ij})$ or, equivalently, that the pair (v, t) satisfies $\delta v = df(t)$. The pair (v, t) is defined up to a choice of the trivializations θ_i or, equivalently, up to an element of the form $(df(w), \delta w)$, $w \in C^0(\mathcal{U}, T_X)$. Similarly, if we replace F by an isomorphic deformation of the form $F' = F\sigma$, where

$$\sigma : X \times \mathrm{Spec}(\mathbf{k}[\epsilon]) \to X \times \mathrm{Spec}(\mathbf{k}[\epsilon])$$

is an automorphism of the trivial deformation of X, we obtain the same element of $D_{X/Y}$.

Conversely, suppose given an element of $D_{X/Y}$, represented by a pair $(v, t) = (\{v_i\}, \{t_{ij}\}) \in C^0(\mathcal{U}, f^*T_Y) \times Z^1(\mathcal{U}, T_X)$ such that $\delta v = df(t)$. The class $\bar{t} \in H^1(X, T_X)$ defines a first-order deformation of X

$$\begin{array}{ccc} X & \to & \mathcal{X} \\ \downarrow & & \downarrow \\ \mathrm{Spec}(\mathbf{k}) & \to & \mathrm{Spec}(\mathbf{k}[\epsilon]) \end{array}$$

which is locally trivial. The 1-cocycle $t = \{t_{ij}\}$ defines local trivializations

$$\theta_i : U_i \times \mathrm{Spec}(\mathbf{k}[\epsilon]) \to \mathcal{X}_{|U_i}$$

and each v_i defines a deformation wfdat F_i of f_i and by composition we get a morphism:

$$\mathcal{X}_{|U_i} \xrightarrow{\theta_i^{-1}} U_i \times \mathrm{Spec}(\mathbf{k}[\epsilon]) \xrightarrow{F_i} Y \times \mathrm{Spec}(\mathbf{k}[\epsilon])$$

By construction $F_{i|U_{ij}}\theta_i^{-1}\theta_j = F_{j|U_{ij}}$ and therefore

$$F_{i|U_{ij}}\theta_i^{-1} = F_{j|U_{ij}}\theta_j^{-1}$$

This means that the morphisms $F_i\theta_i^{-1}$ patch together and define a morphism $F : \mathcal{X} \to Y \times \mathrm{Spec}(\mathbf{k}[\epsilon])$. This obviously gives rise to a first-order deformation of f. It is a straightforward task to verify that the correspondences $F \mapsto \bar{v}$ and $\bar{v} \mapsto F$ are the inverse of each other. Therefore $\mathrm{Def}'_{f/Y}(\mathbf{k}[\epsilon]) \cong D_{X/Y}$ and in particular H_f holds.

Now we want to prove the assertion about obstructions to lifting deformations. We have to show that for every locally trivial infinitesimal deformation

$$\begin{array}{ccc} \mathcal{X} & \xrightarrow{F} & Y \times S \\ & \searrow \quad \swarrow & \\ & S & \end{array}$$

of f with target Y over $S = \mathrm{Spec}(A)$ there is a natural map

$$o_{F/Y} : \mathrm{Ex}_{\mathbf{k}}(A, \mathbf{k}) \to D^1_{X/Y}$$

such that, for a given extension

$$e : 0 \to \mathbf{k} \to \tilde{A} \to A \to 0$$

we have $o_{F/Y}(e) = 0$ if and only if F has a lifting to $\tilde{A}$ which is a locally trivial deformation of f with target Y. Choose $\mathcal{U} = \{U_i\}_{i \in I}$ an affine open cover of X and trivializations

$$\theta_i : U_i \times \mathrm{Spec}(A) \to \mathcal{X}_{|U_i}$$

Since $H^1(U_i, f^*T_Y) = 0$ for each $i \in I$ the morphism

$$F_i := F\theta_i : U_i \times \mathrm{Spec}(A) \to Y \times \mathrm{Spec}(A)$$

has a lifting as a deformation of f_i wfdat:

$$\begin{array}{ccc} U_i \times \mathrm{Spec}(A) & \xrightarrow{F_i} & Y \times \mathrm{Spec}(A) \\ \cap & & \cap \\ U_i \times \mathrm{Spec}(\tilde{A}) & \xrightarrow{\tilde{F}_i} & Y \times \mathrm{Spec}(\tilde{A}) \end{array}$$

If we restrict to U_{ij} we have the identity:

$$F_{i|U_{ij}}\theta_{ij} = F_{j|U_{ij}}$$

where we have written

$$\theta_{ij} := \theta_i^{-1}\theta_j : U_{ij} \times \mathrm{Spec}(A) \to U_{ij} \times \mathrm{Spec}(A)$$

Let

$$\tilde{\theta}_{ij} : U_{ij} \times \mathrm{Spec}(\tilde{A}) \to U_{ij} \times \mathrm{Spec}(\tilde{A})$$

be an automorphism which restricts to θ_{ij} on $U_{ij} \times \mathrm{Spec}(A)$. Then $\tilde{F}_{i|U_{ij}}\tilde{\theta}_{ij}$ and $\tilde{F}_{j|U_{ij}}$ are both liftings of $F_{j|U_{ij}}$. Therefore, since $\Gamma(U_{ij}, f^*T_Y)$ acts faithfully and transitively on the set of such liftings (Proposition 3.4.2 (iii)), there is a $\zeta_{ij} \in \Gamma(U_{ij}, f^*T_Y)$ which carries $\tilde{F}_{j|U_{ij}}$ into $\tilde{F}_{i|U_{ij}}\tilde{\theta}_{ij}$; set $\zeta = \{\zeta_{ij}\} \in C^1(\mathcal{U}, f^*T_Y)$. Now let

$$\tilde{\theta}_{ijk} = \tilde{\theta}_{ij}\tilde{\theta}_{jk}\tilde{\theta}_{ik}^{-1} : U_{ijk} \times \mathrm{Spec}(\tilde{A}) \to U_{ijk} \times \mathrm{Spec}(\tilde{A})$$

Since $\tilde{\theta}_{ijk}$ restricts to the identity on $U_{ijk} \times \mathrm{Spec}(A)$, by Lemma 1.2.6 it corresponds to an $s_{ijk} \in \Gamma(U_{ijk}, T_X)$; by construction $s := \{s_{ijk}\}$ is a 2-cocycle, i.e. it is an element of $\mathcal{Z}^2(\mathcal{U}, T_X)$, and the pair (ζ, s) satisfies

$$\delta(\zeta) = df(s)$$

Define $o_{F/Y}(e) = \overline{(\zeta, s)} \in D^1_{X/Y}$. This definition is well posed because a different choice of the $\tilde{\theta}_{ij}$'s will replace (ζ, s) by $(\zeta + df(u), s + \delta u)$ for some $u \in C^1(\mathcal{U}, T_X)$.

If $o_{F/Y}(e) = 0$ then $(\zeta, s) = (df(u), \delta u)$ for some $u \in C^1(\mathcal{U}, T_X)$. The condition $s = \delta u$ means that for the deformation

$$\xi : \begin{array}{ccc} X & \to & \mathcal{X} \\ \downarrow & & \downarrow \\ \mathrm{Spec}(\mathbf{k}) & \to & \mathrm{Spec}(A) \end{array}$$

we have $o_\xi(e) = \bar{s} = 0 \in H^2(X, T_X)$ so that ξ has a lifting to $\mathrm{Spec}(\tilde{A})$. Such a lifting

$$\tilde{\xi} : \begin{array}{ccccc} X & \to & \mathcal{X} & \to & \tilde{\mathcal{X}} \\ \downarrow & & \downarrow & & \downarrow \\ \mathrm{Spec}(\mathbf{k}) & \to & \mathrm{Spec}(A) & \to & \mathrm{Spec}(\tilde{A}) \end{array}$$

is defined by an appropriate choice of the automorphisms $\tilde{\theta}_{ij}$ such that $\tilde{\theta}_{ij}\tilde{\theta}_{jk} = \tilde{\theta}_{ik}$. The condition $\zeta = df(u)$ means that the liftings $\tilde{F}_i$ and $\tilde{F}_j$ can be chosen so that

$$\tilde{F}_{i|U_{ij}}\tilde{\theta}_{ij} = \tilde{F}_{j|U_{ij}}$$

and this means that they patch together to define a lifting $\tilde{F}$ of F. Conversely, if such a lifting exists then one shows in the same way that $o_{F/Y}(e) = 0$. □

$o_{F/Y}(e)$ is called *the obstruction to lifting F to $\tilde{A}$ as a deformation with target Y*. The notions of *obstructed/unobstructed deformation, obstructed/unobstructed morphism with fixed target* can be given as usual. One can give the notion of *rigid morphism with fixed target* in an obvious way. We have the following:

Corollary 3.4.9. *Under the hypotheses of Theorem 3.4.8 we have:*

(i) If $D_{X/Y} = 0$ then f is rigid as a morphism with fixed target. If $D^1_{X/Y} = 0$ then f is unobstructed as a morphism with fixed target.

(ii) If f is non-degenerate then in the statement of 3.4.8 we can replace $D_{X/Y}$ and $D^1_{X/Y}$ by $H^0(X, N_f)$ and $H^1(X, N_f)$ respectively.

(iii) If f is smooth then in the statement of 3.4.8 we can replace $D_{X/Y}$ and $D^1_{X/Y}$ by $H^1(X, T_{X/Y})$ and $H^2(X, T_{X/Y})$ respectively.

Proof. The proof is immediate in view of Lemma 3.4.7. □

Remarks 3.4.10. (i) If f is a closed embedding of projective nonsingular algebraic schemes then $N_f = N_{X/Y}$ and Theorem 3.4.8 asserts that first-order deformations, resp. obstructions, of f with target Y coincide with first-order deformations, resp. obstructions, of X in Y. This is because, more generally, every infinitesimal deformation of f with target Y is an infinitesimal deformation of X in Y: a proof is given in Note 3.

(ii) If $f : X \to Y$ is a morphism between algebraic varieties with X projective, and if (3.42) is a family of locally trivial deformations of f with target Y, then, using Theorem 3.4.8, we can define a linear map

$$T_{S,s} \to D_{X/Y}$$

which associates to a tangent vector $t : \mathrm{Spec}(\mathbf{k}[\epsilon]) \to S$ at s the element of $D_{X/Y}$ corresponding to the first-order deformation obtained by pulling back F by t. This map is called the *characteristic map* of the family F.

Given a morphism $f : X \to Y$ between algebraic varieties with X projective we have a natural morphism of functors

$$\Phi_f : \mathrm{Def}'_{f/Y} \to \mathrm{Def}'_X$$

called the *forgetful morphism* which associates to a locally trivial deformation (3.42) over $S = \mathrm{Spec}(A)$, A in $\mathrm{ob}(\mathcal{A})$, the family of deformations of X obtained by forgetting the morphism F. This morphism of functors generalizes the analogous forgetful morphism defined for the local Hilbert functor in Subsection 3.2.3. We have the following generalization of Proposition 3.2.9.

Proposition 3.4.11. *Let $f : X \to Y$ be a morphism between algebraic varieties with X projective, and let*

$$\Phi_f : \mathrm{Def}'_{f/Y} \to \mathrm{Def}'_X$$

be the forgetful morphism. Then

(i)

$$d\Phi_f : D_{X/Y} \to H^1(X, T_X)$$

is the map occurring in the exact sequence (3.44)(a).

(ii) The map

$$D^1_{X/Y} \to H^2(X, T_X)$$

occurring in the exact sequence (3.44)(c) is an obstruction map for Φ_f.

(iii) Assume that

$$H^1(X, f^*T_Y) = 0$$

Then Φ_f is smooth.

Proof. The proofs of (i) and (ii) are straightforward.

(iii) The hypothesis and the exact sequences (3.44)(a) and (3.44)(c) imply that

$$D_{X/Y} \to H^1(X, T_X) \qquad \text{is surjective and}$$

$$D^1_{X/Y} \to H^2(X, T_X) \qquad \text{is injective}$$

Now, using parts (i) and (ii), the conclusion follows from Proposition 2.3.6. □

Corollary 3.4.12. *(i) Let $f : X \to Y$ be a non-degenerate morphism of algebraic schemes with X projective. Then*

$$d\Phi_f : H^0(X, N_f) \to H^1(X, T_X)$$

and

$$o(\Phi_f) : H^1(X, N_f) \to H^2(X, T_X)$$

are the coboundary maps coming from the exact sequence

$$0 \to T_X \to f^*T_Y \to N_f \to 0$$

(ii) Let $f : X \to Y$ be a smooth morphism of projective nonsingular algebraic schemes. Then

$$d\Phi_f : H^1(X, T_{X/Y}) \to H^1(X, T_X)$$

and

$$o(\Phi_f) : H^2(X, T_{X/Y}) \to H^2(X, T_X)$$

are the maps induced by the natural inclusion of sheaves $T_{X/Y} \subset T_X$.

Proof. The corollary is just a special case of the proposition. □

Examples 3.4.13. (i) Let $m \geq 0$ be an integer, $F_m = I\!P(\mathcal{F}_m)$ where $\mathcal{F}$ is the locally free rank two sheaf on $I\!P^1$:

$$\mathcal{F}_m = \mathcal{O}_{I\!P^1}(m) \oplus \mathcal{O}_{I\!P^1}$$

and let $\pi : F_m \to I\!P^1$ be the projection. Then we have

$$H^1(F_m, \pi^*T_{I\!P^1}) = 0$$

by an easy calculation using the Leray spectral sequence. Therefore, since π is smooth, we can apply Proposition 3.4.11 to conclude that Φ_π is smooth. Moreover, since F_m is unobstructed as an abstract variety because

$$h^2(F_m, T_{F_m}) = 0$$

(see (B.13)), it follows from Proposition 2.2.5(iii) that π is unobstructed.

We can actually be more precise because we have an exact sequence of locally free sheaves on $I\!P^1$:

$$0 \to \mathcal{O}_{I\!P^1} \to \mathcal{F}_m \otimes \mathcal{F}_m^\vee \to \pi_* T_{F_m/I\!P^1} \to 0$$

which can be deduced easily from the exact sequence (4.28). Since

$$\mathcal{F}_m \otimes \mathcal{F}_m^\vee \cong \mathcal{O}^{\oplus 2} \oplus \mathcal{O}(m) \oplus \mathcal{O}(-m)$$

using the Leray spectral sequence we deduce that $h^1(F_m, T_{F_m/I\!P^1}) = m - 1$ for $m \geq 1$. Therefore, recalling (B.12), we see that the map

$$H^1(F_m, T_{F_m/I\!P^1}) \to H^1(F_m, T_{F_m})$$

is not only surjective but it is actually an isomorphism.

(ii) Let $f : X \to Y$ be a smooth family of projective curves of genus ≥ 2 with X a projective nonsingular surface and Y a projective nonsingular connected curve. Assume that f is non-isotrivial (see Definition 2.6.9). Then $H^1(X, T_{X/Y}) = 0$ and therefore, by (i) and (iii) of Corollary 3.4.9, f is rigid as a morphism with fixed target. This theorem is due to Parshin [139] in char 0. For an exposition we refer the reader to Szpiro [178], where the theorem, and its generalization due to Arakelov [6], is proved without the restriction char$(\mathbf{k}) = 0$.

(iii) Let $f : X \to Y$ be an etale morphism of projective nonsingular schemes of dimension n. Then f is non-degenerate and $df : T_X \to f^*T_Y$ is an isomorphism; therefore $N_f = 0$ and f is rigid as a morphism with fixed target. If we only assume f to be non-degenerate, but not necessarily etale, then df degenerates on a divisor $R \subset X$ (which is the divisor $\overline{D_{n-1}(df)}$ of Example 4.2.8), called the *ramification divisor* of f. N_f is supported on R, and in general f is not rigid as a morphism with fixed target. For the case when X and Y are curves see Subsection 3.4.3.

(iv) Let Y be a projective nonsingular variety, $\gamma \subset Y$ a nonsingular closed subvariety of pure codimension $r \geq 2$ and $\pi : X \to Y$ the blow-up of Y with centre γ. Let $E = \pi^{-1}(\gamma) \subset X$ be the exceptional divisor. Then $E \cong I\!P(N_{\gamma/Y})$ is a projective bundle over γ: let $q : E \to \gamma$ be the structure morphism. Then $N_{E/X} = \mathcal{O}_E(E)$ and it is well known that the restriction of $N_{E/X}$ to each fibre $I\!P$ of q is $\mathcal{O}_{I\!P}(-1)$. Therefore by the Leray spectral sequence of q we immediately deduce that

$$h^i(E, N_{E/X}) = 0 \tag{3.47}$$

for all i. We have $T_{X/Y} = 0$ because $T_{X/Y}$ is a subsheaf of the locally free T_X and is supported on E; therefore π is non-degenerate. Since $\pi_*\mathcal{O}_X = \mathcal{O}_Y$ and $R^i\pi_*\mathcal{O}_X = 0$ for $i \geq 1$, from the Leray spectral sequence we deduce that

$$H^i(X, \pi^*T_Y) = H^i(Y, (\pi_*\pi^*\mathcal{O}_X) \otimes T_Y) = H^i(Y, T_Y), \quad i \geq 0 \tag{3.48}$$

We have an exact and commutative diagram of locally free sheaves on E:

$$\begin{array}{ccccccccc} & & & & 0 & & 0 & & \\ & & & & \downarrow & & \downarrow & & \\ 0 \to & T_{E/\gamma} & \to & & T_E & \to & q^*T_\gamma & \to & 0 \\ & \| & & & \downarrow & & \downarrow & & \\ 0 \to & T_{E/\gamma} & \to & & T_{X|E} & \to & q^*T_Y & \to & N_\pi \to 0 \\ & & & & \downarrow & & \downarrow & & \| \\ & 0 & \to & & N_{E/X} & \to & q^*N_{\gamma/Y} & \to & N_\pi \to 0 \\ & & & & \downarrow & & \downarrow & & \\ & & & & 0 & & 0 & & \end{array} \tag{3.49}$$

In particular, we see that we have an exact sequence of locally free sheaves on E:

$$0 \to N_{E/X} \to q^*N_{\gamma/Y} \to N_\pi \to 0 \tag{3.50}$$

The verification of these facts is straightforward and it is left to the reader.

For example, let $\pi : X = Bl_{[1,0,0]}\mathbb{P}^2 \to \mathbb{P}^2$ be the blow-up of $\mathbb{P}^2$ with centre the point $[1, 0, 0]$. From the exact sequence (3.50) we deduce that $N_\pi = \mathcal{O}_E(1)$. Therefore

$$h^0(X, N_\pi) = 2, \quad h^i(X, N_\pi) = 0, \ i \geq 1$$

In particular, π is unobstructed. Moreover,

$$h^0(X, T_X) = h^0(\mathbb{P}^2, T_{\mathbb{P}^2} \otimes \mathcal{I}_{[1,0,0]}) = 6$$

as can be easily checked using the Euler sequence. Therefore from the exact sequence (3.43) we see that $h^1(X, T_X) = 0$, i.e. X is rigid.

3.4.3 Morphisms from a nonsingular curve with fixed target

Theorem 3.4.8 applies in particular to a morphism

$$\varphi : C \to Y$$

where C and Y are projective and nonsingular, C is a curve, and φ is not constant on each component of C. Consider the exact sequence

$$0 \to T_C \xrightarrow{d\varphi} \varphi^*T_Y \to N_\varphi \to 0$$

The vanishing divisor (see Example 4.2.8 page 200)

$$Z := D_0(d\varphi)$$

of $d\varphi$ is called the *ramification divisor* of φ; the *index of ramification* of φ at $p \in C$ is the coefficient of p in Z. φ is unramified if and only if $Z = 0$. The homomorphism $d\varphi$ extends to a homomorphism

$$T_C(Z) \to \varphi^* T_Y$$

whose cokernel we denote by $\bar{N}_\varphi$; it is locally free. We have

$$\bar{N}_\varphi = N_\varphi / \mathcal{H}_\varphi$$

where $\mathcal{H}_\varphi \subset N_\varphi$ is the torsion subsheaf; it is supported on Z. The following commutative and exact diagram summarizes the situation:

$$\begin{array}{ccccccc} & & & & 0 & & \\ & & & & \downarrow & & \\ & & & & \mathcal{H}_\varphi & & \\ & & & & \downarrow & & \\ 0 \to & T_C & \xrightarrow{d\varphi} & \varphi^* T_Y & \to & N_\varphi & \to 0 \\ & \downarrow & & \| & & \downarrow & \\ 0 \to & T_C(Z) & \to & \varphi^* T_Y & \to & \bar{N}_\varphi & \to 0 \\ & & & & & \downarrow & \\ & & & & & 0 & \end{array} \tag{3.51}$$

We obtain:

$$\chi(N_\varphi) = \chi(\varphi^* T_Y) + 3g - 3 \tag{3.52}$$

Example 3.4.14. Assume that C is connected of genus g, that Y is a projective connected nonsingular curve of genus γ, and that φ has degree d; then

$$\bar{N}_\varphi = 0, \quad N_\varphi = \mathcal{H}_\varphi = \mathcal{O}_Z$$

where $\mathcal{O}(Z) = \varphi^*(T_Y) \otimes K_C$, so that

$$\chi(N_\varphi) = h^0(N_\varphi) = \deg(Z) = 2[g - 1 + (1 - \gamma)d]$$

and φ is unobstructed because $h^1(N_\varphi) = 0$. This corresponds to the fact that the deformations of φ leaving Y fixed are obtained by varying the branch points of φ.

Note that φ is rigid as a morphism with fixed target if $g \geq 2$ and $Z = 0$, i.e. if it is unramified.

* * * * * *

Assume now that

$$\varphi : C \to S$$

is a nonconstant morphism from an irreducible projective nonsingular curve C of genus g to a projective nonsingular surface S and that φ is birational onto its image; let $\Gamma = \varphi(C) \subset S$. Then we have a commutative and exact diagram:

$$\begin{array}{ccccccc} & 0 & & & & & \\ & \downarrow & & & & & \\ 0 \to & T_C & \xrightarrow{d\varphi} & \varphi^* T_S & \to & N_\varphi & \to 0 \\ & \downarrow & & \| & & \downarrow j & \\ 0 \to & \varphi^* T_\Gamma & \to & \varphi^* T_S & \to & \varphi^* N_{\Gamma/S} & \end{array}$$

Since $\varphi^* N_{\Gamma/S}$ is invertible the homomorphism j factors through $\bar{N}_\varphi$ and the above diagram gives rise to the following:

$$\begin{array}{ccccccccc} & & 0 & & & & & & \\ & & \downarrow & & & & & & \\ 0 \to & & T_C(Z) & \longrightarrow & \varphi^* T_S & \to & \bar{N}_\varphi & \to 0 \\ & & \downarrow & & \| & & \downarrow & \\ 0 \to & & \varphi^* T_\Gamma & \to & \varphi^* T_S & \to & \varphi^* N'_{\Gamma/S} & \to 0 \end{array}$$

where Z is the ramification divisor of φ and $N'_{\Gamma/S} = \ker[N_{\Gamma/S} \to T^1_\Gamma]$ is the *equisingular normal sheaf* (see also § 4.7). This diagram implies the following isomorphisms:

$$T_C(Z) \cong \varphi^* T_\Gamma, \quad \bar{N}_\varphi \cong \varphi^* N'_{\Gamma/S}, \quad \mathcal{H}_\varphi \cong \operatorname{coker}[T_C \to \varphi^*(T_\Gamma)] =: N_{\bar{\varphi}}$$

where we have denoted by $\bar{\varphi} : C \to \Gamma$ the morphism induced by φ. In particular, $\varphi^* T_\Gamma$ and $\varphi^* N'_{\Gamma/S}$ are invertible and

$$\varphi_*[T_C(Z)] \cong T_\Gamma \otimes \varphi_* \mathcal{O}_C, \quad \varphi_* \bar{N}_\varphi \cong N'_{\Gamma/S} \otimes \varphi_* \mathcal{O}_C$$

On Γ we have a natural exact sequence:

$$0 \to \mathcal{O}_\Gamma \to \varphi_* \mathcal{O}_C \to \mathbf{t} \to 0$$

where $\mathbf{t}$ is a torsion sheaf supported on the singular locus of Γ. Since $N_{\Gamma/S}$ is invertible the homomorphism

$$N_{\Gamma/S} \to N_{\Gamma/S} \otimes \varphi_* \mathcal{O}_C$$

is injective and it follows that we have an exact sequence

$$\begin{array}{ccccccc} 0 \to & N'_{\Gamma/S} & \to & \varphi_* \bar{N}_\varphi & \to & N'_{\Gamma/S} \otimes \mathbf{t} & \to 0 \\ & & & \| & & & \\ & & & N'_{\Gamma/S} \otimes \varphi_* \mathcal{O}_C & & & \end{array} \tag{3.53}$$

This sequence implies in particular:

$$h^0(N'_{\Gamma/S}) \leq h^0(\bar{N}_\varphi) \leq h^0(N_\varphi)$$

$$h^1(N'_{\Gamma/S}) \geq h^1(\bar{N}_\varphi) = h^1(N_\varphi)$$

Lemma 3.4.15. *If the singularities of Γ are nodes and ordinary cusps then $N'_{\Gamma/S} \otimes \mathbf{t} = 0$; equivalently,*

$$N'_{\Gamma/S} \cong \varphi_* \bar{N}_\varphi$$

In particular, if Γ has only nodes as singularities then

$$N'_{\Gamma/S} \cong \varphi_* N_\varphi$$

Proof. The exact sequence (3.53) can be embedded in the following exact and commutative diagram:

$$\begin{array}{ccccccc}
 & 0 & & 0 & & & \\
 & \downarrow & & \downarrow & & & \\
0\to & N'_{\Gamma/S} & \to & \varphi_*\bar{N}_\varphi & \to & N'_{\Gamma/S}\otimes \mathbf{t} & \to 0 \\
 & \downarrow & & \downarrow a & & \downarrow d & \\
0\to & N_{\Gamma/S} & \to & N_{\Gamma/S}\otimes\varphi_*\mathcal{O}_C & \to & N_{\Gamma/S}\otimes \mathbf{t} & \to 0 \\
 & \downarrow & & \downarrow & & \downarrow c & \\
0\to & T^1_\Gamma & \stackrel{b}{\longrightarrow} & T^1_\Gamma\otimes\varphi_*\mathcal{O}_C & \to & T^1_\Gamma\otimes \mathbf{t} & \to 0 \\
 & \downarrow & & \downarrow & & \downarrow & \\
 & 0 & & 0 & & 0 &
\end{array}$$

The arrow a is injective because it is a nonzero homomorphism of torsion free rank one sheaves. Because of the assumptions made on the singularities, at each singular point $p \in \Gamma$ we have $\mathbf{t}_p = \varphi_*\mathcal{O}_{C,p}/\mathcal{O}_{\Gamma,p} \cong \mathbf{k}$. Therefore the arrow c is an isomorphism because $N_{\Gamma/S}\otimes\mathbf{t} \cong \mathbf{t} \cong T^1_\Gamma\otimes\mathbf{t}$. Thus $d = 0$. The arrow b is injective because at each singular point $p \in \Gamma$ we have

$$(T^1_\Gamma \otimes \varphi_*\mathcal{O}_C)_p = \varphi_*\varphi^*(T^1_{\Gamma,p}) \cong \begin{cases} \mathbf{k}^2 & \text{if } p \text{ is a node} \\ \mathbf{k}^3 & \text{if } p \text{ is a cusp} \end{cases}$$

(proved by easy local computation) while

$$T^1_{\Gamma,p} \cong \begin{cases} \mathbf{k} & \text{if } p \text{ is a node} \\ \mathbf{k}^2 & \text{if } p \text{ is a cusp} \end{cases}$$

(recall Example 3.1.4). The conclusion now follows from the "Snake Lemma". If Γ has only nodes then $\bar{N}_\varphi = N_\varphi$ and we deduce that $N'_{\Gamma/S} \cong \varphi_*N_\varphi$. □

It is possible to show that conversely, if $N'_{\Gamma/S}\otimes\mathbf{t} = 0$ then the singularities of Γ are nodes and ordinary cusps (see [68]).

If $S = \mathbb{P}^2$ then, letting $L = \varphi^*\mathcal{O}(1)$, $d = \deg(L)$, from the Euler sequence restricted to C:

$$0 \to \mathcal{O}_C \to L^{\oplus 3} \to \varphi^*T_{\mathbb{P}^2} \to 0$$

we deduce $\chi(\varphi^*T_{\mathbb{P}^2}) = 3d + 2 - 2g$ and from (3.52)

$$\chi(N_\varphi) = 3d + g - 1 \tag{3.54}$$

By Corollary 3.4.9 the unobstructedness of φ is related to the vanishing of $H^1(C, N_\varphi)$. From (3.51) we see that

$$\deg(N_\varphi) = c_1(\varphi^*T_{\mathbb{P}^2}) - \deg(T_C) = 3d + 2g - 2$$

and that

$$h^1(N_\varphi) = h^1(\bar{N}_\varphi)$$

But

$$\deg(\bar{N}_\varphi) = c_1(N_\varphi) - \deg(Z)$$

and therefore $\bar{N}_\varphi$ is a nonspecial line bundle whenever $\deg(Z) < 3d$. We can therefore state the following result:

Proposition 3.4.16. *Let $\varphi : C \to \mathbb{P}^2$ be a morphism from an irreducible projective nonsingular curve C of genus g, birational onto its image. Let $d = \deg(\varphi^*\mathcal{O}(1))$, and let Z be the ramification divisor of φ. Then*

$$h^0(C, N_\varphi) \geq 3d + g - 1$$

If $\deg(Z) < 3d$ then φ is unobstructed and the above inequality is an equality.

In particular, if $\varphi(C)$ is a plane curve having nodes and cusps as its only singularities and the number κ of cusps satisfies $\kappa < 3d$ then

$$h^0(C, N_\varphi) = 3d + g - 1, \quad h^1(C, N_\varphi) = 0$$

3.4.4 Deformations of a closed embedding

The deformation theory of morphisms is more subtle if we want to allow both the domain and the target to deform nontrivially. In this subsection we will address this case, considering only the simplest situation of a closed embedding.

Let $j : X \subset Y$ be a closed embedding of algebraic schemes. If

$$\begin{array}{ccc} \mathcal{X} & \xrightarrow{J} & \mathcal{Y} \\ & \searrow \quad \swarrow & \\ & S & \end{array}$$

where $S = \mathrm{Spec}(A)$, A in $\mathrm{ob}(\mathcal{A})$, is an infinitesimal deformation of j then J is a closed embedding (see Note 3 at the end of of this section). It is obvious that an infinitesimal deformation of j with fixed target is nothing but a deformation of X in Y. Given another infinitesimal deformation of j:

$$\begin{array}{ccc} \mathcal{X}' & \xrightarrow{J'} & \mathcal{Y}' \\ & \searrow \quad \swarrow & \\ & S & \end{array}$$

over the same $S = \mathrm{Spec}(A)$, an isomorphism between them is a pair of isomorphisms of deformations:

$$\alpha : \mathcal{X} \to \mathcal{X}', \qquad \beta : \mathcal{Y} \to \mathcal{Y}'$$

which make the diagram

$$\begin{array}{ccc} \mathcal{X} & \xrightarrow{J} & \mathcal{Y} \\ \downarrow \alpha & & \downarrow \beta \\ \mathcal{X}' & \xrightarrow{J'} & \mathcal{Y}' \end{array}$$

commutative.

We define

$$\mathrm{Def}_j(A) = \left\{ \begin{array}{c} \text{isomorphism classes of} \\ \text{deformations of } j \text{ over } A \end{array} \right\}$$

$$\mathrm{Def}'_j(A) = \left\{ \begin{array}{c} \text{isomorphism classes of locally} \\ \text{trivial deformations of } j \text{ over } A \end{array} \right\}$$

for each A in $\mathrm{ob}(\mathcal{A})$. These are *the functor of infinitesimal deformations of* j and *of locally trivial infinitesimal deformations of* j respectively.

The locally trivial infinitesimal deformations of a closed embedding are studied by means of a sheaf which we now introduce.

Let's now assume that Y is nonsingular and let $\mathcal{I}_X \subset \mathcal{O}_Y$ be the ideal sheaf of X. Let

$$T_Y\langle X\rangle \subset T_Y$$

be the inverse image of $T_X \subset T_{Y|X}$ under the natural restriction homomorphism $T_Y \to T_{Y|X}$. Then $T_Y\langle X\rangle$ is called *the sheaf of germs of tangent vectors to* Y *which are tangent to* X. We clearly have an inclusion $\mathcal{I}_X T_Y \subset T_Y\langle X\rangle$ such that

$$T_X = T_Y\langle X\rangle/\mathcal{I}_X T_Y$$

and an exact sequence

$$0 \to T_Y\langle X\rangle \to T_Y \to N'_{X/Y} \to 0 \tag{3.55}$$

where $N'_{X/Y} \subset N_{X/Y}$ is the equisingular normal sheaf of X in Y (introduced in Proposition 1.1.9, page 16). From the definition it follows that, for every open set $U \subset Y$, $\Gamma(U, T_Y\langle X\rangle)$ consists of those $\mathbf{k}$-derivations $D \in \Gamma(U, T_Y)$ such that $D(g) \in \Gamma(U, \mathcal{I}_X)$ for every $g \in \Gamma(U, \mathcal{I}_X)$. We also have the following exact commutative diagram:

$$\begin{array}{ccccccccc} & & 0 & & 0 & & & & \\ & & \downarrow & & \downarrow & & & & \\ & & \mathcal{I}_X T_Y & = & \mathcal{I}_X T_Y & & & & \\ & & \downarrow & & \downarrow & & & & \\ 0 & \to & T_Y\langle X\rangle & \to & T_Y & \to & N'_{X/Y} & \to & 0 \\ & & \downarrow & & \downarrow & & \| & & \\ 0 & \to & T_X & \to & T_{Y|X} & \to & N'_{X/Y} & \to & 0 \\ & & \downarrow & & \downarrow & & & & \\ & & 0 & & 0 & & & & \end{array} \tag{3.56}$$

Of course, $N'_{X/Y}$ is replaced by $N_{X/Y}$ in the case where X is nonsingular.

We will describe *locally trivial* infinitesimal deformations of a closed embedding by means of the sheaf $T_Y\langle X\rangle$.

Proposition 3.4.17. *Let $j : X \subset Y$ be a closed embedding of projective algebraic schemes with Y nonsingular. Then* Def'_j *has a formal semiuniversal deformation. Its tangent space is $H^1(Y, T_Y\langle X\rangle)$ and $H^2(Y, T_Y\langle X\rangle)$ is an obstruction space.*

Proof. The proof that Def'_j satisfies Schlessinger's conditions H_0, $\bar{H}$ and H_ϵ is similar to the proof given in Theorem 3.4.8 and will be left to the reader. Since Y is projective the existence of a semiuniversal formal deformation will follow if we will prove the assertion about the tangent space of Def'_j, because $H^1(Y, T_Y\langle X\rangle)$ is finite dimensional.

Let $\mathcal{U} = \{U_i\}_{i\in I}$ be an affine open cover of Y and

$$\mathcal{V} = \{V_i = X \cap U_i\}_{i\in I}$$

the induced affine open cover of X. Every locally trivial first-order deformation of j is obtained by gluing the trivial deformations

$$\begin{array}{ccc} V_i & \subset & V_i \times \mathrm{Spec}(\mathbf{k}[\epsilon]) \\ \cap & & \cap \\ U_i & \subset & U_i \times \mathrm{Spec}(\mathbf{k}[\epsilon]) \end{array}$$

along $V_{ij} \times \mathrm{Spec}(\mathbf{k}[\epsilon])$ and $U_{ij} \times \mathrm{Spec}(\mathbf{k}[\epsilon])$. It is therefore necessary to describe the automorphisms of the trivial deformations

$$\begin{array}{ccc} V_{ij} & \subset & V_{ij} \times \mathrm{Spec}(\mathbf{k}[\epsilon]) \\ \cap & & \cap \\ U_{ij} & \subset & U_{ij} \times \mathrm{Spec}(\mathbf{k}[\epsilon]) \end{array}$$

Every such automorphism A_{ij} consists of a pair $(\theta_{ij}, \Theta_{ij})$ where

$$\theta_{ij} : V_{ij} \times \mathrm{Spec}(\mathbf{k}[\epsilon]) \to V_{ij} \times \mathrm{Spec}(\mathbf{k}[\epsilon])$$

and

$$\Theta_{ij} : U_{ij} \times \mathrm{Spec}(\mathbf{k}[\epsilon]) \to U_{ij} \times \mathrm{Spec}(\mathbf{k}[\epsilon])$$

are automorphisms of deformations such that the following diagram commutes:

$$\begin{array}{ccc} V_{ij} \times \mathrm{Spec}(\mathbf{k}[\epsilon]) & \xrightarrow{\theta_{ij}} & V_{ij} \times \mathrm{Spec}(\mathbf{k}[\epsilon]) \\ \cap & & \cap \\ U_{ij} \times \mathrm{Spec}(\mathbf{k}[\epsilon]) & \xrightarrow{\Theta_{ij}} & U_{ij} \times \mathrm{Spec}(\mathbf{k}[\epsilon]) \end{array}$$

equivalently, such that $\theta_{ij} = \Theta_{ij|V_{ij}}$. According to Lemma 1.2.6, Θ_{ij} and θ_{ij} correspond to sections $D_{ij} \in \Gamma(U_{ij}, T_Y)$ and $d_{ij} \in \Gamma(V_{ij}, T_X)$ respectively such that $D_{ij} \mapsto d_{ij}$ when restricted to X. It follows that $D_{ij} \in \Gamma(U_{ij}, T_Y\langle X\rangle)$ and that to give A_{ij} is the same as to give D_{ij}. This said, the proof of the statement about tangent and obstruction spaces of Def'_j now proceeds in a straightforward way along the lines of the analogous proofs of 1.2.9 and of 1.2.12. We omit the details. □

Also in this case we have the notions of *obstructed* (resp. *unobstructed*) *deformation*, *obstructed* (resp. *unobstructed*) *embedding*, and of *rigid embedding*. It follows from Proposition 3.4.17 that a closed embedding $j : X \subset Y$ of projective nonsingular varieties is rigid if and only if

$$H^1(Y, T_Y\langle X\rangle) = 0$$

Let

$$\begin{array}{ccc} \mathcal{X} & \xrightarrow{J} & \mathcal{Y} \\ & \searrow \quad \swarrow & \\ & S & \end{array}$$

be a locally trivial deformation of $j : X \subset Y$ parametrized by a pointed scheme (S, s). Then we can define a *characteristic map*

$$\chi_J : T_{S,s} \to H^1(Y, T_Y\langle X\rangle)$$

by associating to a tangent vector $t : \mathrm{Spec}(\mathbf{k}[\epsilon]) \to S$ at s the element of $H^1(Y, T_Y\langle X\rangle)$ corresponding to the first-order deformation of f obtained by pulling back J by t.

Remark 3.4.18. Let $j : X \subset Y$ be a closed embedding of projective schemes with Y nonsingular. If

$$H^0(Y, T_Y\langle X\rangle) = 0$$

then Def'_j is prorepresentable. In fact, letting $(R, \hat{u})$ be the formal semiuniversal deformation of Def'_j, one can generalize Theorem 2.6.1 and its corollaries by introducing an automorphism functor

$$\mathrm{Aut}_{\hat{u}} : \hat{\mathcal{A}}_R \to (\mathrm{sets})$$

in an obvious way and proving that it is prorepresentable with tangent space $H^0(Y, T_Y\langle X\rangle)$. This is called the space of *infinitesimal automorphisms of* j. The details of this straightforward generalization are left to the reader.

Example 3.4.19. Let Y be a projective scheme, $p \in Y$ a closed point and $j : \{p\} \subset Y$. Then

$$T_Y(\langle p\rangle) = \mathcal{I}_p T_Y$$

where $\mathcal{I}_p \subset \mathcal{O}_Y$ is the ideal sheaf of p. In this case Def'_j is the functor of locally trivial deformations of the pointed scheme (Y, p). If, for example, Y is a projective nonsingular connected curve of genus g then $T_Y(\langle p\rangle) = T_Y(-p)$ and we get:

$$h^1(Y, T_Y(-p)) = \begin{cases} 0 & \text{if } g = 0 \\ 1 & \text{if } g = 1 \\ 3g-2 & \text{if } g \geq 2 \end{cases}$$

while $h^2(Y, T_Y(-p)) = 0$.

Of course, we can generalize by considering a set of m distinct closed points $\{p_1, \ldots, p_m\}$ of Y, and the inclusion $j : \{p_1, \ldots, p_m\} \to Y$. Then Def'_j is the functor of locally trivial deformations of the m-pointed scheme $(Y; p_1, \ldots, p_m)$.

3.4.5 Stability and costability

Whenever we have a locally trivial infinitesimal deformation

$$\begin{array}{ccc} \mathcal{X} & \rightarrow & \mathcal{Y} \\ & \searrow \quad \swarrow & \\ & \mathrm{Spec}(A) & \end{array}$$

of a closed embedding $j : X \subset Y$ of projective schemes we also have a deformation of X and a deformation of Y (both locally trivial):

$$\xi : \begin{array}{ccc} X & \rightarrow & \mathcal{X} \\ \downarrow & & \downarrow \\ \mathrm{Spec}(\mathbf{k}) & \rightarrow & \mathrm{Spec}(A) \end{array} \qquad \eta : \begin{array}{ccc} Y & \rightarrow & \mathcal{Y} \\ \downarrow & & \downarrow \\ \mathrm{Spec}(\mathbf{k}) & \rightarrow & \mathrm{Spec}(A) \end{array}$$

This means that we have two *forgetful morphisms* of functors:

$$\begin{array}{ccc} \mathrm{Def}'_j & \xrightarrow{\Phi_Y} & \mathrm{Def}'_Y \\ \downarrow \Phi_X & & \\ \mathrm{Def}'_X & & \end{array}$$

The differentials and obstruction maps of these morphisms are described as follows.

Proposition 3.4.20. *If $j : X \rightarrow Y$ is a closed embedding of projective schemes with Y nonsingular then*

(i)

$$d\Phi_Y : H^1(Y, T_Y\langle X\rangle) \rightarrow H^1(Y, T_Y)$$

and

$$o(\Phi_Y) : H^2(Y, T_Y\langle X\rangle) \rightarrow H^2(Y, T_Y)$$

are the maps induced in cohomology by the inclusion $T_Y\langle X\rangle \subset T_Y$.

(ii)

$$d\Phi_X : H^1(Y, T_Y\langle X\rangle) \rightarrow H^1(X, T_X)$$

and

$$o(\Phi_X) : H^2(Y, T_Y\langle X\rangle) \rightarrow H^2(X, T_X)$$

are the maps induced in cohomology by the restriction $T_Y\langle X\rangle \rightarrow T_X$.

Proof. It is a straightforward consequence of the above analysis. Details are left to the reader. □

Remark 3.4.21. Let $j : X \subset Y$ be a closed embedding of projective nonsingular schemes. Then there is a natural morphism of functors

$$H_X^Y \rightarrow \mathrm{Def}_j$$

whose differential is easily seen to be the coboundary map

$$\delta : H^0(X, N_{X/Y}) \to H^1(Y, T_Y\langle X\rangle)$$

determined by the exact sequence (3.55). It follows from the proposition that

$$\ker(d\Phi_Y) = \mathrm{Im}(\delta)$$

as expected, because deformations of X in Y are precisely those deformations of the embedding $X \subset Y$ which induce the trivial deformation of Y.

The proposition implies also that

$$\ker(d\Phi_X) = \mathrm{Im}[H^1(Y, \mathcal{I}_X T_Y) \xrightarrow{\beta} H^1(Y, T_Y\langle X\rangle)]$$

where β is the map induced by the inclusion $\mathcal{I}_X T_Y \subset T_Y\langle X\rangle$. This kernel consists of the first-order deformations of j which induce the trivial deformation of X. If, in particular, $H^0(X, T_X) = 0$ then

$$\ker(d\Phi_X) = H^1(Y, \mathcal{I}_X T_Y)$$

Definition 3.4.22. *If $j : X \subset Y$ is a closed embedding of projective schemes then X is called* stable *in Y if the morphism of functors*

$$\Phi_Y : \mathrm{Def}'_j \to \mathrm{Def}'_Y$$

is smooth. X is called costable *in Y if the morphism of functors*

$$\Phi_X : \mathrm{Def}'_j \to \mathrm{Def}'_X$$

is smooth.

The notion of stability was introduced and studied in [104] for a compact complex submanifold of a complex manifold. As stated in [104], stability means that "no local deformation of Y makes X disappear". Our definition of stability implies that every infinitesimal locally trivial deformation of Y is induced by a locally trivial deformation of j. Costability implies that every infinitesimal locally trivial deformation of X is induced by a locally trivial deformation of j. The notion of costability has been introduced in [88].

Proposition 3.4.23. *Let $j : X \subset Y$ be a closed embedding of projective schemes, Y nonsingular.*

(i) If $H^1(X, N'_{X/Y}) = (0)$ then X is stable in Y.
(ii) If $H^2(Y, \mathcal{I}_X T_Y) = 0$ then X is costable in Y.
(iii) If X is nonsingular and is both stable and costable in Y then X is obstructed if and only if Y is obstructed (as abstract varieties).

Proof. (i) From the exact sequence (3.55) it follows that $d\Phi_Y$ is surjective and that

$$H^2(Y, T_Y\langle X\rangle) \to H^2(Y, T_Y)$$

is injective; but by Proposition 3.4.20(i) this last condition means that Def'_f is less obstructed than Def_Y and the conclusion follows from Proposition 2.3.6.

(ii) The proof is similar using the exact cohomology sequence of the first column of diagram (3.56), Proposition 3.4.20 and Proposition 2.3.6.

(iii) Since the morphisms of functors

$$\mathrm{Def}_X \stackrel{\Phi_X}{\longleftarrow} \mathrm{Def}_j \stackrel{\Phi_Y}{\longrightarrow} \mathrm{Def}_Y$$

are both smooth we deduce that any one of the functors Def_X, Def_j, Def_Y is smooth if and only if the others are. □

Examples 3.4.24. (i) (Kodaira [104], Th. 5) Let Y be a projective nonsingular variety, $\gamma \subset Y$ a nonsingular closed subvariety and $\pi : X \to Y$ the blow-up of Y with centre γ. Let $E = \pi^{-1}(\gamma) \subset X$ be the exceptional divisor; then

$$h^i(E, N_{E/X}) = 0$$

for all i (see (3.47)). From Proposition 3.4.23 we obtain that E is a stable subvariety of X. This is remarkable because γ has not been required to be stable in Y.

(ii) Let X be a projective nonsingular algebraic surface and $Z \subset X$ an irreducible nonsingular rational curve with self-intersection $\nu = Z^2$. Then Z is stable in X if $\nu \geq -1$ because $H^1(Z, N_{Z/X}) = 0$ in this case. On the other hand, if $\nu \leq -2$ then in general Z is not stable in X. An example is provided by the negative section E in the rational ruled surface F_m, for $m \geq 2$. In fact, $E^2 = -m$ and we have seen in Example 1.2.2(ii) that there is a family $f : \mathcal{W} \to \mathbf{A}^1$ of deformations of F_m for which $[E]$ does not extend to the other fibres $\mathcal{W}(t)$, $t \neq 0$, since they are isomorphic to F_n for some $0 \leq n < m$. This implies that E is not stable.

(iii) Assume $\mathbf{k} = \mathbf{C}$. Let C be a projective irreducible nonsingular curve of genus $g \geq 3$, and let $\alpha : C \to JC$ be the Abel–Jacobi embedding of C into its jacobian variety. Then:

- $H^1(JC, \mathcal{O}_{JC}) \cong H^1(C, \mathcal{O}_C)$ ([20], Lemma 11.3.1)
- $T_{JC} = H^1(C, \mathcal{O}_C) \otimes \mathcal{O}_{JC}$.

Thus the restriction $T_{JC} \to T_{JC|C}$ induces isomorphisms

$$\begin{aligned} H^0(JC, T_{JC}) &\cong H^0(C, T_{JC|C}) \cong H^1(C, \mathcal{O}_C) \\ H^1(JC, T_{JC}) &\cong H^1(C, T_{JC|C}) \cong H^1(C, \mathcal{O}_C) \otimes H^1(C, \mathcal{O}_C) \end{aligned} \tag{3.57}$$

In particular, in view of the second column of diagram (3.56), we have

$$H^0(JC, \mathcal{I}_C T_{JC}) = 0 = H^1(JC, \mathcal{I}_C T_{JC})$$

and therefore, taking cohomology of (3.56), we obtain the commutative and exact diagram:

$$\begin{array}{ccccccccc} H^0(N_{C/JC}) & \to & H^1(JC, T_{JC}\langle C\rangle) & \stackrel{d\Phi_{JC}}{\longrightarrow} & H^1(JC, T_{JC}) & \to & H^1(N_{C/JC}) & \to & 0 \\ \| & & \cap & & \| & & \| & & \\ H^0(N_{C/JC}) & \to & H^1(C, T_C) & \stackrel{\sigma}{\longrightarrow} & H^1(C, T_{JC|C}) & \to & H^1(N_{C/JC}) & \to & 0 \end{array} \tag{3.58}$$

which implies that

$$H^1(JC, T_{JC}\langle C\rangle) \cong H^1(C, T_C) \tag{3.59}$$

and

$$H^2(JC, T_{JC}\langle C\rangle) \subset H^2(JC, T_{JC}) \tag{3.60}$$

Since (3.60) is the obstruction map of

$$\Phi_{JC} : \mathrm{Def}_\alpha \to \mathrm{Def}_{JC}$$

from Proposition 2.3.6 we deduce that Def_α is smooth, being less obstructed than Def_{JC} which is smooth (see Example 2.4.11(v), page 74). On the other hand, since (3.59) is the differential of the forgetful morphism

$$\Phi_C : \mathrm{Def}_\alpha \to \mathrm{Def}_C$$

and Def_α is smooth, we deduce by Corollary 2.3.7 that Φ_C is smooth, i.e. C is costable in JC.

Note that $H^0(JC, T_{JC}\langle C\rangle) = 0$ by the first column of diagram (3.56): therefore α has no infinitesimal automorphisms and Def_α is prorepresentable (Remark 3.4.18). It follows that Φ_C is actually an isomorphism of functors (Remark 2.3.8).

We can identify Def_α with Def_C and the differential

$$d\Phi_{JC} : H^1(JC, T_{JC}\langle C\rangle) \to H^1(JC, T_{JC})$$

with the map σ in diagram (3.58). Therefore Φ_{JC} is a closed embedding if and only if σ is injective. In view of the isomorphisms (3.57) σ is Serre-dual to the natural multiplication map:

$$\sigma^\vee : H^0(C, K_C) \otimes H^0(C, K_C) \to H^0(C, 2K_C)$$

This map is surjective if and only if C is non-hyperelliptic: therefore in this case Def_C is a closed smooth subfunctor of Def_{JC}. In the natural decomposition

$$H^1(C, \mathcal{O}_C) \otimes H^1(C, \mathcal{O}_C) = S^2 H^1(\mathcal{O}_C) \oplus \bigwedge^2 H^1(\mathcal{O}_C)$$

we have

$$\mathrm{Im}(\sigma) \subset S^2 H^1(\mathcal{O}_C)$$

because $\bigwedge^2 H^1(\mathcal{O}_C) \subset \ker(\sigma^\vee)$. Therefore $\mathrm{Im}(d\Phi_{JC})$ is contained in $S^2 H^1(\mathcal{O}_C)$ which is the space of first-order deformations of JC preserving the principal polarization (compare with Example 3.3.13, page 151, and observe that in this case $\bar{\Omega} = H^1(\mathcal{O}_C) = V$ by the Hodge decomposition of $H^1(C, \mathbf{C})$).

The validity of the condition "σ injective" is called the *infinitesimal Torelli theorem*: thus it holds if and only if C is non-hyperelliptic. It is not difficult to show that $\alpha(C)$ is unobstructed in JC if and only if C is non-hyperelliptic (see [70], [113]).

The following result, due to Kodaira [104], gives the possibility of relating a local Hilbert functor with the deformation functor of an abstract variety.

Proposition 3.4.25. *Let Y be a projective nonsingular variety, $\gamma \subset Y$ a closed nonsingular subvariety of pure codimension $r \geq 2$, and let $\pi : X \to Y$ be the blow-up of Y with centre γ. Then:*

(i) There is a natural isomorphism of functors

$$B : H_\gamma^Y \to \mathrm{Def}_{\pi/Y}$$

In particular, $\mathrm{Def}_{\pi/Y}$ is prorepresentable.

(ii) Assume that $H^1(Y, T_Y) = 0$, i.e. that Y is rigid. Then the forgetful morphism $\Phi_\pi : \mathrm{Def}_{\pi/Y} \to \mathrm{Def}_X$ is smooth. Therefore we have a smooth morphism of functors $B\Phi_\pi : H_\gamma^Y \to \mathrm{Def}_X$. In particular, if γ is obstructed in Y then X is obstructed as an abstract variety.

Proof. (i) We define

$$B : H_\gamma^Y \to \mathrm{Def}_{\pi/Y}$$

by associating to a family of deformations

$$\begin{array}{ccc} \gamma_A & \subset & Y \times \mathrm{Spec}(A) \\ \downarrow & & \\ \mathrm{Spec}(A) & & \end{array}$$

of γ in Y over A the blow-up

$$\pi_A : X_A := Bl_{\gamma_A}(Y \times \mathrm{Spec}(A)) \to Y \times \mathrm{Spec}(A)$$

of $Y \times \mathrm{Spec}(A)$ along γ_A. Note that, since $\mathcal{O}_{\gamma_A}$ is A-flat and we have the exact sequence on $Y \times \mathrm{Spec}(A)$:

$$0 \to \mathcal{I}_{\gamma_A} \to \mathcal{O}_Y \otimes A \to \mathcal{O}_{\gamma_A} \to 0$$

the sheaf $\mathcal{I}_{\gamma_A}$ is A-flat as well (Proposition A.2(VI)); moreover, $\mathcal{I}_{\gamma_A}^k/\mathcal{I}_{\gamma_A}^{k+1}$ is locally free over $\mathcal{O}_Y \otimes A$ for all $k \geq 1$ because γ_A is regularly embedded in $Y \times \mathrm{Spec}(A)$ by Lemma D.2.3. From this it is easy to deduce that $\mathcal{I}_{\gamma_A}^k$ is A-flat for all $k \geq 1$ and therefore

$$X_A = \mathrm{Proj} \bigoplus_k \mathcal{I}_{\gamma_A}^k$$

is A-flat by Proposition A.2(V). We leave to the reader to check the functoriality of B.

The differential of B is the composition

$$dB : H^0(\gamma, N_{\gamma/Y}) \cong H^0(E, q^* N_{\gamma/Y}) \to H^0(X, N_\pi)$$

where the first map is the obvious isomorphism and the second one comes from the exact sequence (3.50); in a similar way, one describes the obstruction map of B as the one induced by the composition

$$H^1(\gamma, N_{\gamma/Y}) \cong H^1(E, q^*N_{\gamma/Y}) \to H^1(X, N_\pi)$$

deduced from the exact sequence (3.50). These facts can be easily verified by chasing diagram (3.49). Since $H^i(E, N_{E/X}) = 0$ for all i, we see that these maps are both bijective, and the conclusion follows.

(ii) From (3.48) it follows that $H^1(X, \pi^*T_Y) = 0$. The exact sequence

$$0 \to T_X \to \pi^*T_Y \to N_\pi \to 0$$

and Proposition 3.4.20(i) imply that $d\Phi_\pi$ is surjective and $o(\Phi_\pi)$ is injective. The conclusion is a consequence of Proposition 2.3.6. The last assertion is an obvious consequence of the fact that the composition $\Phi_\pi B : H^Y_\gamma \to \mathrm{Def}_X$ is smooth. □

By applying Proposition 3.4.25 to any obstructed nonsingular curve $\gamma \subset \mathbb{P}^3$ (e.g. the curve of degree 14 and genus 24 described in § 4.6) we obtain *an example of obstructed projective variety of dimension 3.* As an application we obtain the following result, which gives examples of obstructed surfaces.

Theorem 3.4.26 (Horikawa [88]). *Let $\gamma \subset \mathbb{P}^3$ be an obstructed nonsingular curve, and X the blow-up of $\mathbb{P}^3$ with centre γ. If $S \subset X$ is a sufficiently ample nonsingular surface then S is obstructed as an abstract variety.*

Proof. By Proposition 3.4.23(iii) it is sufficient to show that S is both stable and costable in X. We have $h^2(X, \mathcal{O}_X) = 0$ and, by the ampleness of S, $h^1(X, \mathcal{O}_X(S)) = 0$ by Serre's vanishing theorem. From the exact sequence

$$0 \to \mathcal{O}_X \to \mathcal{O}_X(S) \to N_{S/X} \to 0$$

we deduce that $h^1(S, N_{S/X}) = 0$ and therefore S is stable in X. On the other hand, we have

$$H^2(X, T_X(-S)) = 0$$

by Serre's vanishing theorem again. Therefore S is costable in X as well. □

We can immediately verify, using the adjunction formula, that the surfaces S constructed in the theorem are regular and with ample canonical class. For related examples see [29].

NOTES

1. The analysis of morphisms from a nonsingular curve is taken from [8]; see also [7]. For Lemma 3.4.15 see [180].

2. For the study of the functors $\mathrm{Def}_{f/Y}$ and Def_f under more general assumptions than those of Theorem 3.4.8 a more careful analysis of first-order deformations and obstructions is needed. We refer the reader to [148] and [150] for more information about this. The deformation theory of closed embeddings is studied in [134] in the analytic category.

3. *Let*

$$\begin{array}{ccc} \mathcal{X} & \xrightarrow{\Phi} & \mathcal{Y} \\ & \searrow\swarrow & \\ & S & \end{array}$$

be a commutative diagram of morphisms of algebraic schemes, with $\mathcal{X}$ and $\mathcal{Y}$ S-flat and Φ projective. Assume that $\Phi_o : \mathcal{X}(o) \to \mathcal{Y}(o)$ is a closed embedding, for some **k***-rational point $o \in S$. Then there is an open neighbourhood $U \subset S$ of o such that the restriction $\Phi(U) : \mathcal{X}(U) \to \mathcal{Y}(U)$ is a closed embedding.*

Proof. Let $\mathcal{K} = \text{coker}[\mathcal{O}_{\mathcal{Y}} \to \Phi_*(\mathcal{O}_{\mathcal{X}})]$. Since Φ is projective, $\Phi_*(\mathcal{O}_{\mathcal{X}})$ is a coherent sheaf and so is $\mathcal{K}$. Moreover, $\mathcal{K}(o) = (0)$ because Φ_o is a closed embedding. It follows that there is an open subset $U \subset S$ containing o such that $\mathcal{K}_{|\mathcal{Y}(U)} = (0)$. Let $\mathcal{Z} = \text{Spec}(\Phi_*(\mathcal{O}_{\mathcal{X}}))$, $h : \mathcal{Z} \to \mathcal{Y}$ the induced S-morphism and

$$\begin{array}{ccc} \mathcal{X} & \xrightarrow{\Phi} & \mathcal{Y} \\ & g \searrow \nearrow h & \\ & \mathcal{Z} & \end{array}$$

the Stein factorization of Φ. Then it follows that $h(U) : \mathcal{Z}(U) \to \mathcal{Y}(U)$ is a closed embedding. Moreover, since g has connected fibres and is bijective over $\mathcal{Z}(o)$, it follows that, modulo shrinking U if necessary, $g(U) : \mathcal{X}(U) \to \mathcal{Z}(U)$ is an isomorphism. The conclusion follows. □

4

The Hilbert schemes and the Quot schemes

Even though this book is centred around the theme of infinitesimal and local deformations, in this chapter we turn our attention to *global deformations*. We will introduce the Hilbert schemes and other related objects, which are important examples of parameter schemes for global families of deformations of algebro-geometric objects. They are used to describe and classify "extrinsic" deformations, i.e. deformations of objects within a given ambient space (e.g. closed subschemes of a given scheme). Their study is preliminary to the construction of "moduli schemes". Moreover, they provide some of the most typical examples of constructions in algebraic geometry by the functorial approach. We will study some of their properties and consider a few applications of the local theory developed so far.

4.1 Castelnuovo–Mumford regularity

In this section we introduce the notion of *m-regularity*, also called *Castelnuovo–Mumford regularity*, and we prove its main properties. They will be needed for the construction of the Hilbert schemes and of the Quot schemes.

Let $m \in \mathbb{Z}$. A coherent sheaf $\mathcal{F}$ on $\mathbb{P}^r$ is *m-regular* if

$$H^i(\mathcal{F}(m-i)) = (0)$$

for all $i \geq 1$.

Because of Serre's vanishing theorem, every coherent sheaf $\mathcal{F}$ on $\mathbb{P}^r$ is m-regular for some $m \in \mathbb{Z}$.

The definition of m-regularity makes sense for a coherent sheaf on any projective scheme X endowed with a very ample line bundle $\mathcal{O}(1)$. For simplicity we will consider the case $X = \mathbb{P}^r$ only, leaving to the reader the obvious modifications of the statements and of the proofs in the general case.

Proposition 4.1.1. *If $\mathcal{F}$ is m-regular then:*

(i) The natural map

$$H^0(\mathcal{F}(k)) \otimes_{\mathbf{k}} H^0(\mathcal{O}(1)) \to H^0(\mathcal{F}(k+1))$$

is surjective for all $k \geq m$.

(ii) $H^i(\mathcal{F}(k)) = (0)$ for all $i \geq 1$ and $k \geq m - i$; in particular, $\mathcal{F}$ is n-regular for all $n \geq m$.

(iii) $\mathcal{F}(m)$, and therefore also $\mathcal{F}(k)$ for all $k \geq m$, is generated by its global sections.

Proof. We prove (i) and (ii) by induction on r. If $r = 0$ there is nothing to prove. Assume $r \geq 1$ and let H be a hyperplane not containing any point of $\mathrm{Ass}(\mathcal{F})$; it exists because $\mathrm{Ass}(\mathcal{F})$ is a finite set. Tensoring by $\mathcal{F}(k)$ the exact sequence:

$$0 \to \mathcal{O}(-H) \to \mathcal{O} \to \mathcal{O}_H \to 0$$

we get an exact sequence:

$$0 \to \mathcal{F}(k-1) \to \mathcal{F}(k) \to \mathcal{F}_H(k) \to 0$$

where $\mathcal{F}_H = \mathcal{F} \otimes \mathcal{O}_H$. For each $i > 0$ we obtain an exact sequence

$$H^i(\mathcal{F}(m-i)) \to H^i(\mathcal{F}_H(m-i)) \to H^{i+1}(\mathcal{F}(m-i-1))$$

which implies that $\mathcal{F}_H$ is m-regular on H. It follows by induction that (i) and (ii) are true for $\mathcal{F}_H$.

Let's consider the exact sequence

$$H^{i+1}(\mathcal{F}(m-i-1)) \to H^{i+1}(\mathcal{F}(m-i)) \to H^{i+1}(\mathcal{F}_H(m-i))$$

If $i \geq 0$ the two extremes are zero (the right one by (ii) for $\mathcal{F}_H$, the left one by the m regularity of $\mathcal{F}$), therefore $\mathcal{F}$ is $(m+1)$-regular. By iteration this proves (ii).

To prove (i) we consider the commutative diagram:

$$\begin{array}{cccc} & H^0(\mathcal{F}(k)) \otimes_{\mathbf{k}} H^0(\mathcal{O}(1)) & \xrightarrow{u} & H^0(\mathcal{F}_H(k)) \otimes_{\mathbf{k}} H^0(\mathcal{O}_H(1)) \\ & \downarrow w & & \downarrow t \\ H^0(\mathcal{F}(k)) \to & H^0(\mathcal{F}(k+1)) & \xrightarrow{v} & H^0(\mathcal{F}_H(k+1)) \end{array}$$

The map u is surjective for $k \geq m$ because $H^1(\mathcal{F}(k-1)) = (0)$; moreover, t is surjective for $k \geq m$ by (i) for $\mathcal{F}_H$. Therefore vw is surjective. It follows that $H^0(\mathcal{F}(k+1))$ is generated by $\mathrm{Im}(w)$ and by $H^0(\mathcal{F}(k))$ for all $k \geq m$. But $H^0(\mathcal{F}(k)) \subset \mathrm{Im}(w)$ because the inclusion $H^0(\mathcal{F}(k)) \subset H^0(\mathcal{F}(k+1))$ is multiplication by H. Therefore w is surjective.

Let's prove (iii). Let $h \gg 0$ be such that $\mathcal{F}(m+h)$ is generated by its global sections. Then the composition

$$H^0(\mathcal{F}(m)) \otimes_{\mathbf{k}} H^0(\mathcal{O}(h)) \otimes_{\mathbf{k}} \mathcal{O} \to H^0(\mathcal{F}(m+h)) \otimes_{\mathbf{k}} \mathcal{O} \to \mathcal{F}(m+h)$$

is surjective because from (i) it follows that the first map is; we deduce that the composition

$$H^0(\mathcal{F}(m)) \otimes_{\mathbf{k}} H^0(\mathcal{O}(h)) \otimes_{\mathbf{k}} \mathcal{O}(-h) \to H^0(\mathcal{F}(m)) \otimes_{\mathbf{k}} \mathcal{O} \to \mathcal{F}(m)$$

is also surjective, hence the second map is surjective too. □

Note that if $\mathcal{F}$ is m-regular then the graded $\mathbf{k}[X_1, \ldots, X_r]$-module

$$\Gamma_*(\mathcal{F}) := \bigoplus_{k \in \mathbb{Z}} H^0(\mathcal{F}(k))$$

can be generated by elements of degree $\leq m$. In fact, this is equivalent to the surjectivity of the multiplication maps

$$H^0(\mathcal{F}(m)) \otimes_{\mathbf{k}} H^0(\mathcal{O}(h)) \to H^0(\mathcal{F}(m+h))$$

for $h \geq 1$, and follows from part (i) of the proposition. In particular, if an ideal sheaf $\mathcal{I} \subset \mathcal{O}_{\mathbb{P}^r}$ is m-regular then the homogeneous ideal

$$I = \Gamma_*(\mathcal{I}) \subset \mathbf{k}[X_0, \ldots, X_r]$$

is generated by elements of degree $\leq m$.

Note also that in proving 4.1.1 we have proved the following:

Proposition 4.1.2. *If $\mathcal{F}$ is m-regular and*

$$0 \to \mathcal{F}(-1) \to \mathcal{F} \to \mathcal{G} \to 0$$

is an exact sequence, then $\mathcal{G}$ is m-regular.

Conversely, we have the following:

Proposition 4.1.3. *Let*

$$0 \to \mathcal{F}(-1) \to \mathcal{F} \to \mathcal{G} \to 0$$

be an exact sequence of coherent sheaves on $\mathbb{P}^r$, and assume that $\mathcal{G}$ is m-regular. Then:

$$(i) \quad H^i(\mathcal{F}(k)) = 0 \qquad \text{for } i \geq 2 \text{ and } k \geq m - i$$

$$(ii) \quad h^1(\mathcal{F}(k-1)) \geq h^1(\mathcal{F}(k)) \text{ for } k \geq m - 1$$

$$(iii) \quad H^1(\mathcal{F}(k)) = 0 \qquad \text{for } k \geq (m-1) + h^1(\mathcal{F}(m-1))$$

In particular, $\mathcal{F}$ is $m + h^1(\mathcal{F}(m-1))$-regular.

Proof. (i) In the exact sequence

$$H^{i-1}(\mathcal{G}(k)) \to H^i(\mathcal{F}(k-1)) \to H^i(\mathcal{F}(k)) \to H^i(\mathcal{G}(k))$$

the first and the last group are zero for $i \geq 2$ and $k \geq m-(i-1)$. Therefore

$$H^i(\mathcal{F}(m-i)) \cong H^i(\mathcal{F}(m-i+1)) \cong H^i(\mathcal{F}(m-i+2)) \cong \cdots$$

From Serre's vanishing theorem we get $H^i(\mathcal{F}(m-i+h)) = 0$ for all $h \gg 0$ and (i) follows.

(ii) For $k \geq m-1$ we have the exact sequence

$$0 \to H^0(\mathcal{F}(k-1)) \to H^0(\mathcal{F}(k)) \xrightarrow{\upsilon_k} H^0(\mathcal{G}(k)) \to H^1(\mathcal{F}(k-1)) \to H^1(\mathcal{F}(k)) \to 0$$

which implies (ii).

(iii) Assume υ_k surjective, and consider the commutative diagram:

$$\begin{array}{ccc} H^0(\mathcal{F}(k)) \otimes H^0(\mathcal{O}(1)) & \xrightarrow{\upsilon_k \otimes id} & H^0(\mathcal{G}(k)) \otimes H^0(\mathcal{O}(1)) \\ \downarrow & & \downarrow w_k \\ H^0(\mathcal{F}(k+1)) & \xrightarrow{\upsilon_{k+1}} & H^0(\mathcal{G}(k+1)) \end{array}$$

Since w_k is surjective for $k \geq m$, we have that υ_{k+1} is surjective too. Therefore

$$H^1(\mathcal{F}(k-1)) \cong H^1(\mathcal{F}(k)) \cong H^1(\mathcal{F}(k+1)) \cong \cdots \cong 0$$

If υ_k is not surjective then $h^1(\mathcal{F}(k-1)) > h^1(\mathcal{F}(k))$. Therefore the function $k \mapsto h^1(\mathcal{F}(k))$ is strictly decreasing for $k \geq m-1$, and this implies (iii). □

The following is a useful characterization of m-regularity.

Theorem 4.1.4. *A coherent sheaf $\mathcal{F}$ on $\mathbb{P}^r$ is m-regular if and only if it has a resolution of the form:*

$$\cdots \to \mathcal{O}(-m-2)^{b_2} \to \mathcal{O}(-m-1)^{b_1} \to \mathcal{O}(-m)^{b_0} \to \mathcal{F} \to 0 \tag{4.1}$$

for some nonnegative $b_0, b_1, b_2, \ldots$.

Proof. Assume that $\mathcal{F}$ has a resolution (4.1) and let

$$\mathcal{R}_1 = \ker\left[\mathcal{O}(-m)^{b_0} \to \mathcal{F}\right]$$

$$\mathcal{R}_j = \ker\left[\mathcal{O}(-m-j+1)^{b_{j-1}} \to \mathcal{O}(-m-j+2)^{b_{j-2}}\right] \qquad j = 2, \ldots, r$$

$$\mathcal{R}_{r+1} = \mathcal{O}(-m-r)^{b_{r+1}}$$

Using the short exact sequences:

$$0 \to \mathcal{R}_1(m-i) \to \mathcal{O}(-i)^{b_0} \to \mathcal{F}(m-i) \to 0$$

$$0 \to \mathcal{R}_j(m-i) \to \mathcal{O}(-i-j+1)^{b_{j-1}} \to \mathcal{R}_{j-1}(m-i) \to 0$$

$$0 \to \mathcal{O}(-i-r-1)^{b_{r+1}} \to \mathcal{O}(-i-r)^{b_r} \to \mathcal{R}_r(m-i) \to 0$$

we see that for all $1 \le i \le r$ we have:

$$H^i(\mathcal{F}(m-i)) \cong H^{i+1}(\mathcal{R}_1(m-i)) \cong \cdots$$

$$\cdots \cong H^r(\mathcal{R}_{r-i}(m-i)) \cong H^{r+1}(\mathcal{R}_{r-i+1}(m-i)) = (0)$$

and $\mathcal{F}$ is m-regular.

Assume conversely that $\mathcal{F}$ is m-regular. By 4.1.1(iii) we have an exact sequence:

$$0 \to \mathcal{R}_1 \to \mathcal{O}(-m)^{b_0} \to \mathcal{F} \to 0$$

with $b_0 = h^0(\mathcal{F}(m))$, which defines $\mathcal{R}_1$.

If $\mathcal{R}_1 = 0$ we are done; if $\mathcal{R}_1 \neq 0$ from the sequences:

$$0 \to \mathcal{R}_1(m-i+1) \to \mathcal{O}(-i+1)^{b_0} \to \mathcal{F}(m-i+1) \to 0$$

we deduce that

$$H^i(\mathcal{R}_1(m-i+1)) \cong H^{i-1}(\mathcal{F}(m-i+1)) \qquad 1 \le i \le r$$

hence $\mathcal{R}_1$ is $(m+1)$-regular. Applying the same argument to $\mathcal{R}_1$ we find an exact sequence:

$$0 \to \mathcal{R}_2 \to \mathcal{O}(-m-1)^{b_1} \to \mathcal{O}(-m)^{b_0} \to \mathcal{F} \to 0$$

with $\mathcal{R}_2$ $(m+2)$-regular. This process can be repeated and gives a resolution as required. □

We will now turn to the problem of finding numerical criteria of m-regularity for a coherent sheaf $\mathcal{F}$ on $I\!P^r$.

Consider a sequence $\sigma_1, \ldots, \sigma_N$ of N sections of $\mathcal{O}_{I\!P^r}(1)$. We will call it $\mathcal{F}$-*regular* if the sequences of sheaf homomorphisms induced by multiplication by $\sigma_1, \ldots, \sigma_N$:

$$0 \to \mathcal{F}(-1) \xrightarrow{\sigma_1} \mathcal{F} \to \mathcal{F}_1 \to 0$$

$$0 \to \mathcal{F}_1(-1) \xrightarrow{\sigma_2} \mathcal{F}_1 \to \mathcal{F}_2 \to 0$$

etc., are exact.

By choosing σ_{i+1} not containing any point of $\mathrm{Ass}(\mathcal{F}_i)$ one shows that $\mathcal{F}$-regular sequences of any length exist. Therefore any general N-tuple $(\sigma_1, \ldots, \sigma_N) \in H^0(\mathcal{O}_{I\!P^r}(1))^N$ is an $\mathcal{F}$-sequence.

Definition 4.1.5. *Let $\mathcal{F}$ be a coherent sheaf on $I\!P^r$, and* $(\mathbf{b}) = (b_0, b_1, \ldots, b_N)$ *a sequence of nonnegative integers such that $N \ge$* $\dim[\mathrm{Supp}(\mathcal{F})]$. *We will call $\mathcal{F}$ a* $(\mathbf{b})$-sheaf *if there exists an $\mathcal{F}$-regular sequence $\sigma_1, \ldots, \sigma_N$ of sections of $\mathcal{O}_{I\!P^r}(1)$ such that $h^0(\mathcal{F}_i(-1)) \le b_i$, $i = 0, \ldots, N$ where $\mathcal{F}_0 = \mathcal{F}$, and $\mathcal{F}_i = \mathcal{F}/(\sigma_1, \ldots, \sigma_i)\mathcal{F}(-1)$, $i \ge 1$.*

Note that from the definition it follows immediately that if $\mathcal{F}$ is a $(\mathbf{b})$-sheaf then $\mathcal{F}_1$ is a $(b_1, \ldots, b_N)$-sheaf.

It is likewise clear that for every coherent sheaf $\mathcal{F}$ on $\mathbb{P}^r$ there is a sequence $(\mathbf{b})$ such that $\mathcal{F}$ is a $(\mathbf{b})$-sheaf. Moreover, a subsheaf of a $(\mathbf{b})$-sheaf is easily seen to be a $(\mathbf{b})$-sheaf.

For example, every ideal sheaf $\mathcal{I} \subset \mathcal{O}_{\mathbb{P}^r}$ is a $(\mathbf{0})$-sheaf, because $\mathcal{O}_{\mathbb{P}^r}$ is clearly a $(\mathbf{0})$-sheaf.

Lemma 4.1.6. *Let*

$$0 \to \mathcal{F}(-1) \to \mathcal{F} \to \mathcal{G} \to 0$$

be an exact sequence of coherent sheaves on $\mathbb{P}^r$. If

$$\chi(\mathcal{F}(k)) = \sum_{i=0}^{r} a_i \binom{k+i}{i}$$

then

$$\chi(\mathcal{G}(k)) = \sum_{i=0}^{r-1} a_{i+1} \binom{k+i}{i}$$

The proof is left to the reader.

Proposition 4.1.7. *Let $\mathcal{F}$ be a $(\mathbf{b})$-sheaf, let $s = \dim[\mathrm{Supp}(\mathcal{F})]$ and*

$$\chi(\mathcal{F}(k)) = \sum_{i=0}^{s} a_i \binom{k+i}{i}$$

Then:

(i) For each $k \geq -1$ we have $h^0(\mathcal{F}(k)) \leq \sum_{i=0}^{s} b_i \binom{k+i}{i}$.
(ii) $a_s \leq b_s$ and $\mathcal{F}$ is also a $(b_0, \dots, b_{s-1}, a_s)$-sheaf.

Proof. (i) By induction on s. If $s = 0$ then $a_0 = h^0(\mathcal{F}) = h^0(\mathcal{F}(-1)) \leq b_0$ and the conclusion is obvious.

Assume $s \geq 1$. We have an exact sequence

$$0 \to \mathcal{F}(-1) \to \mathcal{F} \to \mathcal{F}_1 \to 0$$

with $\mathcal{F}_1$ a $(b_1, \dots, b_N)$-sheaf and $\dim[\mathrm{Supp}(\mathcal{F}_1)] = s - 1$. Then:

$$h^0(\mathcal{F}(k)) - h^0(\mathcal{F}(k-1)) \leq h^0(\mathcal{F}_1(k))$$

and

$$h^0(\mathcal{F}_1(k)) \leq \sum_{i=0}^{s-1} b_{i+1} \binom{k+i}{i}$$

by the inductive hypothesis. Since $h^0(\mathcal{F}(-1)) \leq b_0$ by induction on $k \geq -1$ we get the conclusion.

(ii) By Lemma 4.1.6 and by induction on s we get $a_s \leq b_s$ and $\mathcal{F}_1$ is a $(b_1, \dots, b_{s-1}, a_s)$-sheaf. The conclusion follows. □

Definition 4.1.8. *The following polynomials, defined by induction for each integer* $r \geq -1$*:*

$$\begin{aligned} P_{-1} &= 0 \\ P_r(X_0, \ldots, X_r) &= P_{r-1}(X_1, \ldots, X_r) + \textstyle\sum_{i=0}^r X_i \binom{P_{r-1}(X_1,\ldots,X_r)-1+i}{i} \end{aligned}$$

are called (**b**)-polynomials.

One immediately sees that

$$P_r(X_0, \ldots, X_t, 0, \ldots, 0) = P_t(X_0, \ldots, X_t) \tag{4.2}$$

for each $t < r$.

The following theorem gives a numerical criterion of m-regularity.

Theorem 4.1.9. *Let $\mathcal{F}$ be a* (**b**)*-sheaf on $I\!P^r$, with* (**b**) $= (b_0, b_1, \ldots, b_N)$*, and let*

$$\chi(\mathcal{F}(k)) = \sum_{i=0}^{r} a_i \binom{k+i}{i}$$

be its Hilbert polynomial. Let $(c_0, \ldots, c_r)$ be a sequence of integers such that $c_i \geq b_i - a_i$, for $i = 0, \ldots, r$, and $m = P_r(c_0, \ldots, c_r)$. Then $m \geq 0$ and $\mathcal{F}$ is m-regular. In particular, $\mathcal{F}$ is $P_{s-1}(c_0, \ldots, c_{s-1})$-regular if $s =$ dim[Supp($\mathcal{F}$)].

Proof. By induction on r. If $r = 0$ then $m = 0$ and $\mathcal{F}$ is n-regular for every $n \in \mathbb{Z}$, so the theorem is true in this case. Assume $r \geq 1$. We have an exact sequence:

$$0 \to \mathcal{F}(-1) \to \mathcal{F} \to \mathcal{F}_1 \to 0$$

with $\mathcal{F}_1$ a $(b_1, \ldots, b_N)$-sheaf supported on $I\!P^{r-1}$. From Lemma 4.1.6 and from the inductive hypothesis we deduce that $n \geq 0$ and $\mathcal{F}_1$ is n-regular, where $n = P_{r-1}(c_1, \ldots, c_r)$. From 4.1.3 we deduce that $\mathcal{F}$ is $[n + h^1(\mathcal{F}(n-1)]$-regular and $h^i(\mathcal{F}(n-1)) = 0$ for $i \geq 2$. Therefore:

$$h^1(\mathcal{F}(n-1)) = h^0(\mathcal{F}(n-1)) - \chi(\mathcal{F}(n-1)) \leq \sum_{i=0}^{r} (b_i - a_i) \binom{n-1+i}{i}$$

by 4.1.7(i). It follows that $\mathcal{F}$ is $n + \sum_{i=0}^r c_i \binom{n-1+i}{i}$-regular, by 4.1.1(ii). This proves the first assertion.

The last assertion follows from 4.1.7(ii) and from (4.2). □

Note that the integer m in the statement of the theorem depends on the coefficients of the Hilbert polynomial of $\mathcal{F}$ as well as on the integers b_i. In the special case when $\mathcal{F}$ is a sheaf of ideals we can determine an m for which $\mathcal{F}$ is m-regular which depends only on the Hilbert polynomial of $\mathcal{F}$, as stated in the next corollary.

Corollary 4.1.10. *For each $r \geq 0$ there exists a polynomial $F_r(X_0, \ldots, X_r)$ such that every sheaf of ideals $\mathcal{I} \subset \mathcal{O}_{P^r}$ having the Hilbert polynomial*

$$\chi(\mathcal{I}(k)) = \sum_{i=0}^{r} a_i \binom{k+r}{i}$$

is m-regular, where $m = F_r(a_0, \ldots, a_r)$, and $m \geq 0$.

Proof. It suffices to observe that $\mathcal{I}$ is a $(\mathbf{0})$-sheaf. Therefore the corollary follows from Theorem 4.1.9, taking $F_r(X_0, \ldots, X_r) = P_r(-X_0, \ldots, -X_r)$. □

NOTES

1. Corollary 4.1.10 is in general false for coherent sheaves which are not sheaves of ideals. An example from [130] is

$$\mathcal{F} = \mathcal{O}_{\mathbb{P}^1}(k) \oplus \mathcal{O}_{\mathbb{P}^1}(-k)$$

In fact $\chi(\mathcal{F}) = 2$ is independent of k but the least m such $\mathcal{F}$ is m-regular is $|k|$.

2. If $\mathcal{I}$ is the sheaf of ideals of the closed subscheme $X \subset \mathbb{P}^r$ and $\mathcal{I}$ is m-regular with $m \geq 0$, then $\mathcal{O}_X$ is $(m-1)$-regular. Conversely, if $\mathcal{O}_X$ is $(m-1)$-regular and the restriction map

$$H^0(\mathbb{P}^r, \mathcal{O}_{\mathbb{P}^r}(m-1)) \to H^0(X, \mathcal{O}_X(m-1))$$

is surjective, then $\mathcal{I}$ is m-regular. This follows from the exact sequences

$$0 \to \mathcal{I}(k) \to \mathcal{O}_{\mathbb{P}^r}(k) \to \mathcal{O}_X(k) \to 0$$

$k \geq m-1$.

3. The notion of m-regularity is related to that of *bounded collection of sheaves*, important in moduli theory.

A collection of coherent sheaves $\{F_j\}_{j \in J}$ on a projective scheme X is said to be *bounded* if there is an algebraic scheme S and a coherent sheaf $\mathcal{F}$ on $X \times S$ such that for each $j \in J$ there is a closed point $s \in S$ such that F_j is isomorphic to the sheaf $\mathcal{F}(s) = \mathcal{F}_{|X \times \{s\}}$. One also says that the collection $\{F_j\}_{j \in J}$ is bounded by the sheaf $\mathcal{F}$ on $X \times S$. For details we refer to [98].

4. The notion of Castelnuovo–Mumford regularity has been introduced in [130]. Castelnuovo studied the properties of m-regularity of projective curves in [26], where his upper bound for the genus of projective curves was proved. The treatment of $(\mathbf{b})$-sheaves has been taken from [98].

4.2 Flatness in the projective case

This section is devoted to some properties of flat families of projective schemes which will be needed in this chapter. In particular, we will prove a powerful technical result due to Grothendieck and Mumford, the existence of flattening stratifications, which is a key ingredient in the construction of the Hilbert schemes, of the Quot schemes, and of related schemes like the Severi varieties. The treatment of stratifications closely follows Lecture 8 of [130].

4.2.1 Flatness and Hilbert polynomials

The following result gives the name to the "Hilbert scheme".

Proposition 4.2.1. *(i) Let S be a scheme, $\mathcal{F}$ a coherent sheaf on $\mathbb{P}^r \times S$ and $p : \mathbb{P}^r \times S \to S$ the projection. Then $\mathcal{F}$ is flat over S if and only if $p_*\mathcal{F}(h)$ is locally free on S for all $h \gg 0$.*

(ii) Assume that S is connected. For each $s \in S$ let

$$P_s(t) = \chi(\mathcal{F}(s)(t)) = \sum_i (-1)^i h^i(\mathbb{P}^r(s), \mathcal{F}(s)(t))$$

be the Hilbert polynomial of $\mathcal{F}(s)$. If $\mathcal{F}$ is flat over S then $P_s(t)$ is independent of $s \in S$. Conversely, if S is integral and $P_s(t)$ is independent of s for all $s \in S$, then $\mathcal{F}$ is flat over S. If S is integral and algebraic and $P_s(t)$ is independent of s for all closed $s \in S$, then $\mathcal{F}$ is flat over S.

For the proof of this proposition we refer the reader to [84], Theorem III.9.9.

Corollary 4.2.2. *If*

$$\begin{array}{ccc} \mathcal{X} & \subset & \mathbb{P}^r \times S \\ \downarrow & & \\ S & & \end{array}$$

is a flat family of closed subschemes of $\mathbb{P}^r$ with S connected, then all fibres $\mathcal{X}(s)$ have the same Hilbert polynomial; in particular, they have the same degree.

Proof. It follows from 4.2.1 applied to $\mathcal{F} = \mathcal{O}_{\mathcal{X}}$. □

Examples 4.2.3. (i) Let $U_i = \{(z_0, z_1) \in \mathbb{P}^1 : z_i \neq 0\}$, $U = U_0 \coprod U_1$ and $f : U \to \mathbb{P}^1$ the natural morphism. Then f is flat surjective and quasi-finite. The fibres of f are 0-dimensional, hence projective, but their degree is not constant. This is not a contradiction with Corollary 4.2.2 because the morphism f is not projective, since U is an affine variety.

(ii) In $\mathbb{P}^3$ with homogeneous coordinates $\underline{X} = (X_0, X_1, X_2, X_3)$ consider the curve

$$C_u = \mathrm{Proj}(\mathbf{k}[\underline{X}]/(X_2, X_3)) \cup \mathrm{Proj}(\mathbf{k}[\underline{X}]/(X_1, X_3 - uX_0))$$

for every $u \in \mathbf{A}^1$. If $u \neq 0$ then C_u consists of two disjoint lines, while

$$C_0 = \mathrm{Proj}(\mathbf{k}[\underline{X}]/(X_1X_2, X_3))$$

is a reducible conic in the plane $X_3 = 0$. The Hilbert polynomials are

$$\begin{array}{lll} P_u(t) & = 2t + 2 & u \neq 0 \\ P_0(t) & = 2t + 1 & \end{array}$$

From Corollary 4.2.2 it follows that $\{C_u\}$ cannot be the set of fibres of a flat family of closed subschemes of $\mathbb{P}^3$.

We may try to construct a morphism whose fibres are the C_u's by considering the closed subscheme $\mathcal{X} \subset \mathbb{P}^3 \times \mathbf{A}^1$ defined by the ideal

$$J = (X_2, X_3) \cap (X_1, X_3 - uX_0) = (X_1X_2, X_1X_3, X_2(X_3 - uX_0), X_3(X_3 - uX_0))$$

of $\mathbf{k}[u, X_0, \ldots, X_3]$. From [84], Prop. III.9.7, it follows that $\mathcal{X}$ is flat over $\mathbf{A}^1$. We have:

$$\mathcal{X}(u) = C_u, \qquad u \neq 0$$

$$\mathcal{X}(0) = \mathrm{Proj}(\mathbf{k}[\underline{X}]/(X_1X_2, X_1X_3, X_2X_3, X_3^2))$$

and $\mathcal{X}(0) \neq C_0$: indeed

$$\mathcal{X}(0) = C_0 \cup \mathrm{Proj}(\mathbf{k}[\underline{X}]/(X_1, X_2, X_3^2))$$

is a nonreduced scheme obtained from C_0 by adjoining an embedded point in $(1, 0, 0, 0)$. In particular, we see that $\mathcal{X}(0)$ and C_0 have the same support. Prop. III.9.8 of [84] implies that $\mathcal{X}(0)$ is uniquely determined by the other fibres, i.e. by $\mathcal{X} \cap [\mathbb{P}^3 \times (\mathbf{A}^1 \backslash \{0\})]$.

Fix a scheme S and a coherent sheaf $\mathcal{F}$ on $\mathbb{P}^r \times S$. Consider a morphism $g : T \to S$ and the diagram

$$\begin{array}{ccc} \mathbb{P}^r \times T & \xrightarrow{h} & \mathbb{P}^r \times S \\ \downarrow q & & \downarrow p \\ T & \xrightarrow{g} & S \end{array}$$

where $h = id \times g$. For every open set $U \subset S$ we have homomorphisms

$$H^j(\mathbb{P}^r \times U, \mathcal{F}) \to H^j(\mathbb{P}^r \times g^{-1}(U), h^*\mathcal{F}) \to H^0(g^{-1}(U), R^j q_*(h^*\mathcal{F}))$$

and therefore a homomorphism

$$R^j p_* \mathcal{F} \to g_*[R^j q_*(h^*\mathcal{F})]$$

which corresponds to a homomorphism

$$g^*(R^j p_* \mathcal{F}) \to R^j q_*(h^*\mathcal{F})$$

In the case $j = 0$ we have the following asymptotic result which will be applied later in this section:

Proposition 4.2.4. *For all $m \gg 0$ the homomorphism*

$$g^*(p_*\mathcal{F}(m)) \to q_*(h^*\mathcal{F}(m))$$

is an isomorphism and, if T is noetherian, $R^j q_*(h^*\mathcal{F}(m)) = 0$ all $j \geq 1$.

Proof. We have

$$h^*\mathcal{F} = \Gamma_*(h^*\mathcal{F})^{\sim} := [\oplus_m q_*(h^*\mathcal{F}(m))]^{\sim}$$

Since $\mathcal{F} = \Gamma_*(\mathcal{F})^{\sim}$ we also have

$$h^*\mathcal{F} = h^*[\Gamma_*(\mathcal{F})^{\sim}] = [\oplus_m g^*(p_*\mathcal{F})(m)]^{\sim}$$

and therefore for all $m \gg 0$

$$g^*(p_*\mathcal{F}(m)) \cong q_*(h^*\mathcal{F}(m))$$

For the last assertion cover T by finitely many affine open sets and apply Theorem III.5.2 of [84]. □

The homomorphism of Proposition 4.2.4 is particularly important when g : $\mathrm{Spec}(\mathbf{k}(s)) \to S$ is the inclusion in S of a point $s \in S$; it is denoted by

$$t^j(s) : R^j p_*(\mathcal{F})_s \otimes \mathbf{k}(s) \to H^j(I\!P^r(s), \mathcal{F}(s))$$

The study of these homomorphisms is carried out in [1], Ch. III_2 (see also Chapter III, Section 12, of [84]). Their main properties are summarized in the following theorem and in its corollary.

Theorem 4.2.5. *Let S be a scheme, $\mathcal{F}$ a coherent sheaf on $I\!P^r \times S$, flat over S, $s \in S$ and $j \geq 0$ an integer. Then:*

(i) If $t^j(s)$ is surjective then it is an isomorphism.
(ii) If $t^{j+1}(s)$ is an isomorphism then $R^{j+1}p_*(\mathcal{F})$ is free at s if and only if $t^j(s)$ is an isomorphism.
(iii) If $R^j p_*(\mathcal{F})$ is free at s for all $j \geq j_0 + 1$ then $t^j(s)$ is an isomorphism for all $j \geq j_0$.

Proof. (i) and (ii) are Theorem III.12.11 of [84]. (iii) follows from (i) and (ii) by descending induction on j_0. □

Corollary 4.2.6. *Let $\mathcal{X} \to S$ be a projective morphism, and let $\mathcal{F}$ be a coherent sheaf on $\mathcal{X}$, flat over S. Then:*

(i) If $H^{j+1}(\mathcal{X}(s), \mathcal{F}(s)) = 0$ for some $s \in S$ and $j \geq 0$ then $R^{j+1}p_*(\mathcal{F})_s = (0)$, and

$$t^j(s) : R^j p_*(\mathcal{F})_s \otimes \mathbf{k}(s) \to H^j(\mathcal{X}(s), \mathcal{F}(s))$$

is an isomorphism.
(ii) Let j_0 be an integer such that

$$H^j(\mathcal{X}(s), \mathcal{F}(s)) = 0$$

for all $j \geq j_0 + 1$ and $s \in S$ (e.g. $j_0 = \max_{s \in S}\{\dim[\mathrm{Supp}(\mathcal{F}(s))]\}$). Then $t^{j_0}(s)$ is an isomorphism for all $s \in S$.
(iii) Let $j_0 \geq 0$ be an integer. Then there is a nonempty open set $U \subset S$ such that $t^{j_0}(s)$ is an isomorphism for all $s \in U$.

Proof. (i) follows immediately from 4.2.5. (ii) is a special case of (i).

(iii) It is the open set $U = \bigcap_{j \geq j_0} U_j$, where $U_j = \{s \in S : R^j p_*(\mathcal{F})_s \text{ is free}\}$ (apply 4.2.5(iii)). □

4.2.2 Stratifications

Let S be a scheme. A *stratification* of S consists of a set of finitely many locally closed subschemes $\{S_1, \ldots, S_n\}$ of S, called *strata*, pairwise disjoint and such that

$$S = S_1 \cup \ldots \cup S_n$$

i.e. such that we have a surjective morphism

$$\coprod_i S_i \to S$$

Let $\mathcal{F}$ be a coherent sheaf on S and for each $s \in S$ let

$$e(s) := \dim_{\mathbf{k}(s)}[\mathcal{F}_s \otimes \mathbf{k}(s)]$$

Fix a point $s \in S$, let $e = e(s)$ and let $a_1, \ldots, a_e \in \mathcal{F}_s$ be such that their images in $\mathcal{F}_s \otimes \mathbf{k}(s)$ form a basis. From Nakayama's lemma it follows that the homomorphism $f_s : \mathcal{O}^e_{S,s} \to \mathcal{F}_s$ defined by $a_1, \ldots, a_e$ is surjective; therefore there is an open neighbourhood U of s to which f extends defining a surjective homomorphism $f : \mathcal{O}^e_U \to \mathcal{F}_{|U}$. With a similar argument applied to $\ker(f_s)$ we may find an affine open neighbourhood $U(s)$ of s contained in U and an exact sequence

$$\mathcal{O}^d_{U(s)} \xrightarrow{g} \mathcal{O}^e_{U(s)} \xrightarrow{f} \mathcal{F}_{|U(s)} \to 0 \tag{4.3}$$

It follows that:

(i) $e(s') \leq e(s)$ for all $s' \in U(s)$: therefore $s \mapsto e(s)$ is an upper semicontinuous function from S to $\mathbb{Z}$.
(ii) Let (g_{ij}) be the $e \times d$ matrix with entries in $H^0(U(s), \mathcal{O}_S)$ which defines g. The ideal generated by the g_{ij}'s in $H^0(U(s), \mathcal{O}_S)$ defines a closed subscheme Z_s of $U(s)$ with support equal to $Y_e \cap U(s)$, where for each $e \geq 0$ we have set $Y_e = \{s \in S : e(s) = e\}$. In particular, Y_e is a locally closed subset of S.

Moreover:

(iii) If $q : T \to U(s)$ is a morphism, $q^*(\mathcal{F})$ is locally free of rank e if and only if q factors through the subscheme Z_s.

Proof. q factors through Z_s if and only if all the functions $q^*(g_{ij})$ are zero on T. Since the sequence

$$\mathcal{O}^d_T \xrightarrow{q^*(g)} \mathcal{O}^e_T \xrightarrow{q^*(f)} q^*(\mathcal{F}) \to 0$$

is exact on T, this is equivalent to $q^*(f)$ being an isomorphism and this condition implies that $q^*(\mathcal{F})$ is locally free of rank e. Conversely, if $q^*(\mathcal{F})$ is locally free of rank e, let $\mathcal{G} = \ker[q^*(f)]$. At every point $t \in T$ we have an exact sequence:

$$0 \to \mathcal{G} \otimes \mathbf{k}(t) \to \mathbf{k}(t)^e \to q^*(\mathcal{F}) \otimes \mathbf{k}(t) \to 0$$

Since $q^*(\mathcal{F}) \otimes \mathbf{k}(t)$ is a vector space of dimension e we have $\mathcal{G} \otimes \mathbf{k}(t) = (0)$. By Nakayama's lemma, $\mathcal{G} = (0)$ in a neighbourhood of t and therefore $\mathcal{G} = (0)$ everywhere.

(iv) Since property (iii) characterizes the scheme Z_s and does not depend on the presentation (4.3), for any $s, s' \in S$ the schemes Z_s and $Z_{s'}$ coincide on $U(s) \cap U(s')$; therefore the collection of schemes $\{Z_s : s \in S\}$ defines a locally closed subscheme Z_e of S supported on Y_e. Evidently, $\{Z_e : e \geq 0\}$ is a stratification of S.
(v) Because of (i), for each e the closure of Z_e is contained in $\bigcup_{e' \geq e} Z_{e'}$. In particular, if E is the highest integer such that $Z_E \neq \emptyset$, then Z_E is closed.
(vi) By the right exactness of tensor product, the construction of the schemes Z_e commutes with base change (the proof is similar to that of (iii)). In other words, if $q : T \to S$ is a morphism then

$$\tilde{Z}_e = f^{-1}(Z_e)$$

for all e, where $\tilde{Z}_e \subset T$ is the locally closed stratum associated to the sheaf $q^*\mathcal{F}$.

We have proved the following:

Theorem 4.2.7. *Let S be a scheme and $\mathcal{F}$ a coherent sheaf on S. There is a unique stratification $\{Z_e\}_{e \geq 0}$ of S such that if $q : T \to S$ is a morphism the sheaf $q^*(\mathcal{F})$ is locally free if and only if q factors through the disjoint union of the Z_e's: $T \to \coprod_e Z_e \to S$.*

Moreover, the strata $Z_0, Z_1, \ldots$ are indexed so that for each $e = 0, 1, \ldots$ the restriction of $\mathcal{F}$ to Z_e is locally free of rank e.

For a given e, $\bar{Z}_e \subset \bigcup_{e' \geq e} Z_{e'}$. In particular, if E is the highest integer such that $Z_E \neq \emptyset$, then Z_E is closed.

The stratification $\{Z_e\}_{e \geq 0}$ commutes with base change.

Theorem 4.2.7 describes a natural way to construct stratifications on a scheme. $\{Z_e\}_{e \geq 0}$ is called *the stratification defined by the sheaf $\mathcal{F}$*.

An alternative approach to the construction of this stratification is by the "Fitting ideals" of the sheaf $\mathcal{F}$. Let $k \geq 0$; the *k-th Fitting ideal of $\mathcal{F}$* is the ideal sheaf $Fitt_k^{\mathcal{F}} \subset \mathcal{O}_S$ locally defined by the minors of order $(e - k)$ of the matrix (g_{ij}) in the presentation (4.3). The proof of Theorem 4.2.7 essentially shows that the Fitting ideals are independent of the choice of the presentations (4.3). The closed subscheme of S defined by $Fitt_{k-1}^{\mathcal{F}}$ is denoted by $N_k(\mathcal{F})$. It follows directly from the definition that

$$\mathrm{Supp}(N_k(\mathcal{F})) = \{s \in S : \dim_{\mathbf{k}(s)}(\mathcal{F}_s \otimes \mathbf{k}(s)) \geq k\}$$

and that $N_k(\mathcal{F})$ commutes with base change. Therefore the stratification defined in Theorem 4.2.7 can be also described as follows:

$$Z_e = N_e(\mathcal{F}) \backslash N_{e+1}(\mathcal{F}) \tag{4.4}$$

For details about the properties of the Fitting ideals see [48].

Example 4.2.8. Let $\varphi : \mathbf{A} \to \mathbf{B}$ be a homomorphism of locally free sheaves on the scheme S, of ranks a and b respectively. Applying Theorem 4.2.7 to $\mathrm{coker}(\varphi)$ we obtain a stratification of S with the property that Z_{b-e} is supported on the locus

$$\{s \in S : \mathrm{rk}[\varphi(s) : \mathbf{A}(s) \to \mathbf{B}(s)] = e\}$$

The scheme Z_{b-e} of this stratification will be denoted by $\mathrm{D}_e(\varphi)$. Note that in particular, the subscheme $\mathrm{D}_0(\varphi)$, called the *vanishing scheme* of φ, is closed in S because of (v) above. It has the property that a morphism $f : T \to S$ satisfies $f^*(\varphi) = 0$ if and only if f factors through $D_0(\varphi)$.

The ideal sheaf of $D_0(\varphi)$ is locally generated by the entries of a matrix representing φ. More intrinsically, it can be obtained as follows. Since $\varphi \in \mathrm{Hom}(\mathbf{A}, \mathbf{B})$, it induces by adjunction a homomorphism:

$$Hom(\mathbf{B}, \mathbf{A}) \xrightarrow{\varphi^\vee} \mathcal{O}_S$$

whose image is just the ideal sheaf of $D_0(\varphi)$.

Example 4.2.9. Let $f : X \to S$ be a finite morphism of algebraic schemes. Then $f_*\mathcal{O}_X$ is a coherent sheaf on S. The scheme $N_k(f_*\mathcal{O}_X) \subset S$ is supported on the set of points of S having $\geq k$ preimages (counting multiplicities). It is usually denoted by $N_k(f)$ and it is called *the k-th multiple point scheme of* f. The corresponding stratification is *the multiple point stratification of S relative to* f. There is a vast literature on this stratification. For more about it we refer the reader to [61] and to the literature quoted there.

4.2.3 Flattening stratifications

Definition 4.2.10. *Let S be a scheme and $\mathcal{F}$ a coherent sheaf on $\mathbb{P}^r \times S$. A* flattening stratification *for $\mathcal{F}$ is a stratification $\{S_1, \dots, S_n\}$ of S such that for every morphism $g : T \to S$ the sheaf*

$$\mathcal{F}_g := (1 \times g)^*(\mathcal{F})$$

on $\mathbb{P}^r \times T$ is flat over T if and only if g factors through $\coprod S_i$.

Note that if such a stratification exists it is clearly unique. In the special case $r = 0$ we obtain again the notion of stratification defined by the sheaf $\mathcal{F}$.

The following is a basic technical result.

Theorem 4.2.11. *For every coherent sheaf $\mathcal{F}$ on $\mathbb{P}^r \times S$ the flattening stratification exists.*

Proof. The theorem has already been proved in the case $r = 0$ (Theorem 4.2.7). Therefore we may assume $r \geq 1$. We will proceed in several steps.

Step 1: There are finitely many locally closed subsets $Y^1, \dots, Y^k$ of S such that for each $i = 1, \dots, k$ if we consider on Y^i the reduced scheme structure then $\mathcal{F} \otimes \mathcal{O}_{Y^i \times \mathbb{P}^r}$ is flat over Y^i.

It follows immediately from a repeated use of the fact that there is a nonempty open subset $U \subset S$ such that $\mathcal{F}_{|I\!P^r \times U_{red}}$ is flat over U_{red} (see Note 7).

Step 2: Only finitely many polynomials $P^1, \ldots, P^h$ occur as Hilbert polynomials of the sheaves $\mathcal{F}(s)$, $s \in S$.

In fact from Corollary 4.2.2 it follows that, at most, as many Hilbert polynomials occur as the number of connected components of the sets $Y^1, \ldots, Y^k$.

Step 3: There is an integer N such that for every $m \geq 0$ and for every $s \in S$ we have:

$$H^j(I\!P^r(s), \mathcal{F}(s)(N+m)) = (0)$$

for $j \geq 1$ and the natural map:

$$[p_*\mathcal{F}(N+m)]_s \otimes \mathbf{k}(s) \to H^0(I\!P^r(s), \mathcal{F}(s)(N+m))$$

is an isomorphism, where $p : I\!P^r \times S \to S$ is the projection.

For each $i = 1, \ldots, k$ consider the diagram

$$\begin{array}{rccc} h^i : & I\!P^r \times Y^i & \to & I\!P^r \times S \\ & \downarrow p_i & & \downarrow p \\ & Y^i & \to & S \end{array} \tag{4.5}$$

and let $n_i \gg 0$ be so that $R^j p_{i*}[h^{i*}\mathcal{F}(n_i + m)] = (0)$ for all $m \geq 0$ and all $j \geq 1$ (apply Proposition 4.2.4). Letting

$$N \gg \max\{n_1, \ldots, n_k\}$$

we may apply Proposition 4.2.4 to the diagrams (4.5) and to the sheaf $\mathcal{F}$ and we obtain isomorphisms

$$[p_*\mathcal{F}(N+m)] \otimes \mathcal{O}_{Y^i} \cong p_{i*}[h^{i*}\mathcal{F}(N+m)]$$

for all $s \in Y^i$ and for all $i = 1, \ldots, k$. In particular, we have isomorphisms

$$[p_*\mathcal{F}(N+m)] \otimes \mathbf{k}(s) \cong p_{i*}[h^{i*}\mathcal{F}(N+m)]_s \otimes \mathbf{k}(s) \tag{4.6}$$

for all $s \in Y^i$ and for all $i = 1, \ldots, k$. We may also apply Corollary 4.2.6 to the sheaves $h^{i*}\mathcal{F}$ and to the projections p_i for $j_0 = 0$ to deduce that

$$H^j(I\!P^r(s), \mathcal{F}(s)(N+m)) = (0)$$

for all $s \in S$, $j \geq 1$ and $m \geq 0$, and that

$$p_{i*}[h^{i*}\mathcal{F}(N+m)]_s \otimes \mathbf{k}(s) \cong H^0(I\!P^r(s), \mathcal{F}(s)(N+m)) \tag{4.7}$$

for all $s \in Y^i$ and for all $i = 1, \ldots, k$ and all $m \geq 0$.

Comparing (4.6) and (4.7) we obtain the conclusion.

Step 4: Let N be as in Step 3, and let $g : T \to S$ be a morphism. Then $\mathcal{F}_g$ is flat over T if and only if $g^*[p_*\mathcal{F}(N+m)]$ is locally free for all $m \geq 0$.

Suppose that $\mathcal{F}_g$ is flat over T and let $q : I\!P^r \times T \to T$ be the projection. Since

$$H^j(I\!P^r(t), \mathcal{F}_g(t)(N+m)) = H^j(I\!P^r(g(t)), \mathcal{F}(g(t))(N+m)) = (0)$$

for all $t \in T$, $m \geq 0$ and $j \geq 1$, from Corollary 4.2.6(ii) we deduce that

$$q_*\mathcal{F}_g(N+m)_t \otimes \mathbf{k}(t) \to H^0(I\!P^r(g(t)), \mathcal{F}(g(t))(N+m)) \tag{4.8}$$

is an isomorphism for all $t \in T$. Theorem 4.2.5(ii) applied for $j = -1$ implies that $q_*\mathcal{F}_g(N+m)$ is locally free for all $m \geq 0$. For all $t \in T$ the natural homomorphism

$$\varphi : g^*[p_*\mathcal{F}(N+m)] \to q_*\mathcal{F}_g(N+m)$$

induces an isomorphism:

$$g^*[p_*\mathcal{F}(N+m)]_t \otimes \mathbf{k}(t) \cong q_*\mathcal{F}_g(N+m)_t \otimes \mathbf{k}(t)$$

because both sides are isomorphic to $H^0(I\!P^r(g(t)), \mathcal{F}(g(t))(N+m))$ (the first because of Step 3, the second because of (4.8)). From the fact that $q_*\mathcal{F}_g(N+m)$ is locally free and from Nakayama's lemma it follows that φ is an isomorphism. Therefore $g^*[p_*\mathcal{F}(N+m)]$ is locally free for every $m \geq 0$.

Conversely, suppose that $g^*[p_*\mathcal{F}(N+m)]$ is locally free for all $m \geq 0$. Since for all $m \gg 0$ the natural map φ is an isomorphism (Prop. 4.2.4) it follows that $q_*\mathcal{F}_g(N+m)$ is locally free for all $m \gg 0$: Proposition 4.2.1 implies that $\mathcal{F}_g$ is flat.

Step 5: For every $m \geq 0$ apply Theorem 4.2.7 to the sheaf $p_*\mathcal{F}(N+m)$ and let $Y_{m,j}$ be the component of the corresponding stratification of S where $p_*\mathcal{F}(N+m)$ becomes locally free of rank j. Then for each $j = 1, \ldots, h$ we have the following equality of subsets of S:

$$\bigcap_{m \geq 0} \mathrm{Supp}(Y_{m,P^i(N+m)}) = \bigcap_{m=0,\ldots,r} \mathrm{Supp}(Y_{m,P^i(N+m)})$$

The inclusion $\subset$ is obvious. For $s \in S$ let $P_s(t)$ be the Hilbert polynomial of $\mathcal{F}(s)$. Then $s \in \cap_{m \geq 0}\mathrm{Supp}(Y_{m,P^i(N+m)})$ if and only if

$$P_s(N+m) = h^0(I\!P^r(s), \mathcal{F}(s)(N+m)) = \dim[p_*\mathcal{F}(N+m)_s \otimes \mathbf{k}(s)] = P^i(N+m)$$

for all $m \geq 0$, and this happens if and only if $P_s(t) = P^i(t)$ as polynomials. On the other hand, $s \in \cap_{m=0,\ldots,r}\mathrm{Supp}(Y_{m,P^i(N+m)})$ if and only if $P_s(N+m) = P^i(N+m)$ for $m = 0, \ldots, r$. Since both $P_s(t)$ and $P^i(t)$ have degree $\leq r$, it follows that $P_s(t) = P^i(t)$ and therefore $s \in \cap_{m \geq 0}\mathrm{Supp}(Y_{m,P^i(N+m)})$.

Step 6: Fix i between 1 and h. For each integer $c \geq 0$ the finite intersection

$$\bigcap_{m=0,\dots,c} Y_{m,P^i(N+m)}$$

is a well-defined locally closed subscheme of S. Because of Step 5 the subschemes

$$\bigcap_{m=0,\dots,c} Y_{m,P^i(N+m)}, \qquad c = r, r+1, \dots$$

for a descending chain with fixed support; in particular, they form a descending chain of closed subschemes of a *fixed* open set $V \subset S$, and therefore they stabilize. In other words the intersection

$$Z^i = \bigcap_{m \geq 0} \operatorname{Supp}(Y_{m,P^i(N+m)})$$

is a well-defined locally closed subscheme of S. By Step 5 we have:

$$\operatorname{Supp}(Z^i) = \{s \in S : P_s(t) = P_i(t)\}$$

Step 7: The subschemes $Z^1, \dots, Z^h$ form a stratification of S. It follows immediately from Step 4 that this is the flattening stratification for $\mathcal{F}$. This concludes the proof of Theorem 4.2.11. □

NOTES

1. From the proof of Theorem 4.2.11 it follows that the strata $Z^1, \dots, Z^h$ of the flattening stratification for $\mathcal{F}$ are indexed by the Hilbert polynomials of the sheaves $\mathcal{F}(s)$, $s \in S$.

2. (The see-saw theorem) *Let X be a projective scheme such that $H^0(X, \mathcal{O}_X) = \mathbf{k}$, S an algebraic scheme and $\mathcal{L}$ an invertible sheaf on $X \times S$. Then:*

(i) The locus

$$S_0 = \{s \in S : \mathcal{L}_{|X \times \{s\}} \cong \mathcal{O}_X\}$$

is closed in S.

(ii) Letting $p_0 : X \times S_0 \to S_0$ be the projection, there is an invertible sheaf M on S_0 such that

$$\mathcal{L}_{|X \times S_0} \cong p_0^* M$$

Proof. A line bundle L on X is trivial if and only if it satisfies $h^0(X, L) > 0$ and $h^0(X, L^{-1}) > 0$. In fact, nonzero sections $\sigma \in H^0(X, L)$ and $\tau \in H^0(X, L^{-1})$ correspond to homomorphisms

$$\mathcal{O}_X \xrightarrow{\sigma} L \xrightarrow{\tau^\vee} \mathcal{O}_X$$

whose composition is multiplication by a constant (by the assumption $H^0(X, \mathcal{O}_X) = \mathbf{k}$), so that they are isomorphisms.

Therefore $S_0 = S^+ \cap S^-$ where

$$S^+ = \{s \in S : h^0(X \times \{s\}, \mathcal{L}_{|X\times\{s\}}) > 0\}$$

$$S^- = \{s \in S : h^0(X \times \{s\}, \mathcal{L}^{-1}_{|X\times\{s\}}) > 0\}$$

It follows from the semicontinuity theorem that both S^+ and S^- are closed in S, and (i) follows.

(ii) Since $h^0(X \times \{s\}, \mathcal{L}_{|X\times\{s\}}) = 1$ is constant in $s \in S_0$, from Corollary 12.9 of [84] we deduce that $p_{0*}\mathcal{L}\otimes\mathbf{k}(s) \cong H^0(X\times\{s\}, \mathcal{L}_{|X\times\{s\}})$ for all $s \in S_0$ and in particular, $p_{0*}\mathcal{L} =: M$ is an invertible sheaf. We have a homomorphism of invertible sheaves on $X \times S_0$:

$$p_0^* M = p_0^* p_{0*}\mathcal{L} \to \mathcal{L}_{|X\times S_0}$$

which is surjective because $\mathcal{L}_{|X\times\{s\}}$ is globally generated for all $s \in S_0$. Then it is an isomorphism. □

3. *Let $\mathcal{X} \to S$ be a flat projective morphism of algebraic schemes, and $\mathcal{L}$ an invertible sheaf on $\mathcal{X}$. Assume that for some* **k***-rational point $o \in S$ the sheaf $\mathcal{L}(o)$ is very ample on $\mathcal{X}(o)$ and satisfies $H^1(\mathcal{X}(o), \mathcal{L}(o)) = 0$. Then there is an open neighbourhood $V \subset S$ of o such that $\mathcal{L}_V := \mathcal{L}_{|\mathcal{X}(V)}$ is very ample relative to V. In particular, $\mathcal{L}(s)$ is very ample on $\mathcal{X}(s)$ for every $s \in V$.*

Proof. By Corollary 4.2.6, there is an open neighbourhood $U \subset S$ of o such that $(R^1 f_*\mathcal{L})_{|U} = (0)$ and

$$t^0(u) : (f_*\mathcal{L})_u \otimes \mathbf{k}(u) \to H^0(\mathcal{X}(u), \mathcal{L}(u))$$

is an isomorphism for all $u \in U$. We may even assume that $f_*\mathcal{L}$ is locally free of rank $h^0(\mathcal{X}(o), \mathcal{L}(o))$ on U. From the surjectivity of the map $t^0(o)$ and from the fact that $\mathcal{L}(o)$ is globally generated we deduce that the canonical homomorphism:

$$f^*(f_*\mathcal{L}) \to \mathcal{L}$$

is surjective on $\mathcal{X}(o)$. Since f is projective it follows that there is an open $W \subset U$ containing o such that

$$[f^*(f_*\mathcal{L})]_{|\mathcal{X}(W)} \to \mathcal{L}_W \tag{4.9}$$

is surjective and moreover, $[f^*(f_*\mathcal{L})]_{|\mathcal{X}(W)}$ is locally free. The homomorphism (4.9) defines a W-morphism

$$\begin{array}{ccc} \mathcal{X}(W) & \to & \mathbb{P}(f^*(f_*\mathcal{L}))_{|\mathcal{X}(W)} \\ & \searrow \quad \swarrow & \\ & W & \end{array} \tag{4.10}$$

whose restriction to $\mathcal{X}(o)$ is the embedding defined by the global sections of $\mathcal{L}(o)$. From Note 3 of § 3.4 above it follows that there is an open subset $V \subset W$ containing o such that the restriction of (4.10) to $\mathcal{X}(V)$ is an embedding. This implies the conclusion (see [116], Theorem 1.2.17 for the proof of a more general statement without flatness assumption).

4. *Let $\mathcal{E}$ be a locally free sheaf over $\mathbb{P}^1 \times S$, with S an algebraic integral scheme. Let $o \in S$ be a* **k***-rational point, and $\mathcal{E}(o) \cong \oplus_i \mathcal{O}(n_0^i)$ the fibre over o. Then:*

(i) There is an open set $U \subset S$ such that for each $s \in U$ we have

$$\mathcal{E}(s) \cong \oplus_i \mathcal{O}(n_s^i)$$

with

$$\max_i\{n_s^i\} \leq \max_i\{n_o^i\} \quad and \qquad \min_i\{n_s^i\} \geq \min_i\{n_o^i\}$$

Moreover, if $\mathcal{E}(o)$ is balanced (i.e. $n_o^i = n_o^j$ for all i, j) then $\mathcal{E}(s) \cong \mathcal{E}(o)$ for all $s \in U$.

(ii) For each $s \in S$ we have

$$\sum_i n_s^i = \sum_i n_0^i$$

Proof. ([25]) (i) By the structure theorem for locally free sheaves on $I\!P^1$ (see [136]) we know that for each $s \in S$ we have an isomorphism $\mathcal{E}(s) \cong \oplus_i \mathcal{O}(n_s^i)$ for some integers n_s^i. Let $M_0 = \max_i\{n_0^i\}$ and consider the sheaf $\bar{\mathcal{E}} := \mathcal{E} \otimes p^*\mathcal{O}(-M_0 - 1)$, where $p : I\!P^1 \times S \to I\!P^1$ is the projection. Since $h^0(\bar{\mathcal{E}}(0)) = 0$, from the semicontinuity theorem it follows that there is an open neighbourhood U of 0 such that $h^0(\bar{\mathcal{E}}(s)) = 0$ for all $s \in U$; but this means that $\max_i\{n_s^i\} \leq M_0$ for all $s \in U$, which is the first statement of the proposition. The statement about the minimum is proved similarly after replacing $\mathcal{E}$ by its dual. The last assertion is obvious.

(ii) Applying (i) to $\det(\mathcal{E})$ we find that every point $t \in S$ has an open neighbourhood U_t where $\sum_i n_s^i = \sum_i n_t^i$ for all $s \in U_t$. Since S is connected we deduce that $\sum_i n_s^i$ is constant.

5. Let $\mathcal{X} \to S$ be a flat projective morphism with S an algebraic scheme, and let $o \in S$ be a **k**-rational point. Prove that:

(i) If $\mathcal{X}(o)$ is connected and $\mathcal{X}(s)$ is disconnected for all $s \neq o$ in an open neighbourhood of o then $\mathcal{X}(o)$ is nonreduced.
In particular:

(ii) If $\mathcal{X}(o)$ is connected and reduced then $\mathcal{X}(s)$ is connected for all s in an open neighbourhood of o.

(iii) If $\mathcal{X}(o)$ is disconnected then $\mathcal{X}(s)$ is disconnected for all s in an open neighbourhood of o.

6. *Let $f : X \to Y$ be a proper morphism of algebraic schemes with finite fibres. Let $g : Y' \to Y$ be an arbitrary morphism, $X' = X \times_Y Y'$, $f' : X' \to Y'$ and $g' : X' \to X$ the projections. Then for every quasi-coherent $\mathcal{O}_X$-module $\mathcal{F}$ we have a canonical isomorphism*

$$g^*(f_*\mathcal{F}) \cong f'_*(g'^*\mathcal{F})$$

Proof. Since it is proper and quasi-finite, f is finite, in particular it is affine. The conclusion follows from [1] Ch. II, 1.5.2. □

7. (Generic flatness) *Let $f : \mathcal{X} \to S$ be a morphism of finite type with S integral, and let $\mathcal{F}$ be a coherent sheaf on $\mathcal{X}$. There is a dense open subset $U \subset S$ such that the restriction of $\mathcal{F}$ to $f^{-1}(U)$ is flat over U.*

Proof. See [1], ch. IV, Th. 6.9.1 or [48], Theorem 14.4, p. 308. □

Note that if f is not dominant then $U = S \backslash \overline{f(\mathcal{X})}$ and $f^{-1}(U) = \emptyset$.

4.3 Hilbert schemes

4.3.1 Generalities

Consider a projective scheme Y, and a closed embedding $Y \subset \mathbb{P}^r$. Let's fix a *numerical polynomial* of degree $\leq r$, i.e. a polynomial $P(t) \in \mathbf{Q}[t]$ of the form:

$$P(t) = \sum_{i=0}^{r} a_i \binom{t+r}{i}$$

with $a_i \in \mathbb{Z}$ for all i.

For every scheme S we let:

$$Hilb^Y_{P(t)}(S) = \left\{ \begin{array}{c} \text{flat families } \mathcal{X} \subset Y \times S \text{ of closed subschemes} \\ \text{of } Y \text{ parametrized by } S \text{ with fibres} \\ \text{having Hilbert polynomial } P(t) \end{array} \right\}$$

Since flatness is preserved under base change, this defines a contravariant functor

$$Hilb^Y_{P(t)} : (\text{schemes})^\circ \to (\text{sets})$$

called the *Hilbert functor of Y relative to* $P(t)$.

In the case $Y = \mathbb{P}^r$ we will denote the Hilbert functor with the symbol $Hilb^r_{P(t)}$.

If the functor $Hilb^Y_{P(t)}$ is representable, the scheme representing it will be called *the Hilbert scheme of Y relative to* $P(t)$, and will be denoted by $\mathrm{Hilb}^Y_{P(t)}$ (or $\mathrm{Hilb}^r_{P(t)}$ in case $Y = \mathbb{P}^r$). If $P(t) = n$ a constant polynomial then $\mathrm{Hilb}^Y_{P(t)}$ is usually denoted by $Y^{[n]}$.

If the Hilbert scheme $\mathrm{Hilb}^Y_{P(t)}$ exists then there is a universal element, i.e. there is a flat family of closed subschemes of Y having Hilbert polynomial equal to $P(t)$:

$$\mathcal{W} \subset Y \times \mathrm{Hilb}^Y_{P(t)} \tag{4.11}$$

parametrized by $\mathrm{Hilb}^Y_{P(t)}$ and possessing the following:

Universal property: for each scheme S and for each flat family $\mathcal{X} \subset Y \times S$ of closed subschemes of Y having Hilbert polynomial $P(t)$ there is a unique morphism $S \to \mathrm{Hilb}^Y_{P(t)}$, called the *classifying morphism*, such that

$$\mathcal{X} = S \times_{\mathrm{Hilb}^Y_{P(t)}} \mathcal{W} \subset Y \times S$$

The family (4.11) is called the *universal family*, and the pair $(\mathrm{Hilb}^Y_{P(t)}, \mathcal{W})$ represents the functor $Hilb^Y_{P(t)}$.

The family $\mathcal{W}$ is the universal element of $Hilb^Y_{P(t)}(\mathrm{Hilb}^Y_{P(t)})$, namely the element corresponding to the identity under the identification

$$\mathrm{Hom}(\mathrm{Hilb}^Y_{P(t)}, \mathrm{Hilb}^Y_{P(t)}) = Hilb^Y_{P(t)}(\mathrm{Hilb}^Y_{P(t)})$$

Example 4.3.1. Consider the constant polynomial $P(t) = 1$. Then we have a canonical identification $Y^{[1]} = Y$ and the universal family is the diagonal $\Delta \subset Y \times Y$.

To prove it consider an element of $Y^{[1]}(S)$ for some scheme S:

$$\begin{array}{ccc} \Gamma & \subset & S \times Y \\ \downarrow f & & \\ S & & \end{array}$$

Then f is an isomorphism: in fact it is a one-to-one morphism and $\mathcal{O}_S \to f_*\mathcal{O}_\Gamma$ is an isomorphism since $f_*\mathcal{O}_\Gamma$ is an $\mathcal{O}_S$-algebra which is locally free of rank one over $\mathcal{O}_S$. We therefore have the well-defined morphism $g = qf^{-1} : S \to Y$ where $q : S \times Y \to Y$ is the projection. The morphism

$$(gf, q) : \Gamma \to Y \times Y$$

factors through Δ and induces a commutative diagram

$$\begin{array}{ccc} \Gamma & \to & \Delta \\ \downarrow & & \downarrow \\ S & \xrightarrow[g]{} & Y \end{array}$$

such that $\Gamma \cong g^* \Delta$. Therefore the family Γ is induced by Δ via the morphism g.

Before proving the existence of the Hilbert schemes in general we will consider two important special cases.

4.3.2 Linear systems

If $X \subset I\!P^r$ is a hypersurface of degree d it has Hilbert polynomial

$$h(t) = \binom{t+r}{r} - \binom{t+r-d}{r} = \frac{d}{(r-1)!}\, t^{r-1} + \cdots$$

Conversely, if a closed subscheme Y of $I\!P^r$ has the Hilbert polynomial $h(t)$ then it is a hypersurface of degree d.

In fact, since $h(t)$ has degree $r-1$, Y has dimension $r-1$, so $Y = Y_1 \cup Z$, with Y_1 a hypersurface and $\dim(Z) < r-1$. We have the exact sequence:

$$0 \to \mathcal{I}_{Y_1}/\mathcal{I}_Y \to \mathcal{O}_Y \to \mathcal{O}_{Y_1} \to 0$$

where $\mathcal{I}_Y, \mathcal{I}_{Y_1} \subset \mathcal{O}_{I\!P^r}$ are the ideal sheaves of Y and Y_1. We deduce that

$$h(t) = h_1(t) + k(t)$$

where $h_1(t)$ is the Hilbert polynomial of Y_1 and $k(t)$ is the Hilbert polynomial of $\mathcal{I}_{Y_1}/\mathcal{I}_Y$. Since this sheaf is supported on Z, we have $\deg(k(t)) < r-1$; therefore we see that

$$h_1(t) = \frac{d}{(r-1)!}\, t^{r-1} + \cdots$$

so Y_1 is a hypersurface of degree d, and therefore $h_1(t) = h(t)$. It follows that $k(t) = 0$, i.e. $\mathcal{I}_{Y_1} = \mathcal{I}_Y$; equivalently, $Y = Y_1$.

Therefore $\mathrm{Hilb}^r_{h(t)}$, if it exists, parametrizes a universal family of hypersurfaces of degree d in $\mathbb{P}^r$. To prove its existence let $V := H^0(\mathbb{P}^r, \mathcal{O}(d))$ and in $\mathbb{P}(V)$ take homogeneous coordinates

$$(\ldots, c_{i(0),\ldots,i(r)}, \ldots)_{i(0)+\cdots+i(r)=d}$$

The hypersurface $\mathcal{H} \subset \mathbb{P}^r \times \mathbb{P}(V)$ defined by the equation

$$\sum c_{i(0),\ldots,i(r)} X_0^{i(0)} \cdots X_r^{i(r)} = 0$$

projects onto $\mathbb{P}(V)$ with fibres hypersurfaces of degree d. It follows from Proposition 4.2.1 that $\mathcal{H}$ is flat over $\mathbb{P}(V)$. Let's denote by $p : \mathbb{P}^r \times \mathbb{P}(V) \to \mathbb{P}(V)$ the projection, and let $\mathcal{I}_{\mathcal{H}} \subset \mathcal{O}_{\mathbb{P}^r \times \mathbb{P}(V)}$ be the ideal sheaf of $\mathcal{H}$. For all $x \in \mathbb{P}(V)$ we have

$$1 = h^0(\mathbb{P}^r(x), \mathcal{I}_{\mathcal{H}(x)}(d)) = h^0(\mathbb{P}^r(x), \mathcal{I}_{\mathcal{H}}(d)(x))$$

and

$$0 = h^i(\mathbb{P}^r(x), \mathcal{I}_{\mathcal{H}(x)}(d)) = h^i(\mathbb{P}^r(x), \mathcal{I}_{\mathcal{H}}(d)(x))$$

$$0 = h^i(\mathcal{H}(x), \mathcal{O}_{\mathcal{H}(x)}(d))$$

for all $i \geq 1$. Applying 4.2.5 and 4.2.2 we deduce that:

(a) $R^1 p_* \mathcal{I}_{\mathcal{H}}(d) = 0$;
(b) $p_* \mathcal{I}_{\mathcal{H}}(d)$ is an invertible subsheaf of $p_* \mathcal{O}_{\mathbb{P}^r \times \mathbb{P}(V)}(d) = V \otimes_{\mathbf{k}} \mathcal{O}_{\mathbb{P}(V)}$;
(c) $p_* \mathcal{O}_{\mathbb{P}^r \times \mathbb{P}(V)}(d) / p_* \mathcal{I}_{\mathcal{H}}(d) = p_* \mathcal{O}_{\mathcal{H}}(d)$ is locally free.

It follows that

$$p_* \mathcal{I}_{\mathcal{H}}(d) = \mathcal{O}_{\mathbb{P}(V)}(-1)$$

the tautological invertible sheaf on $\mathbb{P}(V)$, and the natural map

$$p^* p_* \mathcal{I}_{\mathcal{H}}(d) \to \mathcal{I}_{\mathcal{H}}(d)$$

is an isomorphism. Therefore

$$\mathcal{I}_{\mathcal{H}} = [p^* \mathcal{O}_{\mathbb{P}(V)}(-1)](-d)$$

Let's prove that $\mathcal{H} \subset \mathbb{P}^r \times \mathbb{P}(V)$ is a universal family.
Suppose that

$$\begin{array}{ccc} \mathcal{X} & \subset & \mathbb{P}^r \times S \\ \downarrow f & & \\ S & & \end{array}$$

is a flat family of closed subschemes of $\mathbb{P}^r$ with Hilbert polynomial $h(t)$, i.e. hypersurfaces of degree d, and let $\mathcal{I}_{\mathcal{X}} \subset \mathcal{O}_{\mathbb{P}^r \times S}$ be the ideal sheaf of $\mathcal{X}$. Arguing as above we deduce that $f_* \mathcal{I}_{\mathcal{X}}(d)$ is an invertible subsheaf of $V \otimes_{\mathbf{k}} \mathcal{O}_S$ with locally free cokernel $f_* \mathcal{O}_{\mathcal{X}}(d)$, and that

$$\mathcal{I}_{\mathcal{X}} = [f^* f_* \mathcal{I}_{\mathcal{X}}(d)](-d)$$

We have an induced morphism $g : S \to I\!P(V)$ such that

$$g^*[\mathcal{O}_{I\!P(V)}(-1)] = f_* \mathcal{I}_{\mathcal{X}}(d)$$

The subscheme $S \times_{I\!P(V)} \mathcal{H} \subset I\!P^r \times S$ is defined by the ideal sheaf

$$(1 \times g)^* \mathcal{I}_{\mathcal{H}} = (1 \times g)^* [\mathcal{O}_{I\!P(V)}(-1)(-d)]$$

$$= f^*[g^* \mathcal{O}_{I\!P(V)}(-1)](-d) = [f^* f_* \mathcal{I}_{\mathcal{X}}(d)](-d) = \mathcal{I}_{\mathcal{X}}$$

Hence $S \times_{I\!P(V)} \mathcal{H} = \mathcal{X}$. The proof of the uniqueness of g having this property is left to the reader. Therefore we see that $\mathcal{H} \subset I\!P^r \times I\!P(V)$ is a universal family, and $\mathrm{Hilb}^r_{h(t)} = I\!P(V)$.

4.3.3 Grassmannians

The classical grassmannians are special cases of Hilbert schemes, since they parametrize linear spaces, which are the closed subschemes with Hilbert polynomials of the form $\binom{t+n-1}{n-1}$, $n-1$ being their dimension. Let's fix a $\mathbf{k}$-vector space V of dimension N, and let $1 \le n \le N$. Letting

$$\mathbf{G}_{V,n}(S) = \{\text{loc. free rk } n \text{ quotients of the free sheaf } V^\vee \otimes_{\mathbf{k}} \mathcal{O}_S \text{ on } S\}$$

we define a contravariant functor:

$$\mathbf{G}_{V,n} : (\text{schemes})^{\text{o}} \to (\text{sets})$$

called the *Grassmann functor*; we will denote it simply by $\mathbf{G}$ when no confusion is possible.

Theorem 4.3.2. *The Grassmann functor* $\mathbf{G}$ *is represented by a scheme* $G_n(V)$ *together with a locally free quotient of rank n*

$$V^\vee \otimes_{\mathbf{k}} \mathcal{O}_{G_n(V)} \to Q$$

called the universal quotient bundle.

Proof. Given a scheme S and an open cover $\{U_i\}$ of S, to give a locally free rank n quotient of $V^\vee \otimes_{\mathbf{k}} \mathcal{O}_S$ is equivalent to giving one such quotient over each open set U_i so that they patch together on the intersections $U_i \cap U_j$. Therefore $\mathbf{G}$ is a sheaf.

Let's fix a basis $\{e_k\}$ of $V^\vee$ and choose a set J of n distinct indices in $\{1, \ldots, N\}$. We have an induced decomposition $V^\vee = E' \oplus E''$, with E' (resp. E'') a vector subspace of rank n (resp. $N-n$). We can define a subfunctor $\mathbf{G}_J$ of $\mathbf{G}$ letting:

$$\mathbf{G}_J(S) = \left\{ \begin{array}{c} \text{locally free rk } n \text{ quotients } V^\vee \otimes_{\mathbf{k}} \mathcal{O}_S \to \mathcal{F} \\ \text{inducing } E' \otimes_{\mathbf{k}} \mathcal{O}_S \to \mathcal{F} \text{ surjective} \end{array} \right\}$$

Let S be any scheme and $f : \mathrm{Hom}(-, S) \to \mathbf{G}$ a morphism of functors corresponding to a locally free rank n quotient

$$V^\vee \otimes_{\mathbf{k}} \mathcal{O}_S \to \mathcal{F}$$

The fibred product $S_J := \mathrm{Hom}(-, S) \times_{\mathbf{G}} \mathbf{G}_J$ is clearly represented by the open subscheme of S supported on the points where the map $E' \otimes_{\mathbf{k}} \mathcal{O}_S \to \mathcal{F}$ is surjective; this proves that $\mathbf{G}_J$ is an open subfunctor of $\mathbf{G}$. Since clearly the S_J's cover S, we also see that the family of subfunctors $\{\mathbf{G}_J\}$ is an open covering of $\mathbf{G}$.

To prove that $\mathbf{G}_J$ is representable note that if

$$q : V^\vee \otimes_{\mathbf{k}} \mathcal{O}_S \to \mathcal{F}$$

is an element of $\mathbf{G}(S)$ then the induced map

$$\eta : E' \otimes_{\mathbf{k}} \mathcal{O}_S \to \mathcal{F}$$

is surjective if and only if it is an isomorphism; in this case the composition

$$\eta^{-1} \circ q : V^\vee \otimes_{\mathbf{k}} \mathcal{O}_S \to E' \otimes_{\mathbf{k}} \mathcal{O}_S$$

restricts to the identity on $E' \otimes_{\mathbf{k}} \mathcal{O}_S$, hence it is determined by the composition

$$E'' \otimes_{\mathbf{k}} \mathcal{O}_S \to V^\vee \otimes_{\mathbf{k}} \mathcal{O}_S \to E' \otimes_{\mathbf{k}} \mathcal{O}_S$$

It follows that we can identify

$$\mathbf{G}_J(S) = \mathrm{Hom}(E'' \otimes_{\mathbf{k}} \mathcal{O}_S, E' \otimes_{\mathbf{k}} \mathcal{O}_S) = \mathrm{Hom}(E'', E') \otimes_{\mathbf{k}} \mathcal{O}_S$$

This proves that $\mathbf{G}_J$ is isomorphic to $\mathrm{Hom}(-, \mathbf{A}^{n(N-n)})$, hence it is representable. Now the theorem follows from Proposition E.10. □

$G_n(V)$ is called the *grassmannian of n-dimensional subspaces of V*; it is also called the *grassmannian of $(n-1)$-dimensional projective subspaces of $\mathbb{P}(V)$*. When $V = \mathbf{k}^N$ the grassmannian $G_n(\mathbf{k}^N)$ is denoted by $G(n, N)$.

When $n = 1$ the functor $\mathbf{G}_{V,1}$ is represented by

$$G_1(V) = \mathbb{P}(V) = \mathrm{Proj}(\mathrm{Sym}(V^\vee)),$$

the $(N - 1)$-dimensional projective space associated to V. In this case $Q = \mathcal{O}_{\mathbb{P}(V)}(1)$.

From Theorem 4.3.2 it follows that for all schemes S the morphisms $f : S \to G_n(V)$ are in 1–1 correspondence with the locally free rank n quotients

$$V^\vee \otimes_{\mathbf{k}} \mathcal{O}_S \to \mathcal{F}$$

via $f \leftrightarrow f^*Q$. This is the *universal property of* $G_n(V)$.

The universal quotient bundle defines an exact sequence of locally free sheaves on $G_n(V)$:

$$0 \to K \to V^\vee \otimes_{\mathbf{k}} \mathcal{O}_{G_n(V)} \to Q \to 0$$

called the *tautological exact sequence*; K is called the *universal subbundle*.

Let S be a scheme. Associating to every locally free quotient of rank n

$$V^\vee \otimes_{\mathbf{k}} \mathcal{O}_S \to \mathcal{F}$$

the quotient

$$(\wedge^n V^\vee) \otimes_{\mathbf{k}} \mathcal{O}_S \to \wedge^n \mathcal{F}$$

we define a morphism of functors $\mathbf{G}_{V,n} \to \mathbf{G}_{\wedge^n V,1}$, which is induced by a morphism

$$\pi : G_n(V) \to I\!P(\wedge^n V)$$

π is called the *Plücker morphism.*

Proposition 4.3.3. *The Plücker morphism is a closed embedding. In particular, $G_n(V)$ is a projective variety.*

Proof. As in the proof of 4.3.2, we fix a basis of $V^\vee$ and we choose a set J of n distinct indices in $\{1, \dots, N\}$. We obtain a decomposition $V^\vee = E' \oplus E''$ with $\dim(E') = n$, $\dim(E'') = N - n$, and an induced one:

$$\wedge^n V^\vee = \oplus_{i=0}^n (\wedge^{n-i} E') \otimes_{\mathbf{k}} \wedge^i E'' = \wedge^n E' \oplus F$$

where $F = \oplus_{i=1}^n (\wedge^{n-i} E') \otimes_{\mathbf{k}} \wedge^i E''$. For every scheme S let

$$I\!P_J(S) = \left\{ \begin{array}{c} \text{locally free rk 1 quotients } \wedge^n V^\vee \to \mathcal{L} \\ \text{s.t. the induced } \wedge^n E' \to \mathcal{L} \text{ is surjective} \end{array} \right\}$$

We obtain a subfunctor $I\!P_J$ of $\mathbf{G}_{\wedge^n V,1}$. As in the proof of 4.3.2, we see that the $I\!P_J$'s form an open cover of $\mathbf{G}_{\wedge^n V,1}$ by functors representable by affine spaces.

Note that for every locally free rank n quotient $V^\vee \otimes_{\mathbf{k}} \mathcal{O}_S \to \mathcal{F}$ the induced homomorphism:

$$E' \otimes_{\mathbf{k}} \mathcal{O}_S \to \mathcal{F}$$

is surjective if and only if $\wedge^n E' \to \wedge^n \mathcal{F}$ is. Therefore $\mathbf{G}_J = \pi^{-1}(I\!P_J)$ and it suffices to prove that $\pi : \mathbf{G}_J \to I\!P_J$ is a closed embedding.

We may identify $\mathbf{G}_J$ with (the affine space associated to) $\mathrm{Hom}_{\mathbf{k}}(E', E'')$ and $I\!P_J$ with $\mathrm{Hom}_{\mathbf{k}}(F, \wedge^n E')$. Considering that

$$\mathrm{Hom}_{\mathbf{k}}(\wedge^{n-i} E', \wedge^n E') \cong \wedge^i E'$$

canonically via the perfect pairing:

$$\wedge^i E' \times \wedge^{n-i} E' \to \wedge^n E'$$

we have:

$$\begin{aligned} \mathrm{Hom}_{\mathbf{k}}(F, \wedge^n E') &= \oplus_{i=1}^n \mathrm{Hom}_{\mathbf{k}}((\wedge^{n-i} E') \otimes_{\mathbf{k}} \wedge^i E'', \wedge^n E') \\ &= \oplus_{i=1}^n \mathrm{Hom}_{\mathbf{k}}(\wedge^i E'', \mathrm{Hom}_{\mathbf{k}}(\wedge^{n-i} E', \wedge^n E')) \\ &= \oplus_{i=1}^n \mathrm{Hom}_{\mathbf{k}}(\wedge^i E'', \wedge^i E') \end{aligned}$$

and the map

$$\pi : \mathrm{Hom}_{\mathbf{k}}(E'', E') \to \mathrm{Hom}_{\mathbf{k}}(F, \wedge^n E')$$

is

$$\lambda \mapsto (\lambda, \wedge^2\lambda, \ldots, \wedge^n \lambda)$$

This is the graph of a morphism of affine schemes, hence it is a closed embedding. □

For some $1 \leq n \leq r$, let $G = G(n+1, r+1)$ be the grassmannian of n-dimensional projective subspaces of $\mathbb{P}^r$. Consider the projective bundle over G:

$$\mathbf{I} := \mathbb{P}(Q^\vee) = \mathrm{Proj}(\mathrm{Sym}(Q))$$

where Q is the universal quotient bundle on G. Because of the surjection $\mathcal{O}_G^{r+1} \to Q$ we have a closed embedding:

$$\begin{array}{ccc} \mathbf{I} & \subset & \mathbb{P}^r \times G \\ \downarrow p & & \\ G & & \end{array}$$

note that $Q^\vee = p_*\mathcal{I}_{\mathbf{I}}(1) \subset p_*\mathcal{O}_{\mathbb{P}^r \times G}(1) = \mathcal{O}_G^{r+1}$. For every closed point $v \in G$ the fibre $\mathbf{I}(v)$ is the projective subspace $\mathbb{P}(v) \subset \mathbb{P}^r$; for this reason $\mathbf{I}$ is called the *incidence relation*. Since all fibres of p have Hilbert polynomial $\binom{t+n}{n}$, from Proposition 4.2.1 it follows that p is a flat family. Suppose now that

$$\begin{array}{ccc} \Lambda & \subset & \mathbb{P}^r \times S \\ \downarrow q & & \\ S & & \end{array}$$

is another flat family whose fibres have Hilbert polynomial $\binom{t+n}{n}$. We have an inclusion of sheaves on S

$$q_*\mathcal{I}_\Lambda(1) \subset q_*\mathcal{O}_{\mathbb{P}^r \times S}(1) = \mathcal{O}_S^{r+1}$$

which has locally free cokernel $q_*\mathcal{O}_\Lambda(1)$. By the universal property of G the above inclusion induces a unique morphism

$$g : S \to G$$

such that $g^*(Q^\vee) = q_*\mathcal{I}_\Lambda(1)$. Since $\Lambda = \mathbb{P}(q_*\mathcal{I}_\Lambda(1))$ it follows that

$$\Lambda = S \times_G \mathbf{I}$$

namely, the family q is obtained by base change from the incidence relation via the morphism g. This proves that

$$G(n+1, r+1) = \mathrm{Hilb}^r_{\binom{t+n}{n}}$$

4.3.4 Existence

Theorem 4.3.4. *For every projective scheme $Y \subset \mathbb{P}^r$ and every numerical polynomial $P(t)$, the Hilbert scheme $\mathrm{Hilb}^Y_{P(t)}$ exists and is a projective scheme.*

Proof. We will first prove the theorem in the case $Y = \mathbb{P}^r$. From Corollary 4.1.10 it follows that there is an integer m_0 such that for every closed subscheme $X \subset \mathbb{P}^r$ with Hilbert polynomial $P(t)$ the sheaf of ideals $\mathcal{I}_X$ is m_0-regular. It suffices to take

$$m_0 = F_r(-a_0, \dots, -a_{r-1}, 1 - a_r)$$

It follows that for every $k \geq m_0$

$$h^i(\mathbb{P}^r, \mathcal{I}_X(k)) = 0 \tag{4.12}$$

for $i \geq 1$ and

$$h^0(\mathbb{P}^r, \mathcal{I}_X(k)) = \binom{k+r}{r} - P(k)$$

depends only on $P(k)$. Moreover, by Note 2 of Section 4.1 we have

$$h^i(X, \mathcal{O}_X(k)) = 0 \tag{4.13}$$

all $k \geq m_0$ and all $i \geq 1$. Let

$$N = \binom{m_0+r}{r} - P(m_0)$$

$$V = H^0(\mathbb{P}^r, \mathcal{O}_{\mathbb{P}^r}(m_0))$$

Consider the grassmannian $G = G_N(V)$ of N-dimensional vector subspaces of V. Let $V^\vee \otimes_{\mathbf{k}} \mathcal{O}_G \to Q$ be the universal quotient bundle, which is a locally free sheaf of rank N on G, and

$$p : \mathbb{P}^r \times G \to G$$

the projection. We may identify $V \otimes_{\mathbf{k}} \mathcal{O}_G = p_*[\mathcal{O}_{\mathbb{P}^r \times G}(m_0)]$. The image of the composition

$$\begin{array}{ccccc} p^* Q^\vee(-m_0) & \longrightarrow & V \otimes_{\mathbf{k}} \mathcal{O}_{\mathbb{P}^r\times G}(-m_0) & \longrightarrow & \mathcal{O}_{\mathbb{P}^r\times G} \\ & & \| & & \\ & & p^* p_*[\mathcal{O}_{\mathbb{P}^r\times G}(m_0)] \otimes \mathcal{O}_{\mathbb{P}^r\times G}(-m_0) & & \end{array}$$

is a sheaf of ideals: we will denote it by $\mathbf{J}$.

Let $\mathcal{Z} \subset \mathbb{P}^r \times G$ be the closed subscheme defined by $\mathbf{J}$ and denote by $q : \mathcal{Z} \to G$ the restriction of p to $\mathcal{Z}$.

Consider the flattening stratification

$$G^1 \coprod G^2 \coprod \dots \subset G$$

for $\mathcal{O}_{\mathcal{Z}}$ and let H be the stratum relative to the polynomial $P(t)$. We will prove that $H = \text{Hilb}^r_{P(t)}$ and that the universal family is the pullback of q to H:

$$\begin{array}{rccc} \mathcal{W} := & H \times_G \mathcal{Z} & \to & \mathcal{Z} \\ & \downarrow \pi & & \downarrow q \\ & H & \to & G \end{array}$$

By the choice of H, $\mathcal{W}$ defines a flat family of closed subschemes of $\mathbb{P}^r$ with Hilbert polynomial equal to $P(t)$.

Let's prove that $\mathcal{W}$ has the universal property.

Consider a flat family of closed subschemes of $\mathbb{P}^r$ with Hilbert polynomial $P(t)$:

$$\begin{array}{cc} \mathcal{X} \subset & \mathbb{P}^r \times S \\ & \downarrow f \\ & S \end{array}$$

From (4.12) and (4.13) and from Theorem 4.2.5 it follows that

$$R^1 f_* \mathcal{I}_{\mathcal{X}}(m_0) = (0) = R^1 f_* \mathcal{O}_{\mathcal{X}}(m_0)$$

In particular, we have an exact sequence on S:

$$\begin{array}{ccccc} 0 \to f_* \mathcal{I}_{\mathcal{X}}(m_0) \to & f_* \mathcal{O}_{\mathbb{P}^r \times S}(m_0) & \to f_* \mathcal{O}_{\mathcal{X}}(m_0) \to 0 \\ & \| & \\ & V \otimes_{\mathbf{k}} \mathcal{O}_S & \end{array}$$

If we apply Theorem 4.2.5 for $j = -1$ we deduce that $f_* \mathcal{I}_{\mathcal{X}}(m_0)$ and $f_* \mathcal{O}_{\mathcal{X}}(m_0)$ are locally free and $f_* \mathcal{I}_{\mathcal{X}}(m_0)$ has rank N.

From the universal property of G it follows that there exists a unique morphism $g : S \to G$ such that

$$f_* \mathcal{I}_{\mathcal{X}}(m_0) = g^* Q^\vee$$

Claim: For all $m \gg m_0$ we have $f_* \mathcal{O}_{\mathcal{X}}(m) = g^* p_* \mathcal{O}_{\mathcal{Z}}(m)$.

Proof of the Claim: For all $m \gg m_0$ we have exact sequences:

$$0 \to p_* \mathbf{J}(m) \to p_* \mathcal{O}_{\mathbb{P}^r \times G}(m) \to q_* \mathcal{O}_{\mathcal{Z}}(m) \to 0$$

on G and

$$0 \to f_* \mathcal{I}_{\mathcal{X}}(m) \to f_* \mathcal{O}_{\mathbb{P}^r \times S}(m) \to f_* \mathcal{O}_{\mathcal{X}}(m) \to 0$$

on S; since $g^* p_* \mathcal{O}_{\mathbb{P}^r \times G}(m) = f_* \mathcal{O}_{\mathbb{P}^r \times S}(m)$ it suffices to show that:

$$f_* \mathcal{I}_{\mathcal{X}}(m) \cong g^* p_* \mathbf{J}(m)$$

for all $m \gg m_0$. For all such m we have the equality on G:

$$p_* \mathbf{J}(m) = \text{Im}[Q^\vee \otimes p_* \mathcal{O}(m - m_0) \to p_* \mathcal{O}_{\mathbb{P}^r \times G}(m)]$$

induced by the surjections $p^* Q^\vee(m - m_0) \to \mathbf{J}(m)$ of sheaves on $\mathbb{P}^r \times G$. Hence for all $m \gg m_0$ we have:

$$\begin{aligned} g^* p_* \mathbf{J}(m) &= g^* \mathrm{Im}[\mathcal{Q}^\vee \otimes p_* \mathcal{O}_{\mathbb{P}^r \times G}(m - m_0) \to p_* \mathcal{O}_{\mathbb{P}^r \times G}(m)] \\ &= \mathrm{Im}[g^* \mathcal{Q}^\vee \otimes f_* \mathcal{O}_{\mathbb{P}^r \times S}(m - m_0) \to f_* \mathcal{O}_{\mathbb{P}^r \times S}(m)] \\ &= \mathrm{Im}[f_* \mathcal{I}_{\mathcal{X}}(m_0) \otimes f_* \mathcal{O}_{\mathbb{P}^r \times S}(m - m_0) \to f_* \mathcal{O}_{\mathbb{P}^r \times S}(m)] = f_* \mathcal{I}_{\mathcal{X}}(m) \end{aligned}$$

and this proves the Claim.

From the Claim it follows that:

(i) g factors through H.

Indeed, from 4.2.4 it follows that for all $m \gg m_0$:

$$g^* q_* \mathcal{O}_{\mathcal{Z}}(m) = f_*(1 \times g)^* \mathcal{O}_{\mathcal{Z}}(m)$$

Since $g^* q_* \mathcal{O}_{\mathcal{Z}}(m) = f_* \mathcal{O}_{\mathcal{X}}(m)$ is locally free of rank $P(m)$ for all such m Proposition 4.2.1 implies that $(1 \times g)^* \mathcal{O}_{\mathcal{Z}}$ is flat over S with Hilbert polynomial $P(t)$. Hence g factors by the definition of H.

(ii) $\mathcal{X} = S \times_H \mathcal{W}$.

Indeed:

$$\begin{aligned} \mathcal{X} &= \mathrm{Proj}[\bigoplus_{m \gg 0} f_* \mathcal{O}_{\mathcal{X}}(m)] = \mathrm{Proj}[\bigoplus_{m \gg 0} g^* q_* \mathcal{O}_{\mathcal{Z}}(m)] \\ &= \mathrm{Proj}[\bigoplus_{m \gg 0} g^* \pi_* \mathcal{O}_{\mathcal{W}}(m)] = S \times_H \mathrm{Proj}[\bigoplus_{m \gg 0} \pi_* \mathcal{O}_{\mathcal{W}}(m)] = S \times_H \mathcal{W} \end{aligned}$$

Properties (i) and (ii) imply that $H = \mathrm{Hilb}^r_{P(t)}$ and that π is the universal family.

By construction, $\mathrm{Hilb}^r_{P(t)}$ is a quasi-projective scheme. To prove that it is projective it suffices to show that it is proper over $\mathbf{k}$. We will use the valuative criterion of properness. Let A be a discrete valuation $\mathbf{k}$-algebra with quotient field L and residue field K, and let

$$\varphi : \mathrm{Spec}(L) \to \mathrm{Hilb}^r_{P(t)}$$

be any morphism. We must show that φ extends to a morphism

$$\tilde{\varphi} : \mathrm{Spec}(A) \to \mathrm{Hilb}^r_{P(t)}$$

Pulling back the universal family by φ we obtain a flat family

$$\mathcal{X} \subset \mathbb{P}^r \times \mathrm{Spec}(L)$$

of closed subschemes of $\mathbb{P}^r$ with Hilbert polynomial $P(t)$. Since $\mathrm{Spec}(A)$ is nonsingular of dimension one and

$$\mathrm{Spec}(L) = \mathrm{Spec}(A) \backslash \{\text{closed point}\}$$

Proposition III.9.8 of [84] implies the existence of a flat family

$$\mathcal{X}' \subset \mathbb{P}^r \times \mathrm{Spec}(A)$$

which extends $\mathcal{X}$. By the universal property of $\mathrm{Hilb}^r_{P(t)}$ this family corresponds to a morphism $\tilde{\varphi} : \mathrm{Spec}(A) \to \mathrm{Hilb}^r_{P(t)}$ which extends φ. This concludes the proof of the theorem in the case $Y = I\!P^r$.

Let's now assume that Y is an arbitrary closed subscheme of $I\!P^r$. It will suffice to show that the functor $\mathcal{H}ilb^Y_{P(t)}$ is represented by a closed subscheme of $\mathrm{Hilb}^r_{P(t)}$.

Applying Corollary 4.1.10 twice we can find an integer μ such that $\mathcal{I}_Y \subset \mathcal{O}_{I\!P^r}$ is μ-regular and such that for every closed subscheme $X \subset I\!P^r$ with Hilbert polynomial $P(t)$ the ideal sheaf $\mathcal{I}_X \subset \mathcal{O}_{I\!P^r}$ is μ-regular. Let

$$V = H^0(I\!P^r, \mathcal{O}_{I\!P^r}(\mu)), \quad U = H^0(I\!P^r, \mathcal{I}_Y(\mu))$$

It follows from 4.2.5 and 4.2.6 that $\pi_*\mathcal{I}_{\mathcal{W}}(\mu)$ is a locally free subsheaf of $V \otimes_{\mathbf{k}} \mathcal{O}_{\mathrm{Hilb}}$ with locally free cokernel.

On $\mathrm{Hilb}^r_{P(t)}$ consider the composition

$$\Psi : U \otimes_{\mathbf{k}} \mathcal{O}_{\mathrm{Hilb}} \to V \otimes_{\mathbf{k}} \mathcal{O}_{\mathrm{Hilb}} \to V \otimes_{\mathbf{k}} \mathcal{O}_{\mathrm{Hilb}}/\pi_*\mathcal{I}_{\mathcal{W}}(\mu)$$

Let $Z \subset \mathrm{Hilb}^r_{P(t)}$ be the closed subscheme defined by the condition $\Psi = 0$, or equivalently, by the condition

$$U \otimes_{\mathbf{k}} \mathcal{O}_Z \subset \pi_*\mathcal{I}_{\mathcal{W}}(\mu) \otimes \mathcal{O}_Z \tag{4.14}$$

Letting $j : Z \to \mathrm{Hilb}^r_{P(t)}$ be the inclusion, one easily sees that condition (4.14) implies that

$$\mathcal{I}_{Y \times Z} \subset (1 \times j)^*\mathcal{I}_{\mathcal{W}} \subset \mathcal{O}_{I\!P^r \times Z}$$

hence that

$$Z \times_{\mathrm{Hilb}} \mathcal{W} \subset Y \times Z \subset I\!P^r \times Z \tag{4.15}$$

It is straightforward to check that $Z = \mathrm{Hilb}^Y_{P(t)}$ and that (4.15) is the universal family. This concludes the proof of Theorem 4.3.4. □

For any projective scheme $Y \subset I\!P^r$ it is often convenient to consider the functor:

$$\mathcal{H}ilb^Y : (\text{schemes}) \to (\text{sets})$$

defined as:

$$\mathcal{H}ilb^Y(S) = \coprod_{P(t)} \mathcal{H}ilb^Y_{P(t)}(S)$$

This functor is represented by the disjoint union

$$\mathrm{Hilb}^Y = \coprod_{P(t)} \mathrm{Hilb}^Y_{P(t)}$$

which is a scheme locally of finite type (but not of finite type because it has infinitely many connected components unless $\dim(Y) = 0$). It is *the Hilbert scheme of* Y. One convenient feature of Hilb^Y is that it is independent on the projective embedding of Y, even though the indexing of its components $\mathcal{H}ilb^Y_{P(t)}$ by Hilbert polynomials does depend on the embedding. For this reason, when considering Hilb^Y we will not need to specify a projective embedding of Y.

Let's fix a projective scheme Y, and in the Hilbert scheme Hilb^Y let's consider a $\mathbf{k}$-rational point $[X]$ which parametrizes a closed subscheme $X \subset Y$. Denote by $\mathcal{I} \subset \mathcal{O}_Y$ the ideal sheaf of X in Y. The local Hilbert functor H_X^Y is a subfunctor of the restriction to $\mathcal{A}$ of the Hilbert functor; since Hilb^Y represents the Hilbert functor we have, with the notation introduced in § 2.2:

$$H_X^Y(A) = \mathrm{Hom}(\mathrm{Spec}(A), \mathrm{Hilb}^Y)_{[X]}$$

for every A in $\mathrm{ob}(\mathcal{A})$. In particular, H_X^Y is prorepresented by the local ring $\hat{\mathcal{O}}_{\mathrm{Hilb},[X]}$. We can therefore apply the results proved in § 2.4 to obtain information about the local properties of Hilb^Y at $[X]$. In particular, we have the following:

Theorem 4.3.5. *(i) There is a canonical isomorphism of* $\mathbf{k}$*-vector spaces:*

$$T_{[X]}\mathrm{Hilb}^Y \cong H^0(X, N_{X/Y})$$

where $N_{X/Y} = \mathcal{H}om_{\mathcal{O}_X}(\mathcal{I}/\mathcal{I}^2, \mathcal{O}_X)$ *is the normal sheaf of* X *in* Y.

(ii) If $X \subset Y$ *is a regular embedding then the obstruction space of* $\mathcal{O}_{\mathrm{Hilb}^Y,[X]}$ *is a subspace of* $H^1(X, N_{X/Y})$.

The simplest illustration of Theorem 4.3.5 is for $Y^{[1]} = Y$. In this case 4.3.5(i) simply says that $\mathrm{Hom}(m_p/m_p^2, \mathbf{k})$ is the Zariski tangent space of Y at a $\mathbf{k}$-rational point $p \in Y$. The obstruction space is $o(\mathcal{O}_{Y^{[1]},[p]}) = o(\mathcal{O}_{Y,p})$. Of course if p is a singular point then it is not regularly embedded in Y, and $H^1(p, N_{p/Y}) = 0$ is not an obstruction space for the local Hilbert functor.

Consider a flat family of closed subschemes of Y:

$$\begin{array}{ccc} \mathcal{X} \subset & Y \times S \\ & \downarrow f \\ & S \end{array}$$

It induces a functorial morphism $\chi : S \to \mathrm{Hilb}^Y$ (the classifying morphism of the family) whose differential at a $\mathbf{k}$-rational point $s \in S$ is a linear map

$$d\chi_s : T_sS \to H^0(\mathcal{X}(s), N_{\mathcal{X}(s)/Y}) \tag{4.16}$$

called the *characteristic map* of the family f. Obviously, the surjectivity of $d\chi_s$ is a necessary condition for the smoothness of χ at s. We have the following more precise result.

Proposition 4.3.6. *Let* Y *be a projective scheme,*

$$\begin{array}{ccc} \mathcal{X} \subset & Y \times S \\ & \downarrow f \\ & S \end{array}$$

a flat family of closed subschemes of Y with S algebraic, and

$$\chi : S \to \mathrm{Hilb}^Y$$

the classifying map of the family. Then:

(i) If s is a nonsingular point of S and if the characteristic map

$$d\chi_s : T_sS \to H^0(\mathcal{X}(s), N_{\mathcal{X}(s)/Y})$$

is surjective then χ is smooth at s and $\mathcal{X}(s)$ is unobstructed in Y.
(ii) If, moreover, $h^1(\mathcal{X}(s), T_{Y|\mathcal{X}(s)}) = 0$ and $\mathcal{X}(s)$ is nonsingular then f has general moduli at s and $\mathcal{X}(s)$ is unobstructed as an abstract variety.

Proof. (i) The smoothness of χ is a consequence of Theorem 2.1.5 and of the nonsingularity of S at s. The unobstructedness of $\mathcal{X}(s)$ in Y, i.e. the nonsingularity of Hilb^Y at $\chi(s)$, follows from the smoothness of χ and the nonsingularity of S at s.

(ii) The condition $h^1(\mathcal{X}(s), T_{Y|\mathcal{X}(s)}) = 0$ implies that the forgetful morphism $H^Y_{\mathcal{X}(s)} \to \mathrm{Def}_{\mathcal{X}(s)}$ is smooth (Proposition 3.2.9) and therefore the Kodaira–Spencer map of f at s is surjective, being the composition of two surjective maps. We obtain the conclusion from Proposition 2.5.8. □

The following criterion follows at once from Proposition 3.2.9:

Proposition 4.3.7. *Let $X \subset Y$ be a closed embedding of projective nonsingular schemes such that $h^1(X, T_{Y|X}) = 0$. Then the universal family*

$$\begin{array}{ccc} \mathcal{X} & \subset & Y \times \mathrm{Hilb}^Y \\ \downarrow & & \\ \mathrm{Hilb}^Y & & \end{array}$$

has general moduli at the point $[X]$.

NOTES

1. The construction of the grassmannian given here is taken from [97].

2. It is a classical result of Hartshorne that $\mathrm{Hilb}^r_{P(t)}$ is connected for all r and $P(t)$ (see [83] and [140]). For general Y this is no longer true: for example, if $Y \subset \mathbb{P}^3$ is a nonsingular quadric then Hilb^Y_{t+1} has two connected components.

3. Let $X \subset Y$ be a closed embedding of projective schemes. It can be easily verified that for any closed subscheme $Z \subset X$, the induced injective linear map

$$H^0(Z, N_{Z/X}) \to H^0(Z, N_{Z/Y})$$

coincides with the differential at $[Z]$ of the closed embedding

$$\mathrm{Hilb}^X \subset \mathrm{Hilb}^Y$$

4. If $Z \subset Y$ is a closed embedding of projective schemes and $P(t)$ a numerical polynomial, then one can define a functor

$$Hilb(-Z)^Y_{P(t)} : (\text{schemes})^\circ \to (\text{sets})$$

by

$$Hilb(-Z)^Y_{P(t)}(S) = \left\{ \begin{array}{c} \text{flat families } \mathcal{X} \subset Y \times S \text{ with Hilbert polynomial} \\ P(t) \text{ and such that } \mathcal{X}(s) \cap Z \neq \emptyset, \forall s \in S \end{array} \right\}$$

Then $Hilb^Y(-Z)^Y_{P(t)}$ is representable by a closed subscheme

$$\text{Hilb}(-Z)^Y_{P(t)} \subset \text{Hilb}^Y_{P(t)}$$

Let

$$\text{Hilb}^{Y\setminus Z}_{P(t)} := \text{Hilb}^Y_{P(t)} \setminus \text{Hilb}(-Z)^Y_{P(t)} \subset \text{Hilb}^Y_{P(t)}$$

be the corresponding open subscheme and define

$$\text{Hilb}^{Y\setminus Z} := \coprod_{P(t)} \text{Hilb}^{Y\setminus Z}_{P(t)}$$

Then this is a scheme locally of finite type which depends only on $W := Y\setminus Z$ and not on the embedding $W \subset Y$. In the case $P(t) = n$, a constant polynomial, we denote $\text{Hilb}^W_{P(t)}$ by $W^{[n]}$. If W is affine then

$$\text{Hilb}^W = \coprod_n W^{[n]}$$

The proofs of these facts are left to the reader. See also [81].

4.4 Quot schemes

4.4.1 Existence

We will now introduce an important class of schemes, the so-called *Quot schemes*, which generalize the Hilbert schemes. As special cases we will obtain the *relative Hilbert schemes*.

Let $p : X \to S$ be a projective morphism of algebraic schemes, and let $\mathcal{O}_X(1)$ be a line bundle on X very ample with respect to p. Fix a coherent sheaf $\mathcal{H}$ on X and a numerical polynomial $P(t) \in \mathbf{Q}[t]$. We define a functor

$$Quot^{X/S}_{\mathcal{H},P(t)} : (\text{schemes}/S)^\circ \to (\text{sets})$$

called the *Quot functor of X/S relative to $\mathcal{H}$ and $P(t)$*, in the following way:

$$Quot^{X/S}_{\mathcal{H},P(t)}(Z \to S) = \left\{ \begin{array}{c} \text{coherent quotients } \mathcal{H}_Z \to F, \text{ flat over } Z, \text{ with} \\ \text{Hilbert polyn. } P(t) \text{ on the fibres of } X_Z \to Z \end{array} \right\}$$

where $X_Z = Z \times_S X$ and $\mathcal{H}_Z$ denotes the pullback of $\mathcal{H}$ on X_Z, as usual. When $S = \text{Spec}(\mathbf{k})$ we write $Quot^X_{\mathcal{H},P(t)}$ instead of $Quot^{X/\text{Spec}(\mathbf{k})}_{\mathcal{H},P(t)}$.

This definition generalizes the Hilbert functors which are obtained in the case $S = \text{Spec}(\mathbf{k})$ and $\mathcal{H} = \mathcal{O}_X$.

Theorem 4.4.1. *The functor* $\mathcal{Q}uot = \mathcal{Q}uot^{X/S}_{\mathcal{H},P(t)}$ *is represented by a projective* S*-scheme*

$$\mathrm{Quot}^{X/S}_{\mathcal{H},P(t)} \to S$$

Proof. We first consider the case $S = \mathrm{Spec}(\mathbf{k})$ and $X = I\!P^r$. From Theorem 4.1.9 it follows that there is an integer m such that for each scheme Z and for each $(\varphi : \mathcal{H}_Z \to F) \in \mathcal{Q}uot(Z)$, letting $N = \ker(\varphi)$, all the sheaves $N(z), \mathcal{H}(z) = \mathcal{H}, F(z), z \in Z$, are m-regular. Therefore, letting $p_Z : I\!P^r \times Z \to Z$ be the projection, we obtain an exact sequence of locally free sheaves on Z:

$$0 \to p_{Z*}N(m) \to H^0(I\!P^r, \mathcal{H}(m)) \otimes_{\mathbf{k}} \mathcal{O}_Z \to p_{Z*}F(m) \to 0$$

Moreover, for each $m' \geq m$ there is an exact sequence

$$H^0(I\!P^r, \mathcal{O}(m'-m)) \otimes_{\mathbf{k}} p_{Z*}N(m) \to H^0(I\!P^r, \mathcal{H}(m')) \otimes_{\mathbf{k}} \mathcal{O}_Z \to p_{Z*}F(m') \to 0$$

where the first map is given by multiplication of sections. This shows that $p_{Z*}F(m)$ uniquely determines the sheaf of graded $\mathcal{O}_Z[X_0, \ldots, X_r]$-modules $\bigoplus_{k \geq m} p_{Z*}F(k)$, which in turn determines F. Therefore, letting

$$H^0(I\!P^r, \mathcal{H}(m)) = V$$

we have an injective morphism of functors:

$$\mathcal{Q}uot \to \mathbf{G}_{\check{V}, P(m)}$$

given by:

$$\begin{array}{ccc} \mathcal{Q}uot(Z) & \to & \mathbf{G}_{\check{V}, P(m)}(Z) \\ (\mathcal{H}_Z \to F) & \mapsto & [V \otimes_{\mathbf{k}} \mathcal{O}_Z \to p_{Z*}F(m)] \end{array}$$

On $G = G_{P(m)}(\check{V})$ consider the tautological exact sequence

$$0 \to K \to V \otimes_{\mathbf{k}} \mathcal{O}_G \to Q \to 0$$

Let, moreover, $p_2 : I\!P^r \times G \to G$ and $p_1 : I\!P^r \times G \to I\!P^r$ be the projections. On G we have $\Gamma_*(\mathcal{H}) \otimes_{\mathbf{k}} \mathcal{O}_G$, which is a sheaf of graded $\mathcal{O}_G[X_0, \ldots, X_r]$-modules, and determines $p_1^*(\mathcal{H})$. Consider the subsheaf $K\mathcal{O}_G[X_0, \ldots, X_r]$ and the sheaf $\mathcal{F}$ on $G \times I\!P^r$ corresponding to the quotient $\Gamma_*(\mathcal{H}) \otimes_{\mathbf{k}} \mathcal{O}_G / K\mathcal{O}_G[X_0, \ldots, X_r]$, and let $G_P \subset G$ be the stratum corresponding to P of the flattening stratification of $\mathcal{F}$. Then we claim that a morphism of schemes $f : Z \to G$ defines an element of $\mathcal{Q}uot(Z)$ if and only if f factors through G_P, and therefore $\mathcal{Q}uot$ is represented by G_P. The proof of this fact is similar to the one given for the proof of Theorem 4.3.4 and will be left to the reader.

Since G_P is quasi-projective, to prove that it is projective amounts to proving that it is proper over $\mathbf{k}$, and this can be done using the valuative criterion of properness. Let A be a discrete valuation $\mathbf{k}$-algebra with quotient field L and residue field K, and let

$$\varphi : \mathrm{Spec}(L) \to G_P$$

be any morphism. We must show that φ extends to a morphism

$$\tilde{\varphi} : \mathrm{Spec}(A) \to G_P$$

The datum of φ corresponds to an element $(\varphi_L : \mathcal{H}_L \to F_L)$ of $\mathcal{Q}uot(\mathrm{Spec}(L))$. The existence of $\tilde{\varphi}$ will be proved if there is a quotient $\varphi_A : \mathcal{H}_A \to F_A$ on $\mathbb{P}^r \times \mathrm{Spec}(A)$ which is flat over $\mathrm{Spec}(A)$ and which restricts to F_L over $\mathbb{P}^r \times \mathrm{Spec}(L)$. Let $i : \mathbb{P}^r \times \mathrm{Spec}(L) \to \mathbb{P}^r \times \mathrm{Spec}(A)$ be the inclusion, and take $F_A = i_*(F_L)$. Obviously, F_A restricts to F_L. Moreover, if $K_L = \ker(\varphi_L)$, we have $R^1 i_*(K_L) = 0$ and therefore a surjection $\mathcal{H}_A = i_*(\mathcal{H}_L) \to F_A$. We need the following:

Lemma 4.4.2. *Let X be a scheme, U an open subset of X and $i : U \to X$ the inclusion. Then for every coherent sheaf F on U we have*

$$\mathrm{Ass}(i_*(F)) = \mathrm{Ass}(F)$$

Proof. Since $i_*(F)_{|U} = F$ we have $\mathrm{Ass}(i_*(F)) \cap U = \mathrm{Ass}(F)$. Therefore we only need to prove that $\mathrm{Ass}(i_*(F)) \subset U$.

We may assume that $X = \mathrm{Spec}(A)$ and $U = \mathrm{Spec}(B)$ are affine. The inclusion i corresponds to an injective homomorphism $A \to B$ and $F = M^\sim$ for an f.g. B-module M. Let $x \in \mathrm{Ass}(i_*(F))$ and assume that $x \in X \backslash U$. Then the ideal $p_x \subset A$ annihilates an element $m_x \in i_*(F)_x$ which corresponds to a section $m \in \Gamma(V, i_*(F))$ for some open neighbourhood V of x. Up to shrinking X we may assume $V = X$, so that $m \in \Gamma(X, i_*(F)) = \Gamma(U, F) = M$ is annihilated by the ideal $p_x B$. But $p_x B = B$ because $x \notin U$ and therefore $m = 0$: this is a contradiction. The lemma is proved.

From the lemma it follows that $\mathrm{Ass}(F_A) = \mathrm{Ass}(F_L)$: therefore, using the fact that F_L is flat over $\mathrm{Spec}(L)$ and [84], Prop. III.9.7, we deduce that F_A is flat over $\mathrm{Spec}(A)$. This concludes the proof of the theorem in the case $S = \mathrm{Spec}(\mathbf{k})$ and $X = \mathbb{P}^r$.

Assume now that S and X are arbitrary. Consider the closed embedding $j : X \to \mathbb{P}^r \times S$ determined by $\mathcal{O}_X(1)$. Replacing $\mathcal{H}$ by $j_*\mathcal{H}$ we can assume that $X = \mathbb{P}^r \times S$. Let $h, h' \gg 0$ be such that we have an exact sequence:

$$\mathcal{O}_{\mathbb{P}^r \times S}(-h')^{M'} \to \mathcal{O}_{\mathbb{P}^r \times S}(-h)^M \to \mathcal{H} \to 0$$

for some M, M'. Then for each S-scheme $Z \to S$ and for each

$$(\mathcal{H}_Z \to F) \in \mathcal{Q}uot^{\mathbb{P}^r \times S/S}_{\mathcal{H}, P(t)}(Z \to S)$$

we obtain that the composition

$$\mathcal{O}_{\mathbb{P}^r \times Z}(-h)^M \to \mathcal{H}_Z \to F \to 0$$

is a surjection, i.e. is an element of $\mathcal{Q}uot^{\mathbb{P}^r \times S/S}_{\mathcal{O}(-h)^M, P(t)}(Z \to S)$. This proves that the functor $\mathcal{Q}uot^{\mathbb{P}^r \times S/S}_{\mathcal{H}, P(t)}$ is a subfunctor of the functor $\mathcal{Q}uot^{\mathbb{P}^r \times S/S}_{\mathcal{O}(-h)^M, P(t)}$, and this functor is evidently represented by $\mathrm{Quot}^{\mathbb{P}^r}_{\mathcal{O}(-h)^M, P(t)} \times S$.

Conversely, a quotient

$$(\mathcal{O}_{\mathbb{P}^r\times Z}(-h)^M \to F) \in \mathcal{Q}uot^{\mathbb{P}^r\times S/S}_{\mathcal{O}_X(-h)^M, P(t)}(Z \to S)$$

is in $\mathcal{Q}uot^{\mathbb{P}^r\times S/S}_{\mathcal{H},P(t)}(Z \to S)$ if and only if the composition

$$\mathcal{O}_{\mathbb{P}^r\times Z}(-h')^{M'} \to \mathcal{O}_{\mathbb{P}^r\times Z}(-h)^M \to F$$

is zero. This means that the condition for an S-morphism

$$Z \to \mathrm{Quot}^{\mathbb{P}^r\times S/S}_{\mathcal{O}(-h)^M, P(t)} = \mathrm{Quot}^{\mathbb{P}^r/\mathbf{k}}_{\mathcal{O}(-h)^M, P(t)} \times S$$

to define an element of $\mathcal{Q}uot^{\mathbb{P}^r\times S/S}_{\mathcal{H},P(t)}(Z \to S)$ is that it factors through the closed subscheme defined by the entries of the matrix of the homomorphism:

$$\mathcal{O}_{\mathbb{P}^r\times \mathrm{Quot}}(-h')^{M'} \to \mathcal{O}_{\mathbb{P}^r\times \mathrm{Quot}}(-h)^M$$

and this is a closed condition. This proves that $\mathcal{Q}uot^{\mathbb{P}^r\times S/S}_{\mathcal{H},P(t)}$ is represented by a closed subscheme of $\mathrm{Quot}^{\mathbb{P}^r/\mathbf{k}}_{\mathcal{O}(-h)^M, P(t)} \times S$. □

From the fact that $\mathrm{Quot}^{X/S}_{\mathcal{H},P(t)}$ represents the functor $\mathcal{Q}uot^{X/S}_{\mathcal{H},P(t)}$ it follows that there is a *universal quotient*

$$(\mathcal{H}_{\mathrm{Quot}} \to \mathcal{F}) \in \mathcal{Q}uot^{X/S}_{\mathcal{H},P(t)}(\mathrm{Quot}^{X/S}_{\mathcal{H},P(t)})$$

corresponding to the identity morphism under the identification

$$\mathrm{Hom}(\mathrm{Quot}, \mathrm{Quot}) = \mathcal{Q}uot(\mathrm{Quot})$$

In the case $\mathcal{H} = \mathcal{O}_X$ the scheme $\mathrm{Quot}^{X/S}_{\mathcal{O}_X,P(t)}$ is denoted by $\mathrm{Hilb}^{X/S}_{P(t)}$ and called the *relative Hilbert scheme* of X/S with respect to the polynomial $P(t)$.

It will be sometimes convenient to consider the functor

$$\mathcal{Q}uot^{X/S}_{\mathcal{H}} : (\text{schemes}/S)^\circ \to (\text{sets})$$

defined as:

$$\mathcal{Q}uot^{X/S}_{\mathcal{H}}(Z \to S) = \coprod_{P(t)} \mathcal{Q}uot^{X/S}_{\mathcal{H},P(t)}(Z \to S)$$

This functor is represented by the disjoint union

$$\mathrm{Quot}^{X/S}_{\mathcal{H}} = \coprod_{P(t)} \mathrm{Quot}^{X/S}_{\mathcal{H},P(t)}$$

which is a scheme locally of finite type, called *the Quot scheme of X over S relative to $\mathcal{H}$*; it carries a universal quotient $\mathcal{H}_{\mathrm{Quot}} \to \mathcal{F}$.

Similarly, we will consider the *relative Hilbert scheme of X over S*:

$$\mathrm{Hilb}^{X/S} = \coprod_{P(t)} \mathrm{Hilb}^{X/S}_{P(t)}$$

The construction of the Quot scheme commutes with base change; this is a result which follows quite directly from the definition, but it is worth pointing it out:

Proposition 4.4.3 (base change property). *Given a projective morphism $X \to S$, a coherent sheaf $\mathcal{H}$ on X, and a morphism $T \to S$, there is a natural identification:*

$$\mathrm{Quot}^{X_T/T}_{\mathcal{H}_T} = T \times_S \mathrm{Quot}^{X/S}_{\mathcal{H}}$$

Proof. Consider the product diagram

$$\begin{array}{ccc} T \times \mathrm{Quot}^{X/S}_{\mathcal{H}} & \to & \mathrm{Quot}^{X/S}_{\mathcal{H}} \\ \downarrow & & \downarrow \\ T & \to & S \end{array}$$

The universal quotient $\mathcal{H}^{X/S} \to \mathcal{F}$ on $\mathrm{Quot}^{X/S}_{\mathcal{H}}$ pulls back to a quotient $\mathcal{H}^{X_T/T} \to \mathcal{F}_T$ on $T \times \mathrm{Quot}^{X/S}_{\mathcal{H}}$. We have immediately that the T-scheme $T \times \mathrm{Quot}^{X/S}_{\mathcal{H}}$ endowed with this quotient represents the functor $\mathcal{Q}uot^{X_T/T}_{\mathcal{H}_T}$. □

4.4.2 Local properties

Proposition 4.4.4. *Let $X \to S$ be a projective morphism of algebraic schemes, $\mathcal{H}$ a coherent sheaf on X, flat over S, and $\pi : Q = \mathrm{Quot}^{X/S}_{\mathcal{H}} \to S$ the associated Quot scheme over S. Let $s \in S$ be a* **k***-rational point and $q \in \pi^{-1}(s) = Q(s)$ corresponding to a coherent quotient $f : \mathcal{H} \to \mathcal{F}$ with kernel $\mathcal{K}$. Let*

$$f_s : \mathcal{H}(s) \to \mathcal{F}(s)$$

be the restriction of f to the fibre $X(s)$, whose kernel is $\mathcal{K}(s) = \mathcal{K} \otimes \mathcal{O}_{X(s)}$ (by the flatness of $\mathcal{F}$). Then there is an exact sequence

$$0 \to \mathrm{Hom}(\mathcal{K}(s), \mathcal{F}(s)) \to t_q Q \xrightarrow{d\pi_q} t_s S \to \mathrm{Ext}^1_{\mathcal{O}_{X(s)}}(\mathcal{K}(s), \mathcal{F}(s))$$

and an inclusion:

$$\ker[o(\pi_q^\sharp)] \subset \mathrm{Ext}^1_{\mathcal{O}_{X(s)}}(\mathcal{K}(s), \mathcal{F}(s))$$

where $o(\pi_q^\sharp) : o(\mathcal{O}_{Q,q}) \to o(\mathcal{O}_{S,s})$ is the obstruction map of the local homomorphism $\pi_q^\sharp : \mathcal{O}_{S,s} \to \mathcal{O}_{Q,q}$. In particular, π is smooth at q if $\mathrm{Ext}^1_{\mathcal{O}_{X(s)}}(\mathcal{K}(s), \mathcal{F}(s)) = 0$.

Proof. A vector in $\ker(d\pi_q)$ corresponds to a commutative diagram:

$$\begin{array}{ccc} \mathrm{Spec}(\mathbf{k}[\epsilon]) & \to & Q \\ \downarrow & & \downarrow \pi \\ \mathrm{Spec}(\mathbf{k}) & \to & S \end{array}$$

such that the upper horizontal arrow has image $\{q\}$. The above diagram corresponds to an exact and commutative diagram of sheaves on $X(s)$:

$$\begin{array}{ccccccc} & & & 0 & & 0 & \\ & & & \downarrow & & \downarrow & \\ & & & \mathcal{K}(s)_\epsilon & \longrightarrow & \mathcal{K}(s) & \to 0 \\ & & & \downarrow & & \downarrow & \\ 0 \to & \epsilon\mathcal{H}(s) & \to & \mathcal{H}(s)[\epsilon] & \longrightarrow & \mathcal{H}(s) & \to 0 \\ & \downarrow & & \downarrow & & \downarrow & \\ 0 \to & \epsilon\mathcal{F}(s) & \to & \mathcal{F}(s)_\epsilon & \longrightarrow & \mathcal{F}(s) & \to 0 \\ & \downarrow & & \downarrow & & \downarrow & \\ & 0 & & 0 & & 0 & \end{array}$$

where the middle row is exact by the flatness of $\mathcal{H}$. Replacing the middle row by its pushout under $\epsilon\mathcal{H}(s) \to \epsilon\mathcal{F}(s)$ we see that this diagram is equivalent to the following one:

$$\begin{array}{ccccccc} & & & 0 & & 0 & \\ & & & \downarrow & & \downarrow & \\ & & & \mathcal{K}(s) & = & \mathcal{K}(s) & \to 0 \\ & & & \downarrow & & \downarrow & \\ 0 \to & \epsilon\mathcal{F}(s) & \to & \mathcal{P} & \longrightarrow & \mathcal{H}(s) & \to 0 \\ & \| & & \downarrow & & \downarrow & \\ 0 \to & \epsilon\mathcal{F}(s) & \to & \mathcal{F}(s)_\epsilon & \longrightarrow & \mathcal{F}(s) & \to 0 \\ & \downarrow & & \downarrow & & \downarrow & \\ & 0 & & 0 & & 0 & \end{array}$$

and therefore we deduce that $\ker(d\pi_q) = \mathrm{Hom}(\mathcal{K}(s), \mathcal{F}(s))$.

Now consider A in $\mathcal{A}$ and a commutative diagram

$$\begin{array}{ccc} A & \xleftarrow{\varphi} & \mathcal{O}_{Q,q} \\ \uparrow \eta & & \uparrow \pi_q^\sharp \\ B & \underset{\tilde{\varphi}}{\longleftarrow} & \mathcal{O}_{S,s} \end{array} \tag{4.17}$$

where η is a small extension in $\mathcal{A}$. This diagram corresponds to an exact diagram of sheaves on X:

$$
\begin{array}{ccccccccc}
 & & & & & & 0 & & \\
 & & & & & & \downarrow & & \\
 & & & & & & \mathcal{K}_A & & \\
 & & & & & & \downarrow & & \\
\gamma : & 0 \to & \mathcal{H}(s) & \to & \mathcal{H} \otimes_{\mathbf{k}} B & \longrightarrow & \mathcal{H} \otimes_{\mathbf{k}} A & \to 0 & \\
 & & & & & & \downarrow & & \\
 & & & & & & \mathcal{F}_A & & \\
 & & & & & & \downarrow & & \\
 & & & & & & 0 & &
\end{array}
\qquad (4.18)
$$

where the row is exact by the flatness of $\mathcal{H}$ over S. By pushing out by the quotient $f_s : \mathcal{H}(s) \to \mathcal{F}(s)$ and then pulling back by $\alpha : \mathcal{K}_A \to \mathcal{H} \otimes_{\mathbf{k}} A$ we obtain an element

$$[\alpha^* f_{s*}(\gamma)] \in \mathrm{Ext}^1_{\mathcal{O}_X \otimes A}(\mathcal{K}_A, \mathcal{F}(s)) = \mathrm{Ext}^1_{\mathcal{O}_{X(s)}}(\mathcal{K}(s), \mathcal{F}(s))$$

By construction this element vanishes if and only if the previous diagram can be embedded in a commutative diagram with exact rows and columns:

$$
\begin{array}{cccccccc}
 & 0 & & 0 & & 0 & \\
 & \downarrow & & \downarrow & & \downarrow & \\
0 \to & \mathcal{K}(s) & \to & \mathcal{K}_B & \to & \mathcal{K}_A & \to 0 \\
 & \downarrow & & \downarrow & & \downarrow & \\
0 \to & \mathcal{H}(s) & \to & \mathcal{H} \otimes_{\mathbf{k}} B & \longrightarrow & \mathcal{H} \otimes_{\mathbf{k}} A & \to 0 \\
 & \downarrow & & \downarrow & & \downarrow & \\
0 \to & \mathcal{F}(s) & \to & \mathcal{F}_B & \to & \mathcal{F}_A & \\
 & \downarrow & & \downarrow & & \downarrow & \\
 & 0 & & 0 & & 0 &
\end{array}
$$

The middle column of this diagram is an element of $Quot_{\mathcal{H}}^{X/S}(\mathrm{Spec}(B))$, which corresponds to a homomorphism $\varphi' : \mathcal{O}_{Q,q} \to B$ making the diagram

$$
\begin{array}{ccc}
A & \xleftarrow{\varphi} & \mathcal{O}_{Q,q} \\
\uparrow \eta & \swarrow \varphi' & \uparrow \pi_q^{\sharp} \\
B & \xleftarrow[\tilde{\varphi}]{} & \mathcal{O}_{S,s}
\end{array}
$$

commutative. Therefore we have associated an element of $\mathrm{Ext}^1_{\mathcal{O}_{X(s)}}(\mathcal{K}(s), \mathcal{F}(s))$ to each diagram (4.18). It is straightforward to check that this correspondence is linear.

Taking $\eta : \mathbf{k}[\epsilon] \to \mathbf{k}$ we get an inclusion

$$\mathrm{coker}(d\pi_q) \subset \mathrm{Ext}^1_{\mathcal{O}_{X(s)}}(\mathcal{K}(s), \mathcal{F}(s))$$

Taking any small extension η in $\mathcal{A}$ we can apply Proposition 2.1.7 to yield the conclusion. $\square$

Corollary 4.4.5. *Under the assumptions of 4.4.4, if*

$$\mathrm{Ext}^1_{\mathcal{O}_{X(s)}}(\mathcal{K}(s), \mathcal{F}(s)) = 0$$

then $\pi : Q \to S$ *is smooth at* q *of relative dimension* $\dim\big[\mathrm{Hom}(\mathcal{K}(s), \mathcal{F}(s))\big]$.

When $S = \mathrm{Spec}(\mathbf{k})$ we obtain the following "absolute" version of Proposition 4.4.4.

Corollary 4.4.6. *If* X *is a projective scheme,* $\mathcal{H}$ *a coherent sheaf on* X *and* $f : \mathcal{H} \to \mathcal{F}$ *a coherent quotient of* $\mathcal{H}$ *with ker*$(f) = \mathcal{K}$ *then, letting* $Q = \mathrm{Quot}^X_{\mathcal{H}}$, *we have:*

$$T_{[f]}Q = \mathrm{Hom}(\mathcal{K}, \mathcal{F})$$

and the obstruction space of $\mathcal{O}_{Q,[f]}$ *is a subspace of* $\mathrm{Ext}^1(\mathcal{K}, \mathcal{F})$.

In particular, if $\mathrm{Ext}^1(\mathcal{K}, \mathcal{F}) = 0$ *then* Q *is nonsingular of dimension* $\dim\big(\mathrm{Hom}(\mathcal{K}, \mathcal{F})\big)$ *at* $[f]$.

A special case of Proposition 4.4.4 is the following:

Proposition 4.4.7. *Let* $p : \mathcal{X} \to S$ *be a projective flat morphism of algebraic schemes, and* $\pi : \mathrm{Hilb}^{\mathcal{X}/S} \to S$ *the relative Hilbert scheme. For a closed point* $s \in S$ *let* $X = \mathcal{X}(s)$ *be the fibre over* s *and let* $Z \subset X$ *be a closed subscheme with ideal sheaf* $\mathcal{I} \subset \mathcal{O}_X$. *Then there is an exact sequence:*

$$0 \to H^0(Z, N_{Z/X}) \to T_{[Z]}\mathrm{Hilb}^{\mathcal{X}/S} \stackrel{d\pi_{[Z]}}{\longrightarrow} T_sS \to \mathrm{Ext}^1_{\mathcal{O}_X}(\mathcal{I}, \mathcal{O}_Z)$$

If, moreover, $Z \subset X$ *is a regular embedding then the above exact sequence becomes:*

$$0 \to H^0(Z, N_{Z/X}) \to T_{[Z]}\mathrm{Hilb}^{\mathcal{X}/S} \stackrel{d\pi_{[Z]}}{\longrightarrow} T_sS \to H^1(Z, N_{Z/X}) \qquad (4.19)$$

If $\mathrm{Ext}^1_{\mathcal{O}_X}(\mathcal{I}, \mathcal{O}_Z) = (0)$ *(resp.* $H^1(Z, N_{Z/X}) = (0)$ *in the case where* $Z \subset X$ *is a regular embedding) then* π *is smooth at* $[Z]$ *of relative dimension* $h^0(Z, N_{Z/X})$.

NOTES

1. One should compare the statement of 4.4.7 with 3.2.12, since the local relative Hilbert functor of Z in X relative to $X \to S$ is prorepresented by the local ring $\hat{\mathcal{O}}_{\mathrm{Hilb}^{X/S}_{[Z]}}$.

2. Our presentation of the Quot schemes is an adaptation of the one given in [90]. For a description of the sheaf of differentials of the Quot schemes see [117].

3. When $p : \mathcal{X} \to S$ is a projective flat morphism of integral schemes the relative Hilbert scheme $\mathrm{Hilb}^{\mathcal{X}/S}$ has a component isomorphic to S, parametrizing the fibres of p, and a component isomorphic to $\mathcal{X}$, identified with $\mathcal{X}^{[1]}$.

4.5 Flag Hilbert schemes

4.5.1 Existence

Fix an integer $r \geq 1$ and an m-tuple of numerical polynomials

$$\mathbf{P(t)} = (P_1(t), \ldots, P_m(t)), \qquad m \geq 1$$

For every scheme S we let:

$$FH^r_{\mathbf{P(t)}}(S) = \left\{ (\mathcal{X}_1, \ldots, \mathcal{X}_m) : \begin{array}{l} \mathcal{X}_1 \subset \cdots \subset \mathcal{X}_m \subset I\!P^r \times S \\ S\text{-flat closed subschemes} \\ \text{with Hilbert polynomials } \mathbf{P(t)} \end{array} \right\}$$

This clearly defines a contravariant functor:

$$FH^r_{\mathbf{P(t)}} : (\text{schemes})^\circ \to (\text{sets})$$

called the *flag Hilbert functor of* $I\!P^r$ *relative to* $\mathbf{P(t)}$. When $m = 1$ the flag Hilbert functors are just ordinary Hilbert functors.

Theorem 4.5.1. *For every* $r \geq 1$ *and* $\mathbf{P(t)}$ *as above, the flag Hilbert functor* $FH^r_{\mathbf{P(t)}}$ *is represented by a projective scheme* $\mathrm{FH}^r_{\mathbf{P(t)}}$, *called the* flag Hilbert scheme of $I\!P^r$ relative to $\mathbf{P(t)}$, *and by a universal family:*

$$\begin{array}{r} \mathcal{W}_1 \subset \cdots \subset \mathcal{W}_m \subset I\!P^r \times \mathrm{FH}^r_{\mathbf{P(t)}} \\ \downarrow \\ \mathrm{FH}^r_{\mathbf{P(t)}} \end{array}$$

Proof. We will only prove the theorem in the case $m = 2$, leaving to the reader the task of extending the proof to the general case.

By applying Corollary 4.1.10 twice we can find an integer μ such that simultaneously for $i = 1, 2$ we have that for every closed subscheme $\mathcal{X}_i \subset I\!P^r$ with Hilbert polynomial $P_i(t)$ the sheaf of ideals $\mathcal{I}_{\mathcal{X}_i}$ is μ-regular. For every such $\mathcal{X}_1$ and $\mathcal{X}_2$ we thus have:

$$h^0(I\!P^r, \mathcal{I}_{\mathcal{X}_i}(\mu)) = \binom{\mu + r}{r} - P_i(t) =: N_i \qquad i = 1, 2$$

Let $V = H^0(I\!P^r, \mathcal{O}_{I\!P^r}(\mu))$. Consider the Hilbert scheme $\mathrm{H}_i = \mathrm{Hilb}^r_{P_1(t)}$ with universal family $\mathcal{V}_i \subset I\!P^r \times \mathrm{H}_i$, $i = 1, 2$. On the product $\mathrm{H}_1 \times \mathrm{H}_2$ consider the pullback of the two universal families with respect to the projections:

$$\mathcal{V}_i \times_{\mathrm{H}_i} (\mathrm{H}_1 \times \mathrm{H}_2) \subset I\!P^r \times \mathrm{H}_1 \times \mathrm{H}_2, \qquad i = 1, 2$$

and denote by $q : I\!P^r \times \mathrm{H}_1 \times \mathrm{H}_2 \to \mathrm{H}_1 \times \mathrm{H}_2$ the projection. Because of the choice of μ and by Theorem 4.2.5 we have that $q_*\mathcal{I}_i(\mu)$ is a locally free subsheaf of $V \otimes_{\mathbf{k}} \mathcal{O}_{H_1 \times H_2}$ of rank N_i, with locally free cokernel, $i = 1, 2$. Consider the composition

$$\varphi : q_* \mathcal{I}_2(\mu) \subset V \otimes_{\mathbf{k}} \mathcal{O}_{H_1 \times H_2} \to V \otimes_{\mathbf{k}} \mathcal{O}_{H_1 \times H_2} / q_* \mathcal{I}_1(\mu)$$

and let $\mathbf{F} \subset H_1 \times H_2$ be the vanishing scheme of φ (Example 4.2.8). Note that we have

$$q_* \mathcal{I}_2(\mu) \otimes \mathcal{O}_{\mathbf{F}} \subset q_* \mathcal{I}_1(\mu) \otimes \mathcal{O}_{\mathbf{F}} \subset V \otimes \mathcal{O}_{\mathbf{F}} \tag{4.20}$$

We now pull back to $\mathbf{F}$ the two universal families, $i = 1, 2$:

$$\begin{array}{cccc} I\!P^r \times \mathbf{F} & & & I\!P^r \times H_i \\ \cup & & & \cup \\ \mathcal{W}_i & := \mathcal{V}_i \times_{H_i} \mathbf{F} & \to & \mathcal{V}_i \\ & \downarrow k & & \downarrow \\ & \mathbf{F} & \to & H_i \end{array}$$

Claim: $q_* \mathcal{I}_i(\mu) \otimes \mathcal{O}_{\mathbf{F}} = k_* \mathcal{I}_{\mathcal{W}_i}(\mu)$, $i = 1, 2$.

We have natural homomorphisms:

$$\beta_i : q_* \mathcal{I}_i(\mu) \otimes \mathcal{O}_{\mathbf{F}} \to k_* \mathcal{I}_{\mathcal{W}_i}(\mu)$$

Because of the μ-regularity and of Theorem 4.2.5 for every $x \in \mathbf{F}$ we have isomorphisms

$$q_* \mathcal{I}_i(\mu)_x \otimes \mathbf{k}(x) \cong H^0(I\!P^r, \mathcal{I}_{\mathcal{W}_i(x)}(\mu)) \cong k_* \mathcal{I}_{\mathcal{W}_i}(\mu)_x \otimes \mathbf{k}(x)$$

Since the sheaf $k_* \mathcal{I}_{\mathcal{W}_i}(\mu)$ is locally free, from Nakayama's lemma it follows that β_1 and β_2 are isomorphisms, and this proves the claim.

From the claim and from (4.20) we deduce that

$$k_* \mathcal{I}_{\mathcal{W}_2}(\mu) \subset k_* \mathcal{I}_{\mathcal{W}_1}(\mu) \tag{4.21}$$

The μ-regularity and Proposition 4.1.1(iii) imply that the natural homomorphisms

$$k^* k_* \mathcal{I}_{\mathcal{W}_i}(\mu) \to \mathcal{I}_{\mathcal{W}_i}(\mu), \qquad i = 1, 2$$

are surjective; because of (4.21) we deduce that $I_{\mathcal{W}_2} \subset \mathcal{I}_{\mathcal{W}_1}$, hence $\mathcal{W}_1 \subset \mathcal{W}_2$. Therefore $(\mathcal{W}_1, \mathcal{W}_2) \in FH^r_{\mathbf{P(t)}}(\mathbf{F})$.

Claim: The pair $(\mathbf{F}, (\mathcal{W}_1, \mathcal{W}_2))$ represents the functor $FH^r_{\mathbf{P(t)}}$.

Let S be a scheme and let $(\mathcal{X}_1, \mathcal{X}_2) \in FH^r_{\mathbf{P(t)}}(S)$. By definition $\mathcal{X}_1 \subset \mathcal{X}_2 \subset I\!P^r \times S$ are flat over S with Hilbert polynomials $P_1(t)$ and $P_2(t)$ respectively. Let $f : I\!P^r \times S \to S$ be the projection. We have induced classifying morphisms

$$g_1 : S \to H_1, \qquad g_2 : S \to H_2$$

which together define a morphism

$$g : S \to \mathrm{H}_1 \times \mathrm{H}_2$$

Arguing as before, we see that

$$g^* q_* \mathcal{I}_i(\mu) \cong f_* \mathcal{I}_{\mathcal{X}_i}(\mu), \qquad i = 1, 2 \tag{4.22}$$

The fact that $\mathcal{X}_1 \subset \mathcal{X}_2$ implies that $\mathcal{I}_{\mathcal{X}_2}(\mu) \subset \mathcal{I}_{\mathcal{X}_1}(\mu)$, hence that

$$f_* \mathcal{I}_{\mathcal{X}_2}(\mu) \subset f_* \mathcal{I}_{\mathcal{X}_1}(\mu)$$

This, together with (4.22), in turn implies that g factors through $\mathbf{F}$. Since clearly we have

$$\mathcal{X}_1 = S \times_{\mathbf{F}} \mathcal{W}_1, \qquad \mathcal{X}_2 = S \times_{\mathbf{F}} \mathcal{W}_2$$

the claim follows, hence $\mathbf{F} = \mathrm{FH}^r_{\mathbf{P(t)}}$. Moreover, $\mathrm{FH}^r_{\mathbf{P(t)}}$ is projective because it is a closed subscheme of $\mathrm{H}_1 \times \mathrm{H}_2$. □

From the definition it follows that the closed points of $\mathrm{FH}^r_{\mathbf{P(t)}}$ are in 1–1 correspondence with the m-tuples $(X_1, \dots, X_m)$ of closed subschemes of $\mathbb{P}^r$ such that X_i has Hilbert polynomial $P_i(t)$ and

$$X_1 \subset X_2 \subset \cdots \subset X_m$$

Such an m-tuple is called a *flag of closed subschemes* of $\mathbb{P}^r$. We will denote by $[X_1, \dots, X_m]$ the point of $\mathrm{FH}^r_{\mathbf{P(t)}}$ parametrizing such a flag. From the proof of Theorem 4.5.1 it follows that $\mathrm{FH}^r_{\mathbf{P(t)}}$ is a closed subscheme of

$$\prod_{i=1}^{m} \mathrm{Hilb}^r_{P_i(t)}$$

We will denote the projections by

$$pr_i : \mathrm{FH}^r_{\mathbf{P(t)}} \to \mathrm{Hilb}^r_{P_i(t)}, \qquad i = 1, \dots, m$$

For every subset $\mathbf{I} \subset \{1, \dots, m\}$ with cardinality μ we can consider the μ-tuple of polynomials $\mathbf{P_I(t)} = (P_{i_1}, \dots, P_{i_\mu})$ and the flag Hilbert scheme $\mathrm{FH}^r_{\mathbf{P_I(t)}}$. We have natural projection morphisms

$$pr_{\mathbf{I}} : \mathrm{FH}^r_{\mathbf{P(t)}} \to \mathrm{FH}^r_{\mathbf{P_I(t)}}$$

of which the pr_i's are special cases. The flag Hilbert schemes are generalizations of the *flag varieties* (*fibrés en drapeaux* in [77], ch. 1, § 9.9), which parametrize flags of linear subspaces of $\mathbb{P}^r$.

If $Z \subset \mathbb{P}^r$ is a closed subscheme having Hilbert polynomial $Q(t)$, and $\mathbf{P(t)}$ is an m-tuple of numerical polynomials as above, one can define the *flag Hilbert scheme of Z relative to* $\mathbf{P(t)}$

$$FH^Z_{\mathbf{P(t)}} : (\text{schemes})^\circ \to (\text{sets})$$

by an obvious modification of the definition of $FH^r_{\mathbf{P(t)}}$. It is straightforward to prove that $FH^Z_{\mathbf{P(t)}}$ is represented by a closed subscheme $\mathrm{FH}^Z_{\mathbf{P(t)}}$ of the scheme $\mathrm{FH}^r_{\tilde{\mathbf{P}}(\mathbf{t})}$, where

$$\tilde{\mathbf{P}}(\mathbf{t}) = (P_1(t), \ldots, P_m(t), Q(t))$$

Precisely, letting

$$pr_Q : FH^r_{\tilde{\mathbf{P}}(\mathbf{t})} \to \mathrm{Hilb}^r_{Q(t)}$$

be the projection, one has an identification of $\mathrm{FH}^Z_{\mathbf{P(t)}}$ with the scheme-theoretic fibre $pr_Q^{-1}([Z])$ of pr_Q over the point $[Z] \in \mathrm{Hilb}^r_{Q(t)}$. It is convenient to consider the disjoint union

$$\mathrm{FH}^Z = \coprod_{\mathbf{P(t)}} \mathrm{FH}^Z_{\mathbf{P(t)}}$$

which is a scheme locally of finite type, and call it the *flag Hilbert scheme of* Z.

Another variation on the same theme is the following. Given closed subschemes $X \subset Z \subset I\!P^r$, and an m-tuple of numerical polynomials $\mathbf{P(t)}$, one can consider flags of closed subschemes of Z containing X, namely m-tuples $(Y_1, \ldots, Y_m)$ of closed subschemes of Z having Hilbert polynomials $\mathbf{P(t)}$ and such that

$$X \subset Y_1 \subset Y_2 \cdots \subset Y_m \subset Z$$

Again there is an obvious generalization of the definition of the corresponding flag Hilbert functor and of the proof of its representability by a projective scheme. These generalized flag Hilbert schemes can be fruitfully used in concrete geometrical situations, like, for example, in the study of families of pointed closed subschemes, or of reducible closed subschemes, of a given projective scheme.

4.5.2 Local properties

Consider a projective scheme Z and let $(X_1, \ldots, X_m)$ be a flag of closed subschemes of Z; let $\mathcal{I}_i \subset \mathcal{O}_Z$ be the ideal sheaf of X_i. We have inclusions

$$\mathcal{I}_m \subset \mathcal{I}_{m-1} \subset \cdots \subset \mathcal{I}_1$$

and surjections

$$\mathcal{O}_{X_m} \to \mathcal{O}_{X_{m-1}} \to \cdots \to \mathcal{O}_{X_1} \to 0$$

Definition 4.5.2. *The* normal sheaf of $(X_1, \ldots, X_m)$ in Z *is the sheaf of germs of commutative diagrams of homomorphisms of* $\mathcal{O}_Z$*-modules of the following form:*

$$\begin{array}{ccccc} \mathcal{I}_m & \subset & \mathcal{I}_{m-1} & \subset \cdots \subset & \mathcal{I}_1 \\ \downarrow \sigma_m & & \downarrow \sigma_{m-1} & & \downarrow \sigma_1 \\ \mathcal{O}_{X_m} & \to & \mathcal{O}_{X_{m-1}} & \to \cdots \to & \mathcal{O}_{X_1} \end{array}$$

It is denoted by $N_{(X_1,\ldots,X_m)/Z}$.

Note that we have an obvious homomorphism of "projection":

$$N_{(X_1,\ldots,X_m)/Z} \to N_{(X_{i_1},\ldots,X_{i_\mu})/Z}$$

for each choice of a subset $\mathbf{I} = \{i_1, \ldots, i_\mu\} \subset \{1, \ldots, m\}$. In particular, we have homomorphisms:

$$N_{(X_1,\ldots,X_m)/Z} \to N_{X_i/Z}, \qquad i = 1, \ldots, m$$

Proposition 4.5.3. *Let Z be a projective scheme and let $(X_1, \ldots, X_m)$ be a flag of closed subschemes of Z. Then:*

(i) There is a natural identification:

$$T_{[X_1,\ldots,X_m]}\mathrm{FH}^Z = H^0(Z, N_{(X_1,\ldots,X_m)/Z}) \tag{4.23}$$

(ii) If $X_i \subset X_{i+1}$ is a regular embedding for all $i = 1, \ldots, m-1$ and X_m is regularly embedded in Z then the obstruction space of the local ring $\mathcal{O}_{\mathrm{FH}^Z,[X_1,\ldots,X_m]}$ is contained in $H^1(Z, N_{(X_1,\ldots,X_m)/Z})$.

Proof. For simplicity we give the proof in the case $m = 2$, i.e. in the case of a flag (X, Y) of closed subschemes of Z. The general case can be treated similarly.

(i) Let $\mathcal{I}_X \subset \mathcal{O}_Z$ and $\mathcal{I}_Y \subset \mathcal{O}_Z$ be the ideal sheaves of X and of Y. We can represent a first-order deformation $\mathcal{Y} \subset Z \times \mathrm{Spec}(\mathbf{k}[\epsilon])$ of Y in Z by an ideal sheaf $\mathcal{I}_{\mathcal{Y}} \subset \mathcal{O}_Z[\epsilon]$ fitting into the commutative and exact diagram:

$$\begin{array}{ccccccccc}
 & & & & 0 & & 0 & & \\
 & & & & \downarrow & & \downarrow & & \\
 & & & & \mathcal{I}_{\mathcal{Y}} & \to & \mathcal{I}_Y & \to & 0 \\
 & & & & \downarrow & & \downarrow & & \\
0 & \to & \epsilon\mathcal{O}_Z & \to & \mathcal{O}_Z[\epsilon] & \to & \mathcal{O}_Z & \to & 0 \\
 & & \downarrow & & \downarrow & & \downarrow & & \\
0 & \to & \epsilon\mathcal{O}_Y & \to & \mathcal{O}_{\mathcal{Y}} & \to & \mathcal{O}_Y & \to & 0
\end{array}$$

The flatness of $\mathcal{Y}$ over $\mathrm{Spec}(\mathbf{k}[\epsilon])$ follows from the exactness of the last row (Lemma A.9). This diagram is equivalent to the following one, deduced after pushing out the second row by the homomorphism $\epsilon\mathcal{O}_Z \to \epsilon\mathcal{O}_Y$:

$$\begin{array}{ccccccccc}
 & & & & 0 & & 0 & & \\
 & & & & \downarrow & & \downarrow & & \\
 & & & & \mathcal{I}_Y & = & \mathcal{I}_Y & \to & 0 \\
 & & & & \downarrow & & \downarrow & & \\
0 & \to & \epsilon\mathcal{O}_Y & \to & \epsilon\mathcal{O}_Y \oplus \mathcal{O}_Z & \to & \mathcal{O}_Z & \to & 0 \\
 & & \| & & \downarrow & & \downarrow & & \\
0 & \to & \epsilon\mathcal{O}_Y & \to & \mathcal{O}_{\mathcal{Y}} & \to & \mathcal{O}_Y & \to & 0
\end{array}$$

Therefore we have a convenient representation of the given first-order deformation as the middle vertical exact sequence of the previous diagram, which we rewrite:

$$0 \to \mathcal{I}_Y \to \epsilon\mathcal{O}_Y \oplus \mathcal{O}_Z \to \mathcal{O}_{\mathcal{Y}} \to 0$$

Composing the first inclusion with the first projection $\epsilon\mathcal{O}_Y \oplus \mathcal{O}_Z \to \epsilon\mathcal{O}_Y$ we obtain the section $\sigma_{\mathcal{Y}} \in H^0(Y, N_{Y/Z}) = \mathrm{Hom}_{\mathcal{O}_Z}(\mathcal{I}_Y, \mathcal{O}_Y)$ corresponding to $\mathcal{Y}$. Let's assume now that we have an element $(\mathcal{X}, \mathcal{Y}) \in \mathrm{FH}^Z(\mathbf{k}[\epsilon])$. By putting together the above constructions for $\mathcal{X}$ and for $\mathcal{Y}$ we obtain the following commutative diagram:

$$\begin{array}{ccccccccc} 0 \to & \mathcal{I}_Y & \to & \epsilon\mathcal{O}_Y \oplus \mathcal{O}_Z & \to & \mathcal{O}_{\mathcal{Y}} & \to 0 \\ & \cap & & \downarrow & & \downarrow & \\ 0 \to & \mathcal{I}_X & \to & \epsilon\mathcal{O}_X \oplus \mathcal{O}_Z & \to & \mathcal{O}_{\mathcal{X}} & \to 0 \end{array}$$

where the condition $\mathcal{X} \subset \mathcal{Y}$ corresponds to the condition $\mathcal{O}_{\mathcal{Y}} \to \mathcal{O}_{\mathcal{X}}$, and this is in turn equivalent to the condition that composing the inclusions on the left with the first projections in the middle terms we obtain a commutative diagram:

$$\begin{array}{ccc} \mathcal{I}_Y & \subset & \mathcal{I}_X \\ \downarrow \sigma_{\mathcal{Y}} & & \downarrow \sigma_{\mathcal{X}} \\ \epsilon\mathcal{O}_Y & \to & \epsilon\mathcal{O}_X \end{array}$$

This is the global section of $N_{(X,Y)/Z}$ corresponding to $(\mathcal{X}, \mathcal{Y})$ in (4.23). Conversely, given a global section

$$\begin{array}{ccc} \mathcal{I}_Y & \subset & \mathcal{I}_X \\ \downarrow \sigma & & \downarrow \tau \\ \epsilon\mathcal{O}_Y & \to & \epsilon\mathcal{O}_X \end{array} \quad \in H^0(Z, N_{(X,Y)/Z})$$

one finds a first-order deformation of (X, Y) by repeating backwards the above construction.

(ii) Let A be in $\mathrm{ob}(\mathcal{A})$ and let

$$\xi : \qquad \mathcal{X} \subset \mathcal{Y} \subset Z \times \mathrm{Spec}(A)$$

be an infinitesimal deformation of the flag (X, Y). Let

$$\eta : \qquad 0 \to \epsilon\mathbf{k} \to \tilde{A} \to A \to 0$$

define an element $[e] \in \mathrm{Ex}_{\mathbf{k}}(A, \mathbf{k})$, where $m_{\tilde{A}}\epsilon = 0$. Since the embeddings $X \subset Y$ and $Y \subset Z$ are regular we can find an affine open cover $\mathcal{U} = \{U_i = \mathrm{Spec}(P_i)\}_{i \in I}$ of Z such that $Y_i := Y \cap U_i$ is a complete intersection in U_i and $X_i := X \cap U_i$ is a complete intersection in Y_i. Therefore, letting $\mathcal{X}_i = \mathcal{X} \cap U_i$, $\mathcal{Y}_i = \mathcal{Y} \cap U_i$, there are liftings

$$\begin{array}{ccc} \tilde{\mathcal{Y}}_i & \subset & U_i \times \mathrm{Spec}(\tilde{A}) \\ \cup & & \cup \\ \mathcal{Y}_i & \subset & U_i \times \mathrm{Spec}(A) \end{array}$$

and

$$\begin{array}{ccc} \tilde{\mathcal{X}}_i & \subset & \tilde{\mathcal{Y}}_i \\ \cup & & \cup \\ \mathcal{X}_i & \subset & \mathcal{Y}_i \end{array}$$

which together give local liftings of the flags $(\mathcal{X}_i, \mathcal{Y}_i)$:

$$\begin{array}{ccccc} \tilde{\mathcal{X}}_i & \subset & \tilde{\mathcal{Y}}_i & \subset & U_i \times \mathrm{Spec}(\tilde{A}) \\ \cup & & \cup & & \cup \\ \mathcal{X}_i & \subset & \mathcal{Y}_i & \subset & U_i \times \mathrm{Spec}(A) \end{array}$$

In order to find a lifting $(\tilde{\mathcal{X}}, \tilde{\mathcal{Y}})$ of $(\mathcal{X}, \mathcal{Y})$ we must be able to choose the local liftings $(\tilde{\mathcal{X}}_i, \tilde{\mathcal{Y}}_i)$ so that on every $U_{ij} := U_i \cap U_j$ we have

$$(\tilde{\mathcal{X}}_{i|U_{ij}}, \tilde{\mathcal{Y}}_{i|U_{ij}}) = (\tilde{\mathcal{X}}_{j|U_{ij}}, \tilde{\mathcal{Y}}_{j|U_{ij}})$$

At the level of ideals we have:

$$\begin{array}{l} I_{Y_i} \subset I_{X_i} \subset P_i \\ I_{\mathcal{Y}_i} \subset I_{\mathcal{X}_i} \subset P_{iA} := P_i \otimes_{\mathbf{k}} A \\ I_{\tilde{\mathcal{Y}}_i} \subset I_{\tilde{\mathcal{X}}_i} \subset P_{i\tilde{A}} := P_i \otimes_{\mathbf{k}} \tilde{A} \end{array}$$

We have a commutative and exact diagram:

$$\begin{array}{ccccccc} & & & & I_{\tilde{\mathcal{X}}_i} & \rightarrow & I_{\mathcal{X}_i} & \rightarrow 0 \\ & & & & \downarrow & & \downarrow & \\ 0 \rightarrow & \epsilon P_i & \rightarrow & & P_{i\tilde{A}} & \rightarrow & P_{iA} & \rightarrow 0 \\ & \downarrow & & & \downarrow & & \downarrow & \\ 0 \rightarrow & \epsilon P_i / I_{X_i} & \rightarrow & & P_{i\tilde{A}} / I_{\tilde{\mathcal{X}}_i} & \rightarrow & P_{iA} / I_{\mathcal{X}_i} & \rightarrow 0 \end{array}$$

which, after pushing out the middle row by $\epsilon P_i \rightarrow \epsilon P_i / I_{X_i}$, gives the following one:

$$\begin{array}{cccccc} & & & I_{\mathcal{X}_i} & = & I_{\mathcal{X}_i} & \\ & & & \downarrow & & \downarrow & \\ 0 \rightarrow & \epsilon P_i / I_{X_i} & \rightarrow & Q_{iA} & \rightarrow & P_{iA} & \rightarrow 0 \\ & \| & & \downarrow & & \downarrow & \\ 0 \rightarrow & \epsilon P_i / I_{X_i} & \rightarrow & P_{i\tilde{A}} / I_{\tilde{\mathcal{X}}_i} & \rightarrow & P_{iA} / I_{\mathcal{X}_i} & \rightarrow 0 \end{array}$$

where

$$Q_{iA} = (\epsilon P_i / I_{X_i}) \coprod_{\epsilon P_i} P_{iA}$$

therefore the datum of $\tilde{\mathcal{X}}_i$ corresponds to the middle vertical sequence of this diagram:

$$0 \rightarrow I_{\mathcal{X}_i} \rightarrow Q_{iA} \rightarrow P_{i\tilde{A}} / I_{\tilde{\mathcal{X}}_i} \rightarrow 0 \tag{4.24}$$

Repeating the analogous construction for $\tilde{\mathcal{Y}}_i$ we obtain that $\tilde{\mathcal{Y}}_i$ is determined by an exact sequence:

$$0 \to I_{\mathcal{Y}_i} \to Q_{iA} \to P_{i\tilde{A}}/I_{\tilde{\mathcal{Y}}_i} \to 0 \tag{4.25}$$

Since $\tilde{\mathcal{X}}_i \subset \tilde{\mathcal{Y}}_i$ the exact sequences (4.24) and (4.25) fit together:

$$\begin{array}{ccccccc} 0 \to & I_{\mathcal{Y}_i} & \to & Q_{iA} & \to & P_{i\tilde{A}}/I_{\tilde{\mathcal{Y}}_i} & \to 0 \\ & \cap & & \| & & \downarrow & \\ 0 \to & I_{\mathcal{X}_i} & \to & Q_{iA} & \to & P_{i\tilde{A}}/I_{\tilde{\mathcal{X}}_i} & \to 0 \end{array}$$

A different choice of the lifting $\tilde{\mathcal{X}}_i$ corresponds to a different homomorphism $I_{\mathcal{X}_i} \to Q_{iA}$, and they differ by an element

$$\sigma_{\tilde{\mathcal{X}}_i} \in \mathrm{Hom}_{P_{i\tilde{A}}}(I_{\mathcal{X}_i}, \epsilon P_i/I_{X_i}) = \mathrm{Hom}_{P_i}(I_{X_i}, P_i/I_{X_i})$$

Similarly, a different choice of the lifting $\tilde{\mathcal{Y}}_i$ corresponds to an element

$$\sigma_{\tilde{\mathcal{Y}}_i} \in \mathrm{Hom}_{P_i}(I_{Y_i}, P_i/I_{Y_i})$$

The condition that the pair $(\sigma_{\tilde{\mathcal{X}}_i}, \sigma_{\tilde{\mathcal{Y}}_i})$ defines another lifting *of the flag* $(\mathcal{X}_i, \mathcal{Y}_i)$ is that the following diagram

$$\begin{array}{ccc} I_{Y_i} & \subset & I_{X_i} \\ \downarrow \sigma_{\tilde{\mathcal{Y}}_i} & & \downarrow \sigma_{\tilde{\mathcal{X}}_i} \\ P_i/I_{Y_i} & \to & P_i/I_{X_i} \end{array}$$

commutes. This is precisely the condition that $(\sigma_{\tilde{\mathcal{X}}_i}, \sigma_{\tilde{\mathcal{Y}}_i}) \in \Gamma(U_i, N_{(X,Y)/Z})$. Therefore the set of liftings of $(\mathcal{X}_i, \mathcal{Y}_i)$ over $\mathrm{Spec}(\tilde{A})$ is in 1–1 correspondence with $\Gamma(U_i, N_{(X,Y)/Z})$. It follows from this analysis that for all $i, j \in I$ we have a section $(\sigma_{ij}, \tau_{ij}) \in \Gamma(U_{ij}, N_{(X,Y)/Z})$ such that $(\tilde{\mathcal{X}}_{j|U_{ij}}, \tilde{\mathcal{Y}}_{j|U_{ij}})$ is obtained from $(\tilde{\mathcal{X}}_{i|U_{ij}}, \tilde{\mathcal{Y}}_{i|U_{ij}})$ after modifying it by the section (σ_{ij}, τ_{ij}). The collection of these sections is a 1-cocycle

$$\{(\sigma_{ij}, \tau_{ij})\} \in \mathcal{Z}^1(\mathcal{U}, N_{(X,Y)/Z})$$

which defines a cohomology class $o_{(\mathcal{X},\mathcal{Y})}([e])$. It is straightforward to verify that this class is independent of the choices made and that $o_{(\mathcal{X},\mathcal{Y})}([e]) = 0$ if and only if a lifting $(\tilde{\mathcal{X}}, \tilde{\mathcal{Y}})$ exists. □

If a flag $(X_1, \ldots, X_m)$ of closed subschemes of a projective scheme Z satisfies the conditions of Proposition 4.5.3(ii), i.e; if $X_i \subset X_{i+1}$ is a regular embedding for all $i = 1, \ldots, m-1$ and X_m is regularly embedded in Z, we say that the flag $(X_1, \ldots, X_m)$ *is regularly embedded in* Z.

Remarks 4.5.4. (i) One can adapt the above proof to the case $m = 1$ to obtain another, more intrinsic, proof of Propositions 3.2.1(ii) and 3.2.6.

(ii) In the case $m = 2$ considered in the proof of Proposition 4.5.3, to the flag $X \subset Y$ of closed subschemes of Z there is associated a diagram of normal sheaves:

$$\begin{array}{ccccc} & & & & N_{Y/Z} \\ & & & & \downarrow \\ 0 \to N_{X/Y} & \to & N_{X/Z} & \to & N_{Y/Z} \otimes_{\mathcal{O}_Z} \mathcal{O}_X \end{array}$$

and we can immediately check that there is a natural identification

$$H^0(Z, N_{(X,Y)/Z}) = H^0(X, N_{X/Z}) \times_{H^0(X, N_{Y/Z} \otimes_{\mathcal{O}_Z} \mathcal{O}_X)} H^0(Y, N_{Y/Z})$$

(iii) It is an immediate consequence of the definition that, given a flag $(X_1, \ldots, X_m)$ of closed subschemes of a projective scheme Z, the normal sheaf $N_{(X_1,\ldots,X_m)/Z}$ is an $\mathcal{O}_{X_m}$-module. This implies that in the case when we have a regularly embedded flag consisting of 0-dimensional subschemes of Z, we have

$$H^1(Z, N_{(X_1,\ldots,X_m)/Z}) = H^1(X_m, N_{(X_1,\ldots,X_m)/Z}) = 0$$

and from Proposition 4.5.3(ii) we deduce that FH^Z is nonsingular at $[X_1, \ldots, X_m]$. For example, if we consider a projective nonsingular curve Z then it follows that $\mathrm{FH}^Z_{(n_1,\ldots,n_m)}$, which parametrizes flags of effective divisors $(D_1, \ldots, D_m)$ of degrees $n_1 < n_2 < \cdots < n_m$, is nonsingular.

(iv) It is possibile to give a notion of *flag Quot scheme* generalizing the flag Hilbert schemes. This seems not to have been considered in the literature yet.

NOTES

1. The flag Hilbert schemes have been considered in [102]. The proof of Theorem 4.5.1 given here has appeared in [163]. More recent references about flag Hilbert schemes are [30] and [62] (where they are called "nested Hilbert schemes") and [151].

4.6 Examples and applications

4.6.1 Complete intersections

We have already discussed some properties of the local Hilbert functor of a complete intersection $X \subset \mathbb{P}^r$, which of course correspond to local properties of the Hilbert scheme Hilb^r at $[X]$. It is easy to check that, despite the fact that $H^1(X, N_X) \neq (0)$ in general, every complete intersection X is unobstructed in $\mathbb{P}^r$.

We may assume $\dim(X) > 0$. Let's suppose that $X \subset \mathbb{P}^r$, $r \geq 2$, is the complete intersection of $r - n$ hypersurfaces $f_1, \ldots, f_{r-n}$ of degrees $d_1 \leq d_2 \leq \ldots \leq d_{r-n}$ respectively, $n < r$.

Consider a basis $\Phi^{(1)}, \ldots, \Phi^{(m)}$ of $\oplus_j H^0(\mathbb{P}^r, \mathcal{O}(d_j))$ where

$$\Phi^{(h)} = (\phi_1^{(h)}, \ldots, \phi_{r-n}^{(h)})$$

$h = 1, \ldots, m$, and the $\phi_j^{(h)} \in \mathbf{k}[X_0, \ldots, X_r]$. Consider indeterminates $u_1, \ldots, u_m$ and the $(r-n)$-tuple

$$\mathbf{f} + \sum_{h=1}^{m} u_h \Phi^{(h)} = (f_1 + u_1\phi_1^{(1)} + \cdots + u_m\phi_1^{(m)}, \ldots, f_{r-n} + u_1\phi_{r-n}^{(1)} + \cdots + u_m\phi_{r-n}^{(m)}) \tag{4.26}$$

of elements of the polynomial ring $\mathbf{k}[\underline{u}, \underline{x}] = \mathbf{k}[u_1, \ldots, u_m, X_0, \ldots, X_r]$.

Let $K_\bullet(\mathbf{f} + \sum_h u_h \Phi^{(h)})$ be the Koszul complex relative to (4.26) and

$$\Delta := \mathrm{Supp}[H_1(K_\bullet(\mathbf{f} + \sum_h u_h \Phi^{(h)}))] \subset \mathbf{A}^{m+r+1} = \mathrm{Spec}(\mathbf{k}[\underline{u}, \underline{x}])$$

Denoting by $p : \mathbf{A}^{m+r+1} \to \mathbf{A}^m$ the projection, $U := \mathbf{A}^m \backslash p(\Delta)$ is the set of points $\underline{u} \in \mathbf{A}^m$ such that $K_\bullet(\mathbf{f} + \sum_h t_h \Phi^{(h)})$ is exact; U is an open set containing the origin.

In $\mathbb{P}^r \times \mathbf{A}^m$ consider the closed subscheme

$$\mathcal{X} = \mathrm{Proj}\Big(\mathbf{k}[\underline{u}, \underline{x}]/(f_1 + u_1\phi_1^{(1)} + \cdots + u_m\phi_1^{(m)}, \ldots, f_{r-n} + u_1\phi_{r-n}^{(1)} + \cdots + u_m\phi_{r-n}^{(m)})\Big)$$

the projection $\pi : \mathcal{X} \to \mathbf{A}^m$ and its restriction $\pi_U : \mathcal{X}_U \to U$, where $\mathcal{X}_U := \pi^{-1}(U)$. All the fibres of π_U are complete intersections of multidegree $(d_1, \ldots, d_{r-n})$ and $\mathcal{X}(\underline{0}) = X$. The Hilbert polynomial of a complete intersection depends only on its multidegree because it can be computed using the Koszul complex: it follows that all the fibres of π_U have the same Hilbert polynomial $P(t)$ and therefore π_U is a flat family of deformations of X in $\mathbb{P}^r$. In an obvious way the tangent space of U at $\underline{0}$ can be identified with $\oplus_j H^0(\mathbb{P}^r, \mathcal{O}(d_j))$, and the characteristic map with the restriction

$$\varphi : \oplus_j H^0(\mathbb{P}^r, \mathcal{O}(d_j)) \to \oplus_j H^0(X, \mathcal{O}_X(d_j))$$

Since $\dim(X) > 0$ the map φ is surjective, as one easily verifies using the Koszul complex; since, moreover, U is nonsingular at $\underline{0}$ from Proposition 4.3.6 it follows that $\mathrm{Hilb}^r_{P(t)}$ is smooth at $[X]$ and the classifying map is smooth. From this it also follows that complete intersections are parametrized by an open subset of $\mathrm{Hilb}^r_{P(t)}$.

It is interesting to observe that the closure of this open set may contain points parametrizing nonsingular subschemes of $\mathbb{P}^r$ which are not complete intersections. An example of such a subscheme is given by a trigonal canonical curve $C \subset \mathbb{P}^4$: the quadrics containing C intersect in a rational cubic surface S, so it is not a complete intersection since it has degree 8; but $[C]$ is in the closure of the family of complete intersections of three quadrics. It is apparently unknown whether a similar phenomenon may occur in $\mathbb{P}^3$, namely whether there are nonsingular curves in $\mathbb{P}^3$ which are flat limits of complete intersections without being complete intersections. See [51] for more about this.

The Kodaira–Spencer map of the families π_U has been studied in [161] in the case of complete intersections of dimension ≥ 2: π_U has general moduli except for surfaces of multidegrees (4), (2, 3), (2, 2, 2) (respectively in $\mathbb{P}^3$, in $\mathbb{P}^4$ and in $\mathbb{P}^5$), i.e. for complete intersection K3-surfaces. The special case of hypersurfaces had already been considered in [107] (see Example 3.2.11(i), page 134).

4.6.2 An obstructed nonsingular curve in $\mathbb{P}^3$

We will prove that the Hilbert scheme $\text{Hilb}^{\mathbb{P}^3}$ has an everywhere nonreduced component Σ whose general point parametrizes a nonsingular curve of degree 14 and genus 24. It will follow that every curve parametrized by a general point of Σ is obstructed in $\mathbb{P}^3$. This example is due to Mumford ([129]).

A general element of Σ is constructed as follows. Let $F \subset \mathbb{P}^3$ be a nonsingular cubic surface, $E, H \subset F$ respectively a line and a plane section in F. Let $C \subset F$ be a general member of the linear system $|4H+2E|$. Using Bertini's theorem one easily checks that C is irreducible and nonsingular; its degree and genus are $(C \cdot H) = 14$ and $\frac{1}{2}(C - H \cdot C) + 1 = 24$. From the exact sequence:

$$0 \to \mathcal{K}_C(H) \to N_C \to \mathcal{O}_C(3H) \to 0$$

we see that

$$h^1(C, N_C) = h^1(C, \mathcal{O}_C(3H)) = h^0(C, \mathcal{K}_C(-3H)) = h^0(C, \mathcal{O}_C(2E)) = 1$$

where the last equality follows easily from the exact sequence

$$0 \to \mathcal{O}_F(-4H) \to \mathcal{O}_F(2E) \to \mathcal{O}_C(2E) \to 0$$

and from $h^0(\mathcal{O}_F(-4H)) = 0 = h^1(\mathcal{O}_F(-4H))$ and $h^0(\mathcal{O}_F(2E)) = h^0(\mathcal{O}_F(E)) = 1$. Moreover, the linear system $|C| = |4H + 2E|$ has dimension

$$\dim(|C|) = 1 + \dim(|C|_C) = h^0(C, \mathcal{K}_C(H)) = 37$$

and therefore, since every curve C is contained in a unique cubic surface (because $9 < 14$), the dimension of the family W of all curves C we are considering is $19 + 37 = 56 = 4 \cdot 14$ but they satisfy $h^0(C, N_C) = 56 + h^1(C, N_C) = 57$. We will prove that W is an open set of a component Σ of $\text{Hilb}^{\mathbb{P}^3}$ and this will imply that Σ is everywhere nonreduced. Our assertion will be proved if we show that our curves C are not contained in a family whose general member is a curve D not contained in a cubic surface. But on every such D the line bundle $\mathcal{O}_D(4)$ is nonspecial and therefore, by Riemann-Roch, $h^0(D, \mathcal{O}_D(4)) = 33$, hence D is contained in a pencil of quartic surfaces. Let G_1, G_2 be two linearly independent quartics containing D: they are both irreducible because otherwise D would be either contained in a plane or in a quadric, which is not the case because there are no nonsingular curves of degree 14 and genus 24 on such surfaces. We have $G_1 \cap G_2 = D \cup q$ where q is a conic; since q has at most double points D has at most triple points and therefore G_1 and G_2 cannot be simultaneously singular at any point of D, thus the general quartic surface G containing D is nonsingular along D. By applying Riemann-Roch on G we obtain $\dim(|D|_G) = 24$. Therefore, since G is not a general quartic surface (because D is not a complete intersection), we see that the family of pairs (D, G) has dimension $\leq 33 + 24 = 57$ so that the family Z of curves D has dimension ≤ 56. This shows that the family W, which has dimension 56, cannot be in the closure of Z and this proves the assertion.

It is instructive to observe that we can write the linear system $|C|$ on a nonsingular cubic surface F as $|4H+2E| = |6H-2(H-E)|$ and this means that we can find a sextic surface F_6 such that $F \cap F_6 = C \cup q_1 \cup q_2$ where q_1 and q_2 are disjoint conics; if $[C] \in \Sigma$ is general then one can show that q_1, q_2 and F_6 can be chosen to be nonsingular.

There is another component R of $\mathrm{Hilb}^{\mathbb{P}^3}$ whose general point parametrizes a nonsingular curve C' of degree 14 and genus 24 such that

$$C' \cup E \cup \Gamma = F_3 \cap F_6$$

where E is a line and Γ is a rational normal cubic which are disjoint. We have in this case $|C'| = |6H - E - \Gamma|$ and

$$\begin{aligned} & h^1(C', N_{C'}) = h^1(C', \mathcal{O}_{C'}(3H)) = h^0(C', K_{C'}(-3H)) \\ &= h^0(C', \mathcal{O}_{C'}(2H - E - \Gamma)) = h^0(F_3, \mathcal{O}_{F_3}(2H - E - \Gamma)) = 0 \end{aligned}$$

Thus C' is unobstructed.

We refer the reader to [41] for another point of view about this example. This is the first published example of an obstructed space curve. Many others have since appeared in the literature (see [78], [162], [79], [50], [103], [187], [22], [125], [80]). A final word on the search for pathologies of this kind is contained in [183], where it is shown that virtually every singularity can appear as a point of $\mathrm{Hilb}^{\mathbb{P}^3}$ parametrizing a nonsingular curve. For examples of obstructed nonsingular curves in higher dimensional projective spaces see [35], [55].

4.6.3 An obstructed (nonreduced) scheme

In $\mathbb{P}^3$ consider the scheme

$$X = \mathrm{Proj}\Big(\mathbf{k}[X_0, \ldots, X_3]/J\Big)$$

where

$$J = (X_1X_2, X_1X_3, X_2X_3, X_3^2)$$

X is supported on the reducible conic defined by the equations

$$X_1X_2 = 0, \quad X_3 = 0$$

has an embedded point at $(1,0,0,0)$ and has Hilbert polynomial $2(t+1)$ (see Example 4.2.3(ii)). As in 4.2.3(ii) we consider the flat family parametrized by $\mathbf{A}^1$:

$$\mathcal{X} = \mathrm{Proj}\Big(\mathbf{k}[u, \underline{X}]/(X_1X_2, X_1X_3, X_2(X_3 - uX_0), X_3(X_3 - uX_0))\Big) \subset \mathbb{P}^3 \times \mathbf{A}^1$$

where $\mathbf{k}[u, \underline{X}] = \mathbf{k}[u, X_0, \ldots, X_3]$. We have $X = \mathcal{X}(0)$. If $u \neq 0$ then $\mathcal{X}(u)$ is a pair of disjoint lines. Let

$$g : \mathbf{A}^1 \to \mathrm{Hilb}^3_{2(t+1)}$$

be the classifying map. If $u \neq 0$ we have

$$h^1(\mathcal{X}(u), N_{\mathcal{X}(u)}) = 0; \quad h^0(\mathcal{X}(u), N_{\mathcal{X}(u)}) = 8$$

Therefore $g(u)$ is a smooth point and the tangent space has dimension 8.

In order to show that X is obstructed it suffices to show that

$$h^0(X, N_X) > 8 \tag{4.27}$$

because $g(0)$ and $g(u)$ belong to the same irreducible component of $\mathrm{Hilb}^3_{2(t+1)}$.

Consider the surjection

$$\mathbf{f} : \mathcal{O}_{I\!P^3}(-2)^{\oplus 4} \to \mathcal{I}_X \to 0$$

determined by the four equations of degree 2 which define X. Elementary computations, based on the fact that the generators of the ideal are monomials, lead to the following resolution of $\mathcal{O}_X$ which extends $\mathbf{f}$:

$$0 \to \mathcal{O}_{I\!P^3}(-4) \xrightarrow{B} \mathcal{O}_{I\!P^3}(-3)^{\oplus 4} \xrightarrow{A} \mathcal{O}_{I\!P^3}(-2)^{\oplus 4} \xrightarrow{\mathbf{f}} \mathcal{O}_{I\!P^3} \to \mathcal{O}_X \to 0 \tag{4.28}$$

A and B being given by the following matrices:

$$B = \begin{pmatrix} X_3 \\ -X_3 \\ X_2 \\ -X_1 \end{pmatrix} \qquad A = \begin{pmatrix} X_3 & X_3 & 0 & 0 \\ -X_2 & 0 & X_3 & 0 \\ 0 & -X_1 & 0 & X_3 \\ 0 & 0 & -X_1 & -X_2 \end{pmatrix}$$

By taking $Hom(-, \mathcal{O}_X)$ we obtain the following exact sequence:

$$0 \to N_X \to \mathcal{O}_X(2)^{\oplus 4} \xrightarrow{{}^tA} \mathcal{O}_X(3)^{\oplus 4}$$

from which we deduce that

$$H^0(X, N_X) = \ker[H^0(\mathcal{O}_X(2))^{\oplus 4} \xrightarrow{{}^tA} H^0(X, \mathcal{O}_X(3))^{\oplus 4}] \tag{4.29}$$

Using resolution (4.28) it is easy to show that the restriction maps

$$\varphi_n : H^0(I\!P^3, \mathcal{O}(n)) \to H^0(X, \mathcal{O}_X(n))$$

are surjective if $n \geq 2$. This allows us to identify $H^0(X, \mathcal{O}_X(2))$ and $H^0(X, \mathcal{O}_X(3))$ with the homogeneous parts of degree 2 and 3 respectively of $\mathbf{k}[X_0, \ldots, X_3]/J$. Hence using (4.29) we can represent $H^0(X, N_X)$ by 4-tuples of polynomials. Precisely, $H^0(X, N_X)$ is, modulo J, the vector space of 4-tuples

$$\underline{q} = (q_1, q_2, q_3, q_4)$$

of homogeneous polynomials of degree 2 such that $A\,{}^t\underline{q} \in (J_3)^4$. It is easy to find all of them because J is generated by monomials. Computing, one finds that a basis of $H^0(X, N_X)$ is defined by the following column vectors:

$$\begin{matrix} X_1^2 & X_1X_0 & X_2^2 & X_2X_0 & X_3X_0 & 0 & 0 & 0 & 0 & 0 & 0 & 0 \\ 0 & 0 & 0 & 0 & 0 & X_1^2 & X_1X_0 & X_3X_0 & 0 & 0 & 0 & 0 \\ 0 & 0 & 0 & 0 & 0 & 0 & 0 & 0 & X_2^2 & X_2X_0 & X_3X_0 & 0 \\ 0 & 0 & 0 & 0 & 0 & 0 & 0 & 0 & 0 & 0 & 0 & X_3X_0 \end{matrix}$$

In particular, we see that $h^0(X, N_X) = 12$, and this proves (4.27).

A little extra work shows that $[X] = g(0)$ belongs to two irreducible components of $\mathrm{Hilb}^3_{2(t+1)}$. We already know one of them of dimension 8: it contains $g(u)$, $u \neq 0$, and a general point of it parametrizes a pair of disjoint lines.

The other component has dimension 11 and a general point of it parametrizes the disjoint union $Y = Q \cup \{p\}$ of a conic Q and a point p. Note that

$$h^0(Y, N_Y) = h^0(Q, N_Q) + h^0(p, N_p) = 8 + 3 = 11$$

and $h^1(Y, N_Y) = 0$. Hence Y is a smooth point of a component of dimension 11 of $\mathrm{Hilb}^3_{2(t+1)}$. Therefore it suffices to produce a flat family parametrized by an irreducible curve, e.g. $\mathbf{A}^1$,

$$\mathcal{Y} \subset I\!P^3 \times \mathbf{A}^1$$

such that $\mathcal{Y}(0) = X$, $\mathcal{Y}(1) = Y$. Here it is:

$$\mathcal{Y} = \mathrm{Proj}\Big(\mathbf{k}[\upsilon, X_0, X_1, X_2, X_3]/\mathbf{I}\Big)$$

where

$$\mathbf{I} = (X_1X_2,\ X_1X_3 + \upsilon X_1X_0,\ X_2X_3 + \upsilon X_2X_0,\ X_3^2 - \upsilon^2X_0^2)$$

Clearly $\mathcal{Y}(0) = X$; since

$$\mathbf{I} = (X_1, X_2, X_3 - \upsilon X_0) \cap (X_3 + \upsilon X_0, X_1X_2)$$

it follows that for all $\upsilon \neq 0$, $\mathcal{Y}(\upsilon)$ is the disjoint union of a conic and a point. The flatness of $\mathcal{Y}$ follows from [84], Prop. III.9.8.

This example shows that in general the Hilbert schemes are reducible and not equidimensional. For a description of Hilb^3_{3t+1}, which presents several analogies with the one given here of $\mathrm{Hilb}^3_{2(t+1)}$, we refer to [142].

4.6.4 Relative grassmannians and projective bundles

Consider a coherent sheaf $\mathcal{E}$ on an algebraic scheme S, and let $P(t) = n$, where n is a positive integer, be a constant polynomial. Then $\mathrm{Quot}^{S/S}_{\mathcal{E},n}$ is a projective S-scheme which will be denoted by $\mathrm{Quot}_n(\mathcal{E})$ in what follows. We will denote by

$$\rho : \mathrm{Quot}_n(\mathcal{E}) \to S$$

the structural projective morphism and by

$$\rho^*\mathcal{E} \to \mathcal{Q}$$

the universal quotient bundle; $\mathcal{Q}$ is locally free of rank n. If $n = 1$ then $\mathcal{Q}$ is an invertible sheaf: it will be denoted by $\mathcal{O}_{\mathrm{Quot}_1(\mathcal{E})}(1)$, or simply by $\mathcal{O}(1)$ if no confusion arises. The pair $(\mathrm{Quot}_n(\mathcal{E})/S, \mathcal{Q})$ represents the functor

$$\mathcal{Q}uot_n(\mathcal{E}) : (\text{schemes}/S)^\circ \to (\text{sets})$$

defined by:

$$\mathcal{Q}uot_n(\mathcal{E})(f : T \to S) = \{\text{locally free rk } n \text{ quotients } f^*\mathcal{E} \to \mathcal{F} \text{ on } T\}$$

On $\mathrm{Quot}_n(\mathcal{E})$ we have a *tautological exact sequence*

$$0 \to \mathcal{K} \to \rho^*(\mathcal{E}) \to \mathcal{Q} \to 0$$

If $\mathcal{E}$ is locally free then $\mathcal{K}$ is locally free as well and it is called the *universal subbundle*.

If $\mathcal{E}$ is locally free we define

$$G_n(\mathcal{E}) := \mathrm{Quot}_n(\mathcal{E}^\vee)$$

and call it the *grassmannian bundle* of subbundles of rank n of $\mathcal{E}$; for $n = 1$ we have

$$G_1(\mathcal{E}) = \mathbb{P}(\mathcal{E})$$

the *projective bundle* associated to $\mathcal{E}$, according to our definition (which differs from the one adopted in [84], p. 162). The tautological exact sequence on $\mathbb{P}(\mathcal{E})$ is:

$$0 \to \mathcal{K} \to \rho^*(\mathcal{E}^\vee) \to \mathcal{O}_{\mathbb{P}(\mathcal{E})}(1) \to 0$$

In particular, for a finite-dimensional **k**-vector space V we have

$$\mathbb{P}(V \otimes_{\mathbf{k}} \mathcal{O}_S) = \mathbb{P}(V) \times S$$

and more generally,

$$G_n(V \otimes_{\mathbf{k}} \mathcal{O}_S) = G_n(V) \times S$$

Therefore, if $\mathcal{E}$ is locally free on S then $\mathrm{Quot}_n(\mathcal{E})$ is locally the product of S by a grassmannian; in particular, the projection $\rho : \mathrm{Quot}_n(\mathcal{E}) \to S$ is a smooth morphism.

Proposition 4.6.1. *Let $\mathcal{E}$ be a locally free sheaf on the algebraic scheme S, and let*

$$0 \to \mathcal{K} \to \rho^*(\mathcal{E}) \to \mathcal{Q} \to 0 \tag{4.30}$$

be the tautological exact sequence on $\mathrm{Quot}_n(\mathcal{E})$ *for some* $1 \le n \le \mathrm{rk}(\mathcal{E})$. *Then there is a natural isomorphism*

$$\Omega^1_{\mathrm{Quot}_n(\mathcal{E})/S} \cong \mathcal{H}om(\mathcal{Q}, \mathcal{K})$$

and therefore

$$T_{\mathrm{Quot}_n(\mathcal{E})/S} \cong \mathcal{H}om(\mathcal{K}, \mathcal{Q})$$

Proof. Letting $B = \mathrm{Quot}_n(\mathcal{E})$ consider the product $B \times_S B$ with projections $\mathrm{pr}_i : B \times_S B \to B$, $i = 1, 2$, and let $\mathcal{E}_{B\times_S B}$ be the pullback of $\mathcal{E}$ on $B \times_S B$. Denote by $\mathcal{I}_\Delta \subset \mathcal{O}_{B\times_S B}$ the ideal sheaf of the diagonal $\Delta \subset B \times_S B$. The tautological exact sequence (4.30) pulls back to two exact sequences:

$$0 \to \mathrm{pr}_i^*\mathcal{K} \to \mathcal{E}^\vee_{B\times_S B} \to \mathrm{pr}_i^*\mathcal{Q} \to 0$$

on $B \times_S B$ whose restrictions to Δ coincide, and Δ is characterized by this property. This can be also expressed by saying that Δ is the vanishing scheme of the composition

$$\mathrm{pr}_1^*\mathcal{K} \to \mathcal{E}^\vee_{B\times_S B} \to \mathrm{pr}_2^*\mathcal{Q}$$

Therefore we have a surjective homomorphism:

$$Hom(\mathrm{pr}_2^*\mathcal{Q}, \mathrm{pr}_1^*\mathcal{K}) \to \mathcal{I}_\Delta$$

(see 4.2.8) which, restricted to Δ, gives a surjective homomorphism:

$$Hom(\mathcal{Q}, \mathcal{K}) \to \mathcal{I}_\Delta/\mathcal{I}_\Delta^2 = \Omega^1_{B/S}$$

which has to be an isomorphism since both sheaves are locally free and have the same rank. □

Proposition 4.6.2. *Let*

$$0 \to \mathcal{E} \xrightarrow{\alpha} \mathcal{F} \xrightarrow{\beta} \mathcal{G} \to 0$$

be an exact sequence of locally free sheaves on the algebraic scheme S, and $n \geq 1$ an integer. Then there is a closed regular embedding

$$\mathrm{Quot}_n(\mathcal{G}) \subset \mathrm{Quot}_n(\mathcal{F})$$

and a natural identification:

$$N_{\mathrm{Quot}_n(\mathcal{G})/\mathrm{Quot}_n(\mathcal{F})} = \rho^*\mathcal{E}^\vee \otimes \mathcal{Q} \otimes \mathcal{O}_{\mathrm{Quot}_n(\mathcal{G})}$$

where $\rho : \mathrm{Quot}_n(\mathcal{F}) \to S$ is the structure morphism and $\rho^*\mathcal{F} \to \mathcal{Q}$ is the universal quotient.

In particular, if $n = 1$ we have

$$N_{\mathrm{Quot}_1(\mathcal{G})/\mathrm{Quot}_1(\mathcal{F})} = \rho^*\mathcal{E}^\vee \otimes \mathcal{O}(1) \otimes \mathcal{O}_{\mathrm{Quot}_1(\mathcal{G})}$$

Proof. Let $f : T \to S$ be a morphism. For every locally free rank n quotient

$$(f^*\mathcal{G} \to \mathcal{H}) \in Quot_n(\mathcal{G})(T)$$

there is associated, by composition with the surjective homomorphism $f^*(\beta) : f^*\mathcal{F} \to f^*\mathcal{G}$, an element

$$(f^*\mathcal{F} \to \mathcal{H}) \in Quot_n(\mathcal{F})(T)$$

Therefore $Quot_n(\mathcal{G})$ is a subfunctor of $Quot_n(\mathcal{F})$. Consider the diagram of homomorphisms on $\mathrm{Quot}_n(\mathcal{F})$:

$$\begin{array}{ccc} \rho^*(\mathcal{E}) & & \\ \downarrow \rho^*(\alpha) & & \\ \rho^*(\mathcal{F}) & \xrightarrow{\gamma} & \mathcal{Q} \to 0 \\ \downarrow \rho^*(\beta) & & \\ \rho^*(\mathcal{G}) & & \\ \downarrow & & \\ 0 & & \end{array}$$

Given a morphism $f : T \to S$, an element of

$$Quot_n(\mathcal{F})(T) = \mathrm{Hom}_S(T, \mathrm{Quot}_n(\mathcal{F}))$$

belongs to $Quot_n(\mathcal{G})(T)$ if and only if it factors through the closed subscheme $\mathrm{D}_0(\gamma\rho^*(\alpha))$ of $\mathrm{Quot}_n(\mathcal{F})$. This proves that $Quot_n(\mathcal{G})$ is a closed subfunctor of $Quot_n(\mathcal{F})$, and therefore the embedding $\mathrm{Quot}_n(\mathcal{G}) \subset \mathrm{Quot}_n(\mathcal{F})$ is closed. More precisely, this analysis shows that $\mathrm{Quot}_n(\mathcal{G}) = \mathrm{D}_0(\gamma\rho^*(\alpha))$. Since $\mathrm{Quot}_n(\mathcal{G})$ has codimension $\mathrm{rk}(\mathcal{E})n$ in $\mathrm{Quot}_n(\mathcal{F})$ it follows that it is regularly embedded. According to Example 4.2.8 we have a surjective homomorphism:

$$Hom(\mathcal{Q}, \rho^*(\mathcal{E})) \to \mathcal{I}$$

where $\mathcal{I} \subset \mathcal{O}_{\mathrm{Quot}_n(\mathcal{F})}$ is the ideal sheaf of $\mathrm{Quot}_n(\mathcal{G})$. By restricting to $\mathrm{Quot}_n(\mathcal{G})$ we obtain a surjective homomorphism:

$$Hom(\mathcal{Q}, \rho^*\mathcal{E}) \otimes \mathcal{O}_{\mathrm{Quot}_n(\mathcal{G})} \to \mathcal{I}/\mathcal{I}^2 \to 0$$

which is an isomorphism because both are locally free and of the same rank. □

Corollary 4.6.3. *Let*

$$0 \to \mathcal{E} \to \mathcal{F} \to \mathcal{G} \to 0 \tag{4.31}$$

be an exact sequence of locally free sheaves on the algebraic scheme S. Then there is a closed immersion

$$\mathbb{P}(\mathcal{E}) \subset \mathbb{P}(\mathcal{F})$$

and a natural identification:

$$N_{\mathbb{P}(\mathcal{E})/\mathbb{P}(\mathcal{F})} = \rho^*\mathcal{G} \otimes \mathcal{O}_{\mathbb{P}(\mathcal{E})}(1)$$

In particular, to every exact sequence (4.31) with $\mathcal{E}$ an invertible sheaf there corresponds a section

$$\sigma : S \to \mathbb{P}(\mathcal{E}) \subset \mathbb{P}(\mathcal{F})$$

of the projective bundle $\mathbb{P}(\mathcal{F}) \to S$ whose normal bundle is $\mathcal{G} \otimes \mathcal{E}^\vee$.

Proof. Only the last assertion requires a proof. It follows by observing that

$$\mathcal{O}_{\mathbb{P}(\mathcal{F})}(1) \otimes \mathcal{O}_{\mathbb{P}(\mathcal{E})} = \mathcal{O}_{\mathbb{P}(\mathcal{E})}(1) = \mathcal{E}^\vee$$

and therefore the formula for the normal bundle is a direct consequence of Proposition 4.6.2. □

Remark 4.6.4. Let $\mathcal{E}$ be a locally free sheaf on an algebraic scheme S, and let

$$0 \to \mathcal{Q}^\vee \to \rho^*(\mathcal{E}) \to \mathcal{K}^\vee \to 0 \tag{4.32}$$

be the dual of the tautological exact sequence (4.30) on $G_n(\mathcal{E}) = \mathrm{Quot}_n(\mathcal{E}^\vee)$. Tensoring with $\mathcal{Q}$ we obtain the exact sequence:

$$\begin{array}{ccccccc} 0 \to & \mathcal{Q}^\vee \otimes \mathcal{Q} & \to & \rho^*(\mathcal{E}) \otimes \mathcal{Q} & \to & \mathcal{K}^\vee \otimes \mathcal{Q} & \to 0 \\ & & & & & \| & \\ & & & & & T_{G_n(\mathcal{E})/S} & \end{array} \tag{4.33}$$

In the case $n = 1$ and $S = \mathrm{Spec}(\mathbf{k})$ we have $\mathcal{E} = V$ a vector space and $G_1(V) = \mathbb{P}(V) = \mathbb{P}$; the dual of the tautological sequence is

$$0 \to \mathcal{O}_{\mathbb{P}}(-1) \to V \otimes \mathcal{O}_{\mathbb{P}} \to T_{\mathbb{P}}(-1) \to 0 \tag{4.34}$$

and the sequence (4.33) is the Euler sequence

$$0 \to \mathcal{O}_{\mathbb{P}} \to V \otimes \mathcal{O}_{\mathbb{P}}(1) \to T_{\mathbb{P}} \to 0$$

Therefore (4.33) is a generalization of the Euler sequence. Dualizing (4.34) we get an inclusion

$$\mathbf{I} = \mathbb{P}(\Omega^1_{\mathbb{P}}) \subset \mathbb{P} \times \mathbb{P}^\vee$$

where $\mathbf{I} := \{(x, H) : x \in H\}$ is the incidence relation (see Note 2 of Appendix B). From Corollary 4.6.3 we obtain

$$N_{\mathbf{I}/\mathbb{P}\times\mathbb{P}^\vee} = p_1^*\mathcal{O}_{\mathbb{P}}(1) \otimes p_2^*\mathcal{O}_{\mathbb{P}^\vee}(1) \otimes \mathcal{O}_{\mathbf{I}}$$

where

$$\mathbb{P} \xleftarrow{p_1} \mathbb{P} \times \mathbb{P}^\vee \xrightarrow{p_2} \mathbb{P}^\vee$$

are the projections.

Example 4.6.5. Let X be a projective nonsingular variety and L an invertible sheaf on X. Consider the Atiyah extension

$$0 \to \mathcal{O}_X \to \mathcal{E}_L \to T_X \to 0$$

(see page 145). Then $\rho : \mathbb{P}(\mathcal{E}_L) \to X$ is a $\mathbb{P}^r$-bundle, $r = \dim(X)$, and $\mathbb{P}(\mathcal{O}_X) \subset \mathbb{P}(\mathcal{E}_L)$ is a section of ρ with normal bundle T_X. In the case $X = \mathbb{P} := \mathbb{P}(V)$, where V is a finite-dimensional vector space, and $L = \mathcal{O}(1)$ the Atiyah extension coincides with the Euler sequence (Remark 3.3.10) so that we have $\mathbb{P}(\mathcal{E}_L) = \mathbb{P} \times \mathbb{P}$ and

$$\mathbb{P} = \mathbb{P}(\mathcal{O}_{\mathbb{P}}) \subset \mathbb{P}(\mathcal{E}_L) = \mathbb{P} \times \mathbb{P}$$

is the diagonal embedding.

Example 4.6.6. Let V be a finite-dimensional $\mathbf{k}$-vector space, $X \subset \mathbb{P} := \mathbb{P}(V)$ a projective irreducible nonsingular variety and $\mathcal{I} \subset \mathcal{O}_{\mathbb{P}}$ its ideal sheaf. Then we have inclusions of locally free sheaves on X:

$$N^{\vee}_{X/\mathbb{P}} = \mathcal{I}/\mathcal{I}^2 \subset \Omega^1_{\mathbb{P}|X} \subset V^{\vee} \otimes \mathcal{O}_X(-1)$$

which induce closed embeddings of projective bundles:

$$\begin{array}{ccccc} \mathbb{P}(N^{\vee}_{X/\mathbb{P}}) \subset & \mathbb{P}(\Omega^1_{\mathbb{P}|X}) & \subset X \times \mathbb{P}^{\vee} \\ & \| & \\ & \{(x, H) : x \in H\} & \end{array}$$

Recalling the exact sequence

$$0 \to N^{\vee}_{X/\mathbb{P}} \to \Omega^1_{\mathbb{P}|X} \to \Omega^1_X \to 0 \tag{4.35}$$

we see that we have an identification:

$$\mathbb{P}(N^{\vee}_{X/\mathbb{P}}) = \{(x, H) : T_x X \subset H\} \subset X \times \mathbb{P}^{\vee} \tag{4.36}$$

Letting

$$\rho : \mathbb{P}(N^{\vee}_{X/\mathbb{P}}) \to X$$

be the structural morphism, $n = \dim(X)$, and $r = \dim(\mathbb{P})$, from the description (4.36) we deduce that for every $x \in X$, the fibre of ρ is identified with the $(r-n-1)$-dimensional linear system of hyperplanes which are tangent to X at x.

Let's consider the special case when X is a rational normal curve in $\mathbb{P} = \mathbb{P}^r$, $r \geq 2$. Denote by λ the invertible sheaf of degree 1 on X, so that $\mathcal{O}_X(1) = \lambda^r$. We have a morphism

$$q : \mathbb{P}(N^{\vee}_{X/\mathbb{P}}) \to |\lambda^{r-2}| \cong \mathbb{P}^{r-2}$$

which associates to (x, H) the divisor $q(H)$ of degree $r-2$ such that

$$H \cdot X = 2x + q(H)$$

The existence of q implies that $\mathbb{P}(N^{\vee}_{X/\mathbb{P}}) \cong X \times \mathbb{P}^{r-2}$ and therefore

$$N^{\vee}_{X/\mathbb{P}} = \underbrace{\lambda^s \oplus \cdots \oplus \lambda^s}_{r-1}$$

for some $s \in \mathbb{Z}$. From the exact sequence (4.35) one immediately computes that

$$(r-1)s = (r+2)(1-r)$$

i.e. $s = -(r+2)$. The conclusion is the following formula for *the normal bundle of a rational normal curve $X \subset \mathbb{P}^r$*:

$$N_{X/\mathbb{P}^r} \cong \underbrace{\lambda^{r+2} \oplus \cdots \oplus \lambda^{r+2}}_{r-1} \tag{4.37}$$

For another approach to the same computation see [131]. The normal bundles of rational curves of any degree in $\mathbb{P}^3$ (resp. in $\mathbb{P}^r$) have been computed in [63] and [49] (resp. in [153]).

Example 4.6.7 (Zappa [188]). Let C be a projective nonsingular connected curve of genus 1 and let

$$0 \to \mathcal{O}_C \to \mathcal{E} \to \mathcal{O}_C \to 0 \tag{4.38}$$

be an exact sequence of locally free sheaves corresponding to a nonzero element of $\mathrm{Ext}^1_{\mathcal{O}_C}(\mathcal{O}_C, \mathcal{O}_C) = H^1(C, \mathcal{O}_C)$; in particular, (4.38) does not split. Consider the ruled surface $\rho : \mathbb{P}(\mathcal{E}) \to C$ and let $C' := \mathbb{P}(\mathcal{O}_C) \subset \mathbb{P}(\mathcal{E})$ be the section of ρ corresponding to the subsheaf $\mathcal{O}_C \subset \mathcal{E}$ in (4.38). By Corollary 4.6.3 we have $N := N_{C'/\mathbb{P}(\mathcal{E})} = \mathcal{O}_{C'}$ and therefore $h^0(C', N) = 1$. If $C'' \subset \mathbb{P}(\mathcal{E})$ is a curve such that $[C']$ and $[C'']$ belong to the same component of $\mathrm{Hilb}^{\mathbb{P}(\mathcal{E})}$ then, since N is the trivial sheaf, either $C' = C''$ or $C'' \cap C' = \emptyset$. But both C' and C'' are sections of ρ and the second possibility implies that the exact sequence (4.38) splits. This is a contradiction, and we conclude that $[C']$ is isolated in $\mathrm{Hilb}^{\mathbb{P}(\mathcal{E})}$, with a one-dimensional Zariski tangent space. Therefore C' is obstructed in $\mathbb{P}(\mathcal{E})$.

* * * * * *

Let $\mathcal{H}$ be a coherent sheaf on the projective scheme X, and let $P(t) = n$ be a constant polynomial, where n is a positive integer. Then we have two different Quot schemes associated to these data.

The first one is $\mathrm{Quot}^{X/X}_{\mathcal{H},n}$, whose $\mathbf{k}$-rational points are quotients $\mathcal{H} \to \mathcal{F}$ which are locally free of rank n: it has just been considered.

The other one is $\mathrm{Quot}^{X}_{\mathcal{H},n}$. A $\mathbf{k}$-rational point of this scheme is nothing but a quotient $\mathcal{H} \to \mathcal{F}$ such that $\mathcal{F}$ is a torsion sheaf with finite support and $h^0(\mathcal{F}) = n$. When $n = 1$ then $\mathcal{F} \cong \mathbf{k}(x)$ for some closed point $x \in X$: therefore we have a natural morphism

$$\begin{array}{cccc} q: & \mathrm{Quot}^{X}_{\mathcal{H},1} & \to & X \\ & (\mathcal{H} \to \mathcal{F}) & \mapsto & \mathrm{Supp}(\mathcal{F}) \end{array}$$

and $\mathrm{Quot}^{X}_{\mathcal{H},1}$ is a scheme over X. Let $\mathcal{H} \to \mathcal{F}$ be a $\mathbf{k}$-rational point of $\mathrm{Quot}^{X}_{\mathcal{H},1}$; then $\ker[\mathcal{H} \to \mathcal{F}]$ is called an *elementary transform* of $\mathcal{H}$. The process of passing from $\mathcal{H}$ to $\ker[\mathcal{H} \to \mathcal{F}]$ is called an *elementary transformation* centred at x. This construction is classical when X is a projective nonsingular curve and $\mathcal{H}$ is locally free. For generalizations of it see [126]. $\mathrm{Quot}^{X}_{\mathcal{H},1}$ is *the scheme of elementary transformations of* $\mathcal{H}$.

Proposition 4.6.8. *Assume that* $\mathrm{Supp}(\mathcal{H})$ *is connected. Then* $\mathrm{Quot}^{X}_{\mathcal{H},1}$ *is connected.*

Proof. The natural morphism

$$\begin{array}{cccc} q: & \mathrm{Quot}^{X}_{\mathcal{H},1} & \to & X \\ & (\mathcal{H} \to \mathcal{F}) & \mapsto & \mathrm{Supp}(\mathcal{F}) \end{array}$$

has image $\mathrm{Supp}(\mathcal{H})$. Every $\mathcal{H} \to \mathcal{F}$ factors as

$$\begin{array}{ccc} \mathcal{H} & \rightarrow & \mathcal{H}\otimes \mathcal{O}_x \\ & & \downarrow \\ & & \mathcal{F} \quad \cong \mathcal{O}_x \\ & & \downarrow \\ & & 0 \end{array}$$

and therefore the fibre $q^{-1}(x)$ is identified with $I\!P(H^0(\mathcal{H}\otimes\mathcal{O}_x))^\vee$ which is connected. The conclusion follows. □

4.6.5 Hilbert schemes of points

The Hilbert schemes parametrizing 0-dimensional subschemes of a given scheme are inexpectedly complicated and have a variety of applications. Since it is beyond our scope to give an exhaustive overview of the subject, we will only explain some of the basic facts. The reader is referred to [91], [92], [133] for more details. See also [81] for another approach to Hilbert schemes of points.

Consider a projective scheme Y and, for a positive integer n, the Hilbert scheme $Y^{[n]}$. We have already seen in § 4.3 that $Y^{[1]}\cong Y$: we will therefore assume $n\geq 2$.

Let $Z\subset Y$ be a closed subscheme of length n. Then

$$H^0(Z,N_Z)=\bigoplus_{y\in \mathrm{Supp}(Z)}\mathrm{Hom}(\mathcal{I}_{Z_y},\mathcal{O}_{Z_y})$$

and

$$H^1(Z,N_Z)=(0)$$

Moreover, from the spectral sequence of Exts (see [64], p. 265) we see that

$$\mathrm{Ext}^1(\mathcal{I}_Z,\mathcal{O}_Z)=H^0(Y,Ext^1(\mathcal{I}_Z,\mathcal{O}_Z))=\bigoplus_{y\in\mathrm{Supp}(Z)}\mathrm{Ext}^1(\mathcal{I}_{Z_y},\mathcal{O}_{Z_y})$$

It follows that the local properties of $Y^{[n]}$ at $[Z]$ are determined by the independent contributions from each of the points of Supp$([Z])$. We immediately have the following properties:

(a) *If Z is reduced and supported at n distinct points of Y then $[Z]$ is a nonsingular point of $Y^{[n]}$ if and only if it is supported at nonsingular points of Y.*
(b) *If Y is reduced then the set of $[Z]$'s with Z supported at n nonsingular points of Y is an open set of dimension $n\,\dim(Y)$ contained in the nonsingular locus of $Y^{[n]}$.*

Another important property is:

(c) [58] *If Y is connected then $Y^{[n]}$ is also connected.*

Proof. Let $n \geq 1$ and let $\mathcal{I} \subset \mathcal{O}_{Y \times Y^{[n]}}$ be the ideal sheaf of the universal family in $Y \times Y^{[n]}$. Then we have a diagram of morphisms:

$$\begin{array}{ccc} & \mathrm{Quot}_{\mathcal{I},1}^{Y \times Y^{[n]}} & \\ \swarrow p & & \searrow q \\ Y^{[n+1]} & & Y \times Y^{[n]} \end{array}$$

where q is the natural morphism, which is surjective because $\mathrm{Supp}(\mathcal{I}) = Y \times Y^{[n]}$. The morphism p is defined as follows.

Let $(y, [Z]) \in Y \times Y^{[n]}$ be a $\mathbf{k}$-rational point and let $\gamma : \mathcal{I} \to \mathbf{k}(y, [Z])$ be a quotient, which is a $\mathbf{k}$-rational point of $\mathrm{Quot}_{\mathcal{I},1}^{Y \times Y^{[n]}}$. Then $\ker(\gamma) \subset \mathcal{O}_{Y \times Y^{[n]}}$ is an ideal sheaf such that $\ker(\gamma)\mathcal{O}_{Y \times [Z]}$ has colength 1 in $\mathcal{I} \otimes \mathcal{O}_{Y \times [Z]}$. Therefore $\ker(\gamma)\mathcal{O}_{Y \times [Z]}$ defines a subscheme $W \subset Y$ of length $n + 1$ containing Z and x; we define $p(\gamma) = [W]$. The morphism p is clearly surjective. Since $Y \times Y^{[n]}$ is connected by induction, we conclude that $Y^{[n+1]}$ is connected by Proposition 4.6.8. □

In general, $Y^{[n]}$ is singular and reducible even if Y is nonsingular connected. Notable exceptions are the cases $\dim(Y) = 1, 2$.

If C is a projective nonsingular curve and $n \geq 1$ an integer, every closed subscheme $D \subset C$ of length n is a Cartier divisor, therefore regularly embedded in C. It follows that $C^{[n]}$ is nonsingular of dimension

$$h^0(D, N_D) = h^0(D, \mathcal{O}_D) = n$$

because $H^1(D, N_D) = (0)$ is an obstruction space for $\mathcal{O}_{C^{[n]},[D]}$ (Theorem 4.3.5(ii)). Actually, $C^{[n]}$ is naturally isomorphic to $C^{(n)}$, the n-th symmetric product of C, the isomorphism being given by the *cycle map*

$$C^{[n]} \to C^{(n)}$$

which maps a closed $D \subset C$ to the associated Weil divisor.

The case of surfaces is more subtle.

Theorem 4.6.9 (Fogarty [58]). *If Y is a projective nonsingular connected surface then $Y^{[n]}$ is nonsingular connected of dimension $2n$.*

Proof. Let $[Z] \in Y^{[n]}$. We then have:

$$\mathrm{Ext}^i(\mathcal{I}_Z, \mathcal{O}_Z) = (0) \qquad i \geq 3$$

Moreover, from the exact sequence:

$$0 \to \mathcal{I}_Z \to \mathcal{O}_Y \to \mathcal{O}_Z \to 0$$

we obtain the sequence:

$$\begin{array}{ll} 0 & \to \mathrm{Hom}(\mathcal{O}_Z, \mathcal{O}_Z) \to \mathrm{Hom}(\mathcal{O}_Y, \mathcal{O}_Z) \to \mathrm{Hom}(\mathcal{I}_Z, \mathcal{O}_Z) \\ & \to \mathrm{Ext}^1(\mathcal{O}_Z, \mathcal{O}_Z) \to \mathrm{Ext}^1(\mathcal{O}_Y, \mathcal{O}_Z) \to \mathrm{Ext}^1(\mathcal{I}_Z, \mathcal{O}_Z) \\ & \to \mathrm{Ext}^2(\mathcal{O}_Z, \mathcal{O}_Z) \to \mathrm{Ext}^2(\mathcal{O}_Y, \mathcal{O}_Z) \to \mathrm{Ext}^2(\mathcal{I}_Z, \mathcal{O}_Z) \to \quad 0 \end{array}$$

Since $\mathrm{Ext}^i(\mathcal{O}_Y, \mathcal{O}_Z) = H^i(Y, \mathcal{O}_Z) = (0)$ for $i \geq 1$ we see that

$$\mathrm{Ext}^2(\mathcal{I}_Z, \mathcal{O}_Z) = (0)$$

and

$$\begin{aligned} \mathrm{Ext}^1(\mathcal{I}_Z, \mathcal{O}_Z) \cong \mathrm{Ext}^2(\mathcal{O}_Z, \mathcal{O}_Z) &= \mathrm{Hom}(\mathcal{O}_Z, \mathcal{O}_Z \otimes \omega_Y)^\vee \\ &= \mathrm{Hom}(\mathcal{O}_Z, \mathcal{O}_Z)^\vee = H^0(Z, \mathcal{O}_Z)^\vee \end{aligned}$$

Therefore

$$\sum_{i=0}^{2} (-1)^i \dim[\mathrm{Ext}^i(\mathcal{I}_Z, \mathcal{O}_Z)] = h^0(Z, N_Z) - h^0(Z, \mathcal{O}_Z) = h^0(Z, N_Z) - n$$

Since the left-hand side is independent of Z, it follows that $h^0(Z, N_Z)$ is also independent of Z. But $Y^{[n]}$ is connected and has an open set which is nonsingular and of dimension $2n$: the conclusion follows. □

To see that $Y^{[n]}$ is singular if $\dim(Y) = 3$ consider $I\!P^3$ with homogeneous coordinates $X_0, \ldots, X_3$ and the subscheme

$$Z = V(X_1^2, X_2^2, X_3^2, X_1X_2, X_1X_3, X_2X_3)$$

Then $[Z] \in (I\!P^3)^{[4]}$. A computation similar to that of the example of Subsection 4.6.3 shows that the Zariski tangent space of $(I\!P^3)^{[4]}$ at $[Z]$ has dimension 18. But $(I\!P^3)^{[4]}$ is connected and has a component which is nonsingular of dimension 12 at its general point: it follows that $(I\!P^3)^{[4]}$ is singular. For more examples and for a useful detailed introduction to punctual Hilbert schemes we refer the reader to [91].

4.6.6 Schemes of morphisms

Let X and Y be schemes, with X projective and Y quasi-projective. For every scheme S let:

$$F(S) = \mathrm{Hom}_S(X \times S, Y \times S)$$

This defines a contravariant functor:

$$F : (\text{schemes})^\circ \to (\text{sets})$$

called *the functor of morphisms from X to Y*.

For every $\Phi \in F(S)$ let $\Gamma_\Phi \subset X \times Y \times S$ be its graph. Then $\Gamma_\Phi \cong X \times S$ is flat over S and therefore defines a flat family of closed subschemes of $X \times Y$ parametrized by S. This means that F is a subfunctor of $Hilb^{X \times Y}$.

If $G \subset X \times Y \times S$ is a flat family of closed subschemes of $X \times Y$, proper over S, then the projection $\pi : G \to X \times S$ is a family of morphisms into X and the locus of points $s \in S$ such that $\pi(s)$ is an isomorphism is open (Note 2 of § 4.2). This means that F is an open subfunctor of $Hilb^{X\times Y}$, represented by an open subscheme of $\mathrm{Hilb}^{X\times Y}$, which we denote by $\underline{\mathrm{Hom}}(X, Y)$. It is called *the scheme of morphisms from X to Y*.

The scheme $\underline{\mathrm{Hom}}(X, Y)$ contains an open and closed subscheme isomorphic to Y and consisting of the constant morphisms. In particular, if the only morphisms $f : X \to Y$ are the constant ones then

$$\underline{\mathrm{Hom}}(X, Y) \cong Y$$

Let X and Y be as above, and consider the contravariant functor

$$G : (\text{schemes})^{\circ} \to (\text{sets})$$

defined as follows:

$$G(S) = \{S\text{-isomorphisms } X \times S \to Y \times S\}$$

Clearly, G is a subfunctor of F. It is easy to prove that G is represented by an open subscheme $\underline{\mathrm{Isom}}(X, Y)$ of $\underline{\mathrm{Hom}}(X, Y)$, called *the scheme of isomorphisms from X to Y*. When $X = Y$ it is denoted by $\underline{\mathrm{Aut}}(X)$ and called *the scheme of automorphisms of X*. It is a group scheme. The following result follows immediately from Proposition 3.4.2 and Corollary 3.4.3:

Proposition 4.6.10. *Let $f : X \to Y$ be a morphism of algebraic schemes, with X reduced and projective and Y nonsingular and quasiprojective. Then*

$$T_{[f]}(\underline{\mathrm{Hom}}(X, Y)) \cong H^0(X, f^*T_Y)$$

and the obstruction space of $\underline{\mathrm{Hom}}(X, Y)$ *at* $[f]$ *is contained in* $H^1(X, f^*T_Y)$.

If X is nonsingular then the tangent space to $\underline{\mathrm{Aut}}(X)$ *at* 1_X *is* $H^0(X, T_X)$.

Let $j : X \subset Y$ be a closed embedding of projective nonsingular schemes. Then j induces an inclusion

$$J : \underline{\mathrm{Aut}}(X) \subset \underline{\mathrm{Hom}}(X, Y)$$

such that $J(1_X) = j$ and which is induced by the closed embedding

$$1_X \times j : X \times X \subset X \times Y.$$

It follows that J is a closed embedding. Its differential at 1_X is the injective linear map

$$H^0(T_X) \to H^0(T_{Y|X})$$

coming from the natural inclusion $T_X \subset T_{Y|X}$. In fact, from the diagram of inclusions:

$$\begin{array}{ccc} X \times X & \subset & X \times Y \\ \cup & & \cup \\ \Delta & \cong & \Gamma_j \end{array}$$

we deduce the commutative diagram:

$$\begin{array}{ccc} N_{\Delta/X\times X} & \subset & N_{\Gamma_j/X\times Y} \\ \| & & \| \\ T_X & \subset & T_{Y|X} \end{array}$$

and we conclude according to § 4.3, Note 3.

Example 4.6.11. Consider $X = I\!P^1$, $Y = I\!P^r$ and $j : I\!P^1 \to I\!P^r$ the r-th Veronese embedding. Locally around $[j]$ we have a well-defined morphism

$$M : \underline{\mathrm{Hom}}(I\!P^1, I\!P^r) \to \mathrm{Hilb}^{I\!P^r}$$

sending $[j] \mapsto [j(I\!P^1)]$ with fibre $M^{-1}([j(I\!P^1)])$ an open neighbourhood of the identity in $\mathrm{Aut}(I\!P^1)$. Consider the following diagram consisting of two exact sequences:

$$\begin{array}{ccccccc} & & & 0 & & & \\ & & & \uparrow & & & \\ 0 \to & T_{I\!P^1} & \to & j^*T_{I\!P^r} & \to & N_j & \to 0 \\ & & & \uparrow & & & \\ & & & \mathcal{O}_{I\!P^1}(r)^{r+1} & & & \\ & & & \uparrow & & & \\ & & & \mathcal{O}_{I\!P^1} & & & \\ & & & \uparrow & & & \\ & & & 0 & & & \end{array}$$

From the vertical sequence (the Euler sequence restricted to $I\!P^1$) we get

$$h^1(j^*T_{I\!P^r}) = 0, \quad h^0(j^*T_{I\!P^r}) = r(r+2)$$

Since $h^1(T_{I\!P^1}) = 0$ from the other sequence we obtain $h^1(N_j) = 0$ and the exact sequence

$$\begin{array}{ccccccc} 0 \to & H^0(T_{I\!P^1}) & \to & H^0(j^*T_{I\!P^r}) & \xrightarrow{q} & H^0(N_j) & \to 0 \\ & \| & & \| & & \| & \\ & T_{1_{I\!P^1}}\mathrm{Aut}(I\!P^1) & & T_{[j]}\underline{\mathrm{Hom}}(I\!P^1, I\!P^r) & & T_{[j(I\!P^1)]}\mathrm{Hilb}^{I\!P^r} & \end{array}$$

Since the map q can be identified with $dM_{[j]}$ we see that M and $\underline{\mathrm{Hom}}(I\!P^1, I\!P^r)$ are smooth at $[j]$ and $\mathrm{Hilb}^{I\!P^r}$ is smooth at $[j(I\!P^1)]$; moreover,

$$\dim_{[j]}(\underline{\mathrm{Hom}}(I\!P^1, I\!P^r) = r(r+2) = \dim_{j(I\!P^1)}(\mathrm{Hilb}^{I\!P^r}) + 3$$

For more on the schemes $\underline{\mathrm{Hom}}(I\!P^1, X)$ and applications to uniruledness see [42].

4.6.7 Focal loci

We assume $\mathrm{char}(\mathbf{k}) = 0$. Consider a flat family of closed subschemes of a projective scheme Y:

$$\begin{array}{c} \Xi \subset Y \times B \xrightarrow{q_1} Y \\ \downarrow q_2 \\ B \end{array} \tag{4.39}$$

parametrized by a scheme B. Let

$$f : \Xi \subset Y \times B \to Y$$

be the composition of the inclusion (4.39) with the projection q_1. Denote by

$$\mathcal{N} := N_{\Xi/Y\times B}$$

the normal sheaf of Ξ in $Y \times B$, and let

$$(q_2^* T_B)_{|\Xi}$$

be the restriction to Ξ of the tangent sheaf along the fibres of q_1. The *global characteristic map* of the family (4.39) is the homomorphism:

$$\chi : (q_2^* T_B)_{|\Xi} \to \mathcal{N}$$

defined by the following exact and commutative diagram:

$$\begin{array}{ccccccccc} & & & & 0 & & & & \\ & & & & \downarrow & & & & \\ & & & & (q_2^* T_B)_{|\Xi} & \xrightarrow{\chi} & \mathcal{N} & & \\ & & & & \downarrow & & \| & & \\ 0 & \to & T_\Xi & \to & T_{Y\times B|\Xi} & \to & \mathcal{N} & \to & 0 \\ & & \downarrow df & & \downarrow & & & & \\ & & f^* T_Y & = & f^* T_Y & & & & \end{array} \tag{4.40}$$

For each $b \in B$ the homomorphism χ induces a homomorphism

$$\chi_b : T_{B,b} \otimes \mathcal{O}_{\Xi(b)} \to N_{\Xi(b)/Y}$$

called the *local characteristic map* of the family (4.39) at the point b.

Let

$$\varphi : B \to Hilb_Y$$

be the classifying morphism induced by the flat family (4.39). Then if $\Xi(b)$ is connected, the linear map:

$$H^0(\chi_b) : T_{B,b} \to H^0(N_{\Xi(b)/Y})$$

is $d\varphi_b$, the characteristic map of (4.39) at the point b (see page 217).

Assume that Y, B and the family (4.39) are smooth. In this case all the sheaves in (4.40) are locally free. From a diagram-chasing it follows that

$$\ker(\chi) = \ker(df)$$

and therefore:

Proposition 4.6.12.

$$\dim[f(\Xi)] = \dim(\Xi) - \mathrm{rk}[\ker(\chi)]$$

Let's denote by $V(\chi)$ the closed subscheme of Ξ defined by the condition:

$$\mathrm{rk}(\chi) < \min\{\mathrm{rk}[(q_2^* T_B)_{|\Xi}], \mathrm{rk}(\mathcal{N})\} = \min\{\dim(B), \mathrm{codim}_{Y\times B}(\Xi)\}$$

We will call the points of $V(\chi)$ *first-order foci* of the family (4.39). $V(\chi)$ is *the scheme of first-order foci*, and the fibre of $V(\chi)$ over a point $b \in B$:

$$V(\chi)_b = V(\chi_b) \subset \Xi(b)$$

is *the scheme of first-order foci at b*.

If χ has maximal rank, i.e. if χ is injective or has torsion cokernel, then $V(\chi)$ is a proper closed subscheme of Ξ. If χ does not have maximal rank then $V(\chi) = \Xi$.

The definition of first-order foci at a point b depends only on the geometry of the family (4.39) in a neighbourhood of b. A focus $y \in V(\chi)_b$ is a point where there is an intersection between the fibre $\Xi(b)$ and the infinitesimally near ones. One defines higher-order foci inductively: second-order foci are the first-order foci of the family of first-order foci, and so on.

Focal loci have been studied classically in the case of families of linear spaces (see e.g. [158]). Recently, they have been applied to the geometry of the theta divisor of an algebraic curve in [37] and [38]. Related work is [31], [33], [39].

NOTES

1. Let $Q \subset \mathbb{P}^3$ be a quadric cone with vertex v, and $L \subset Q$ a line. Then

$$N_{L/Q} = \mathcal{O}_L(1) \subset N_{L/\mathbb{P}^3} = \mathcal{O}_L(1) \oplus \mathcal{O}_L(1)$$

(see Example 3.2.8(iii), page 131); in particular, $H^0(L, N_{L/Q}) = 2$ and $H^1(L, N_{L/Q}) = 0$. On the other hand, the Hilbert scheme Hilb^Q is one-dimensional at the point $[L]$ since L moves in a one-dimensional family. It follows that L is obstructed in Q (see [44] for generalizations of this example).

2. Another example of reducible Hilbert scheme is the following, which appears in [170]. Let $Y = C \times C'$ where C and C' are projective nonsingular connected curves of genera g and g' respectively and let

$$\begin{array}{ccc} Y & \xrightarrow{p'} & C' \\ \downarrow p & & \\ C & & \end{array}$$

be the projections; assume $g, g' \geq 2$. Consider an effective divisor $D = x_1 + \cdots + x_g$ of degree g on C, and an effective divisor $D' = x'_1 + \cdots + x'_{g'}$ of degree g' on C', both consisting of distinct points, and let

$$\Gamma = p^{-1}(D) + p'^{-1}(D') = C'_{x_1} + \cdots + C'_{x_g} + C_{x'_1} + \cdots + C_{x'_{g'}}$$

where $C'_x = p^{-1}(x)$ and $C_{x'} = p'^{-1}(x')$. Γ is a reduced divisor, has gg' nodes and no other singularity. If either D or D' is nonspecial the curve Γ belongs to an irreducible component H_1 of Hilb^Y of dimension $g + g'$ generically consisting of curves of the same form, obtained by moving D and D'. When both D and D' are special divisors the curve Γ belongs to a linear system of dimension ≥ 3 whose general member is a nonsingular curve and therefore belongs to another irreducible component H_2 of Hilb^Y which has dimension $g + g' - 1$. The intersection $H_1 \cap H_2$ is irreducible of dimension $g + g' - 2$.

4.7 Plane curves

An important refinement of the Hilbert functors derives from the consideration of flat families of closed subschemes of a projective scheme Y having prescribed singularities, i.e. of families all of whose members have the same type of singularity in some specified sense. This leads to the notion of *equisingularity* and to a related vast area of research. In this section we will only concentrate on the specific case of families of plane curves with assigned number of nodes and cusps: we will show how to construct universal families of such curves, whose parameter schemes are called *Severi varieties* for historical reasons. This is a subject with a long history and a wealth of important results, both classical and modern. Here we will limit ourselves to proving a few basic results and to indicating some of their generalizations and the main references in the literature. We will assume char($\mathbf{k}$) $= 0$ in this section.

4.7.1 Equisingular infinitesimal deformations

Let Y be a projective nonsingular variety and $X \subset Y$ a closed subscheme with ideal sheaf $\mathcal{I}_X \subset \mathcal{O}_Y$. Recall ((1.3)) that on X we have an exact sequence of coherent sheaves:

$$0 \to T_X \to T_{Y\ X} \to N_{X/Y} \to T^1_X \to 0$$

Recall that the sheaf

$$N'_{X/Y} := \ker[N_{X/Y} \to T^1_X]$$

is called the *equisingular normal sheaf* of X in Y (see Proposition 1.1.9, page 16). Clearly, $N'_{X/Y} = N_{X/Y}$ if X is nonsingular. By definition, sections of the equisingular normal sheaf parametrize first-order deformations of X in Y which are locally trivial, because they induce trivial deformations around every point of X.

We recall that an alternative description of the equisingular normal sheaf can be given by means of the sheaf of germs of tangent vectors to Y which are tangent to X

$$T_Y\langle X\rangle \subset T_Y$$

introduced in § 3.4. In fact, there is an exact sequence (see (3.55))

$$0 \to T_Y\langle X\rangle \to T_Y \to N'_{X/Y} \to 0 \tag{4.41}$$

From the definition it follows that, for every open set $U \subset Y$, $\Gamma(U, T_Y\langle X\rangle)$ consists of those $\mathbf{k}$-derivations $D \in \Gamma(U, T_Y)$ such that $D(g) \in \Gamma(U, \mathcal{I})$ for every $g \in \Gamma(U, \mathcal{I})$.

Examples 4.7.1. (i) Assume that X is a hypersurface in Y; then $N_{X/Y} \cong \mathcal{O}_X(X)$. Locally on an affine open set $U \subset Y$ we have $(N_{X/Y})_{|U\cap X} \cong \mathcal{O}_{U\cap X}$. Assume that X can be represented by an equation $f(x_1, \ldots, x_n) = 0$ in local coordinates on U; then from the definition of T^1_X it follows that $(N'_{X/Y})_{|U\cap X} \subset \mathcal{O}_{U\cap X}$ is the image of the ideal sheaf $(\partial f/\partial x_1, \ldots, \partial f/\partial x_n) \subset \mathcal{O}_U$. We deduce that an equisingular first-order deformation of X in Y corresponding to a local section $\bar{g}$ of $N'_{X/Y}$ can be written locally as

$$f(\underline{x}) + \epsilon g(\underline{x}) = 0$$

where

$$g(\underline{x}) = a_1(\underline{x})\frac{\partial f}{\partial x_1} + \cdots + a_n(\underline{x})\frac{\partial f}{\partial x_n}$$

restricts to $\bar{g}$. Therefore if $Y = \mathbb{P}^n$ and X is a hypersurface of degree d we have an exact sequence

$$0 \to \mathcal{O}_{\mathbb{P}^n} \to \mathcal{I}(d) \to N'_{X/\mathbb{P}^n} \to 0$$

where $\mathcal{I} \subset \mathcal{O}_{\mathbb{P}^n}$ is the ideal sheaf locally generated by the partial derivatives of a local equation of X. In the special case of a curve X in a surface Y assume that $p \in X$ is a singular point and let $f(x, y) = 0$ be a local equation of X around p. If p is a node then $(\partial f/\partial x, \partial f/\partial y) = m_p$ is just the maximal ideal of p; if p is an ordinary cusp with principal tangent say $y = 0$ then $(\partial f/\partial x, \partial f/\partial y) = (x, y^2)$ is an ideal of colength 2.

(ii) Let $X \subset \mathbb{P}^2$ be a (possibly reducible) plane curve of degree d of equation $F(X_0, X_1, X_2) = 0$, having δ nodes $p_1, \ldots, p_\delta$ and no other singularity. This case is important because every nonsingular projective curve is birationally equivalent to a nodal plane curve. Denote by $\Delta = \{p_1, \ldots, p_\delta\} \subset \mathbb{P}^2$ the 0-dimensional reduced scheme of the nodes of X and by $\nu : C \to X$ the normalization map. The above analysis shows that sections of $H^0(\mathcal{I}_\Delta(d))$, i.e. curves of degree d which are adjoint to X, cut on X sections of $N'_{X/\mathbb{P}^2}$. This means that

$$\nu^*(N'_{X/\mathbb{P}^2}) = \nu^*[\mathcal{O}_X(d) \otimes \mathcal{I}_\Delta]$$

$$= \nu^*\mathcal{O}_X(d)(-p'_1 - p''_1 - \cdots - p'_\delta - p''_\delta) = \omega_C \otimes \nu^*\mathcal{O}(3)$$

where $\nu^{-1}(p_i) = \{p'_i, p''_i\}$, $i = 1, \ldots, \delta$, and therefore we have

$$h^0(C, \nu^*(N'_{X/\mathbb{P}^2})) = 3d + g - 1, \quad h^1(C, \nu^*(N'_{X/\mathbb{P}^2})) = 0$$

where g is the geometric genus of X. Moreover, since

$$\nu_*[\nu^*(N'_{X/\mathbb{P}^2})] = \nu_*[\nu^*(\mathcal{O}_X(d) \otimes \mathcal{I}_\Delta]$$

$$= \mathcal{O}_X(d) \otimes \nu_*\mathcal{O}_C(-p'_1 - p''_1 - \cdots - p'_\delta - p''_\delta) = \mathcal{O}_X(d) \otimes \mathcal{I}_\Delta = N'_{X/\mathbb{P}^2}$$

we have

$$\begin{aligned} h^0(X, N'_{X/\mathbb{P}^2}) &= h^0(C, \nu^*(N'_{X/\mathbb{P}^2})) = 3d + g - 1 = \tbinom{d+2}{2} - \delta - 1 \\ h^1(X, N'_{X/\mathbb{P}^2}) &= h^1(C, \nu^*(N'_{X/\mathbb{P}^2})) = 0 \end{aligned} \tag{4.42}$$

Finally, since

$$h^0(\mathcal{I}_\Delta(d)) \geq \binom{d+2}{2} - \delta = 3d + g$$

and $H^0(\mathcal{I}_\Delta(d))/(F) \subset H^0(N'_{X/\mathbb{P}^2})$, comparing with (4.42) we see that Δ imposes independent conditions to curves of degree d and that

$$H^0(N'_{X/\mathbb{P}^2}) = H^0(\mathcal{I}_\Delta(d))/(F)$$

or, equivalently, the restriction map

$$H^0(\mathcal{I}_\Delta(d)) \to H^0(N'_{X/\mathbb{P}^2})$$

is surjective. Note that $N'_{X/\mathbb{P}^2}$ is a *noninvertible* subsheaf of $N_{X/\mathbb{P}^2}$.

(iii) Another interesting case is obtained by taking an irreducible curve $X \subset \mathbb{P}^2$ of degree d having δ nodes $p_1, \dots, p_\delta$ and κ ordinary cusps $q_1, \dots, q_\kappa$ as its only singularities. This case is important because branch curves of generic projection on $\mathbb{P}^2$ of projective nonsingular surfaces are curves of this type.

Let $\nu : C \to X$ be the normalization map. Letting $\bar{q}_j = \nu^{-1}(q_j)$, $j = 1, \dots, \kappa$, we have in this case, according to the above description

$$\begin{aligned} \nu^*(N'_{X/P^2}) &= \mathcal{O}_C(d)(-p'_1 - p''_1 - \cdots - p'_\delta - p''_\delta - 3\bar{q}_1 - \cdots - 3\bar{q}_\kappa) \\ &= \omega_C \otimes \nu^*\mathcal{O}_X(3)(-\bar{q}_1 - \cdots - \bar{q}_\kappa) \end{aligned}$$

As before, one shows that $\nu_*[\nu^*(N'_{X/P^2})] = N'_{X/P^2}$ and therefore

$$h^0(X, N'_{X/\mathbb{P}^2}) = h^0(C, \omega_C \otimes \nu^*\mathcal{O}_X(3)(-\bar{q}_1 - \cdots - \bar{q}_\kappa)) \geq \binom{d+2}{2} - \delta - 2\kappa - 1 \tag{4.43}$$

and in general we may have strict inequality and $h^1(X, N'_{X/\mathbb{P}^2}) \neq 0$ because the invertible sheaf $\omega_C \otimes \nu^*\mathcal{O}_X(3)(-\bar{q}_1 - \cdots - \bar{q}_\kappa)$ can be special. But if $\kappa < 3d$ then it is certainly nonspecial and therefore in such a case we have

$$h^0(X, N'_{X/\mathbb{P}^2}) = \tbinom{d+2}{2} - \delta - 2\kappa - 1 = 3d + g - 1 - \kappa$$

$$h^1(X, N'_{X/\mathbb{P}^2}) = 0$$

4.7.2 The Severi varieties

Given an integer $d > 0$, consider the complete linear system $|\mathcal{O}(d)|$ of plane curves of degree d. It is a flat family of closed subschemes of $\mathbb{P}^2$ parametrized by the projective space $\Sigma_d = \mathbb{P}[H^0(\mathbb{P}^2, \mathcal{O}(d))]$:

$$\begin{array}{rl} & \mathcal{H} \subset \mathbb{P}^2 \times \Sigma_d \\ |\mathcal{O}(d)| : & \downarrow \\ & \Sigma_d \end{array}$$

The linear system $|\mathcal{O}(d)|$ has a universal property with respect to families of plane curves of degree d because the pair $(\Sigma_d, |\mathcal{O}(d)|)$ represents the Hilbert functor:

$$\Lambda_d : (\text{algschemes})^\circ \to (\text{sets})$$

given by:

$$\Lambda_d(S) = \left\{ \begin{array}{c} \text{flat families } \mathcal{C} \subset I\!P^2 \times S \text{ of plane} \\ \text{curves of degree } d \text{ parametrized by } S \end{array} \right\}$$

(see § 4.3). In this subsection we want to consider the problem of constructing a universal family of reduced curves in $I\!P^2$ having degree d, an assigned number δ of nodes and κ of ordinary cusps and no other singularity. If such a universal family exists it is parametrized by a scheme which we denote by $\mathcal{V}_d^{\delta,\kappa}$. These schemes have been studied classically: the foundations of their theory are given in [171] and they are therefore called *Severi schemes* or *Severi varieties*.

If the Severi scheme $\mathcal{V}_d^{\delta,\kappa}$ exists then, by the universal property, there is a functorially defined morphism

$$\mathcal{V}_d^{\delta,\kappa} \to \Sigma_d \tag{4.44}$$

We start from the definition of the functor we want to represent.

Definition 4.7.2. *Let d, δ, κ as above. Then*

$$\mathbf{V}_d^{\delta,\kappa} : (\text{algschemes})^\circ \to (\text{sets})$$

is defined as follows. For each algebraic scheme S

$$\mathbf{V}_d^{\delta,\kappa}(S) = \left\{ \begin{array}{c} \text{flat families } \mathcal{C} \subset I\!P^2 \times S \text{ of plane curves of deg. } d \text{ formally} \\ \text{locally trivial at each } \mathbf{k}\text{-rational } s \in S \text{ whose geometric} \\ \text{fibres are curves with } \delta \text{ nodes and } \kappa \text{ cusps as singularities} \end{array} \right\}$$

(see § 2.5 for the definition of formal local triviality). Obviously, $\mathbf{V}_d^{\delta,\kappa}$ is a subfunctor of Λ_d. The main result about $\mathbf{V}_d^{\delta,\kappa}$ is the following:

Theorem 4.7.3. *For each d, δ, κ as above, the functor $\mathbf{V}_d^{\delta,\kappa}$ is represented by an algebraic scheme $\mathcal{V}_d^{\delta,\kappa}$ which is a (possibly empty) locally closed subscheme of Σ_d.*

In the case $\kappa = 0$ we write $\mathcal{V}_d^{\delta}$ instead of $\mathcal{V}_d^{\delta,0}$. The first published proof of this result is in [186]. We will not reproduce it in full generality here, but we will only consider the case $\kappa = 0$, i.e. the case of nodal curves. This assumption allows a technically simpler argument without changing the structure of the original proof. We need some lemmas.

Lemma 4.7.4. *Let $p \in I\!P^2$, $\mathcal{O}$ the local ring of $I\!P^2$ at p, $B_0 = \mathcal{O}/(f_0)$ the local ring of a plane curve having a node at p. Assume that for some A in* ob$(\mathcal{A})$ *we have a deformation $A \to B$ of B_0 over A such that $T^1_{B/A}$ is A-flat. Then B is trivial and $T^1_{B/A} \cong A$.*

Proof. By induction on $\dim_{\mathbf{k}}(A)$. The case $A = \mathbf{k}$ is trivial because

$$T^1_{B_0} = \mathcal{O}/(f_0, f_{0X}, f_{0Y}) = \mathcal{O}/(f_{0X}, f_{0Y}) \cong \mathbf{k}$$

(see 3.1.4). In the general case consider a small extension

$$0 \to (\epsilon) \to A \to A' \to 0$$

and the induced deformation $A' \to B'$. We have $B = (A \otimes_{\mathbf{k}} \mathcal{O})/(f)$ for some f which reduces to f_0 modulo m_A, and $T^1_{B/A} = B/(f_X, f_Y)$. Therefore $B' = (A' \otimes_{\mathbf{k}} \mathcal{O})/(f')$ where f' is obtained from f by reducing the coefficients to A', and

$$T^1_{B'/A'} = B'/(f'_X, f'_Y) = B/(f_X, f_Y) \otimes_A A' = T^1_{B/A} \otimes_A A'$$

It follows that $T^1_{B'/A'}$ is A'-flat and, by induction, we have

$$B' = (A' \otimes_{\mathbf{k}} \mathcal{O})/(f_0)$$

and

$$T^1_{B'/A'} = A' \otimes_{\mathbf{k}} [\mathcal{O}/(f_0, f_{0X}, f_{0Y})] = A'$$

Thus $f = f_0 + \epsilon g$ where $g \in \mathbf{k}$. We have:

$$T^1_{B/A} = (A \otimes_{\mathbf{k}} \mathcal{O})/(f_0 + \epsilon g, f_{0X}, f_{0Y}) = A/(\epsilon g)$$

where the last equality follows from the fact that $f_0 \in (f_{0X}, f_{0Y})$. Since $A/(\epsilon g)$ is A-flat if and only if $g = 0$ it follows that $B = (A \otimes_{\mathbf{k}} \mathcal{O})/(f_0) = A \otimes_{\mathbf{k}} (\mathcal{O}/(f_0))$ is the trivial deformation and $T^1_{B/A} = A$. □

Lemma 4.7.5. *Let $f : X \to S$ be a flat morphism of algebraic schemes which factors as*

$$\begin{array}{ccc} X & \xrightarrow{j} & Y \\ & \searrow & \downarrow q \\ & & S \end{array}$$

where j is a regular embedding of codimension 1 and q is smooth. Then for every morphism of algebraic schemes $\varphi : S' \to S$ we have

$$T^1_{S' \times_S X/S'} \cong \Phi^* T^1_{X/S}$$

where $\Phi : S' \times_S X \to X$ is the projection (i.e. $T^1_{X/S}$ commutes with base change).

Proof. Since the question is local we may reduce to a diagram of $\mathbf{k}$-algebras of the form:

$$\begin{array}{ccc} B = P/(f) & \to & B' = P_{A'}/(f') \\ \uparrow & & \uparrow \\ A & \to & A' \end{array}$$

where P is a smooth A-algebra, $f \in P$ is a regular element, and f' is the image of f in $P_{A'} = P \otimes_A A'$. Then we have

$$\begin{array}{ccc} (f)/(f^2) \cong B & \xrightarrow{\delta} & \Omega_{P/A} \otimes_P B \\ \bar{f} & \mapsto & df \otimes 1 \end{array}$$

and

$$\begin{array}{ccc} (f')/(f'^2) \cong B' & \xrightarrow{\delta'} & \Omega_{P'/A'} \otimes_{P'} B' \\ \bar{f}' & \mapsto & df' \otimes 1 \end{array}$$

But since

$$\Omega_{P'/A'} \otimes_{P'} B' = (\Omega_{P/A} \otimes_P B) \otimes_B B'$$

we have $\delta' = \delta \otimes_B B'$ and

$$T^1_{B/A} \otimes_B B' = \operatorname{coker}(\delta^\vee) \otimes_B B' = \operatorname{coker}(\delta'^\vee) = T^1_{B'/A'}$$ □

Lemma 4.7.6. *Let S be an algebraic scheme and $\mathcal{C} \subset \mathbb{P}^2 \times S$ a flat family of plane curves of degree d. Let $s \in S$ be a* **k***-rational point such that the fibre $\mathcal{C}(s)$ is a curve having at most nodes as singularities. Then $T^1_{\mathcal{C}/S}$ is S-flat at a point $p \in \mathcal{C}(s)$ if and only if the family is formally locally trivial at s around p.*

Proof. If p is a nonsingular point of $\mathcal{C}(s)$ then $\mathcal{C} \to S$ is smooth at p, hence $T^1(\mathcal{C}/S, \mathcal{O}_\mathcal{C})_p = 0$ and the assertion is obvious.

Let's assume that p is a node of $\mathcal{C}(s)$. Let $A = \mathcal{O}_{S,s}$, $A_\alpha := A/m^\alpha$, $S_\alpha = \operatorname{Spec}(A_\alpha)$ and $\mathcal{C}_\alpha \to S_\alpha$ the induced infinitesimal deformation of $\mathcal{C}_1 = \mathcal{C}(s)$; let $B = \mathcal{O}_{\mathcal{C},p}$ and $B_\alpha = \mathcal{O}_{\mathcal{C}_\alpha,p}$. By Lemma 4.7.5 we have:

$$T^1_{\mathcal{C}_\alpha/S_\alpha,p} = T^1_{B_\alpha/A_\alpha} \cong T^1(\mathcal{C}/S, \mathcal{O}_\mathcal{C})_p \otimes_{\mathcal{O}_{\mathcal{C},p}} B_\alpha = T^1_{B/A} \otimes_B B_\alpha \tag{4.45}$$

Assume that $T^1(\mathcal{C}/S, \mathcal{O}_\mathcal{C})_p = T^1_{B/A}$ is A-flat. Then $T^1_{B_\alpha/A_\alpha}$ is A_α-flat by (4.45) and therefore B_α is the trivial deformation of $B_0 = B/m_A B = \mathcal{O}_{\mathcal{C}(s),p}$, by Lemma 4.7.4. Since this is true for every α we conclude that the family $\mathcal{C} \to S$ is formally locally trivial at p.

Conversely, assume that $\mathcal{C} \to S$ is formally locally trivial at p. Then $B_\alpha \cong B_0 \otimes_\mathbf{k} A_\alpha$ for all α and $T^1_{B_\alpha/A_\alpha}$ is A_α-flat by Lemma 4.7.4. But by (4.45) we have

$$T^1_{B_\alpha/A_\alpha} = T^1_{B/A} \otimes_B B_\alpha = T^1_{B/A} \otimes_A A_\alpha$$

and from the local criterion of flatness we deduce that $T^1_{B/A} = T^1(\mathcal{C}/S, \mathcal{O}_\mathcal{C})_p$ is A-flat. □

Proof of Theorem 4.7.3. Consider the universal family $\mathcal{H} \subset \mathbb{P}^2 \times \Sigma_d$ of plane curves of degree d and let $\{W_i\}$ be the flattening stratification of $T^1_{\mathcal{H}/\Sigma_d}$. Let $W = W_i$ be a stratum containing a **k**-rational point $s \in \Sigma_d$ parametrizing a reduced curve $\mathcal{H}(s)$ having δ nodes and no other singularity, and let $\mathcal{H}' \subset \mathbb{P}^2 \times W$ be the induced family of degree d curves. By Lemma 4.7.5 we have

$$T^1_{\mathcal{H}'/W} \cong T^1_{\mathcal{H}/\Sigma_d} \otimes \mathcal{O}_{\mathcal{H}'}$$

and therefore, by construction, $T^1_{\mathcal{H}'/W}$ is flat over W. Moreover, since $\mathcal{H}' \subset I\!P^2 \times W$ is a regular embedding of codimension 1, $T^1_{\mathcal{H}'/W}$ is of the form $\mathcal{O}_V$ for some closed subscheme $V \subset I\!P^2 \times W$. By applying Lemma 4.7.5 again we deduce

$$\mathcal{O}_{V(s)} = T^1_{\mathcal{H}'/W} \otimes \mathbf{k} \cong T^1_{\mathcal{H}(s)}$$

which is a reduced scheme of length δ supported at $\mathrm{Sing}(\mathcal{H}(s))$. This implies that $V \to W$ is etale at the δ points of $V(s)$. Therefore there is an open neighbourhood U of $s \in W$ such that $V(U) \to U$ is etale of degree δ. If $u \in U$ is a $\mathbf{k}$-rational point then $\mathcal{H}(u)$ is a curve such that $T^1_{\mathcal{H}(u)} \cong V(u)$, hence $\mathcal{H}(u)$ has δ singular points $p_1, \ldots, p_\delta$ and no other singularity, such that $T^1_{\mathcal{O}_{p_j}} \cong \mathbf{k}$. From Proposition 3.1.5 it follows that $\mathcal{H}(u)$ is a δ-nodal curve. Therefore by applying Lemma 4.7.6 we see that the family $\mathcal{H}'(U) \to U$ is an element of $\mathbf{V}_d^{\delta,0}(U)$.

Putting together all these open sets we obtain a locally closed subset $U_i \subset W_i$ such that the induced family $\mathcal{H}'_i \subset I\!P^2 \times U_i$ defines an element of $\mathbf{V}_d^{\delta,0}(U_i)$. Now let

$$\mathcal{V}_d^\delta = \bigcup_i U_i$$

and let $\overline{\mathcal{H}} \subset I\!P^2 \times \mathcal{V}_d^\delta$ be the induced family. If $\mathcal{C} \subset I\!P^2 \times S$ is an element of $\mathbf{V}_d^{\delta,0}(S)$ for an algebraic scheme S then by the universal property of $|\mathcal{O}(d)|$ we obtain a unique morphism $S \to \Sigma_d$ inducing the given family by pullback. By Lemma 4.7.6 and the defining property of the flattening stratification this morphism factors through $\mathcal{V}_d^\delta$. Thus $(\mathcal{V}_d^\delta, \overline{\mathcal{H}})$ represents the functor $\mathbf{V}_d^{\delta,0}$. □

We now consider the local properties of the Severi varieties.

Proposition 4.7.7. *Let $C \subset I\!P^2$ be a reduced curve having degree d, δ nodes and κ ordinary cusps and no other singularity. Let $[C] \in \mathcal{V} = \mathcal{V}_d^{\delta,\kappa}$ be the point parametrizing C. Then there is a natural identification:*

$$T_{[C]}\mathcal{V} = H^0(C, N'_{C/I\!P^2})$$

and $H^1(C, N'_{C/I\!P^2})$ is an obstruction space for $\mathcal{O}_{\mathcal{V},[C]}$.

Proof. $T_{[C]}\mathcal{V}$ is the subspace of $T_{[C]}\Sigma_d = H^0(C, \mathcal{O}_C(d))$ corresponding to locally trivial first-order deformations, and these are the elements of $H^0(C, N'_{C/I\!P^2})$ by the very definition of $N'_{C/I\!P^2}$. From the proof of Proposition 3.2.6 it is obvious that obstructions to deforming locally trivial deformations lie in the space $H^1(C, N'_{C/I\!P^2})$. □

According to the classical terminology, we call $\mathcal{V}_d^{\delta,\kappa}$ *regular* at a point $[C]$ if $H^1(C, N'_{C/I\!P^2}) = 0$; otherwise $\mathcal{V}_d^{\delta,\kappa}$ is called *superabundant* at $[C]$. An irreducible component W of $\mathcal{V}_d^{\delta,\kappa}$ is called regular (resp. superabundant) if it is regular (resp.

superabundant) on a nonempty open subset. $\mathcal{V}_d^{\delta,\kappa}$ is called regular if all its components are regular; otherwise it is called superabundant. From 4.7.7 and from Example 4.7.1(iii) it follows that if a component W of $\mathcal{V}_d^{\delta,\kappa}$ is regular then it is generically nonsingular of dimension $3d + g - 1 - \kappa$, where g is the geometric genus of C, i.e. of pure codimension $\delta + 2\kappa$ in Σ_d.

Corollary 4.7.8. *If $\kappa < 3d$ then $\mathcal{V}_d^{\delta,\kappa}$ is regular at every point. In particular, $\mathcal{V}_d^{\delta}$ is regular at every point, thus it is nonsingular of pure dimension*

$$3d + g - 1 = \binom{d+2}{2} - 1 - \delta$$

if it is nonempty.

Proof. It follows from Proposition 4.7.7 and from Example 4.7.1(iii). □

Remark 4.7.9. Note that the corollary does not claim that $\mathcal{V}_d^{\delta,\kappa} \neq \emptyset$. The nonemptiness of the Severi varieties will be discussed in the next subsection. Corollary 4.7.8 follows also from Proposition 3.4.16 recalling that, by Lemma 3.4.15, we have

$$H^i(C, N'_{C/\mathbb{P}^2}) \cong H^i(\tilde{C}, N_\varphi), \quad i = 0, 1$$

where $\varphi : \tilde{C} \to C$ is the normalization map. The description of the deformations of C given by Proposition 3.4.16 can be considered as the "parametric" counterpart of the "cartesian" point of view of Corollary 4.7.8.

The Severi varieties $\mathcal{V}_d^{\delta,\kappa}$ may have a complicated structure. If there are too many cusps then in general a $[C] \in \mathcal{V}_d^{\delta,\kappa}$ satisfies $H^1(C, N'_{C/\mathbb{P}^2}) \neq (0)$ (see Example 4.7.10 below) and in fact, $\mathcal{V}_d^{\delta,\kappa}$ can be singular at such a $[C]$. To decide whether this effectively happens has been a long-standing classical problem (see [189], ch. VIII, where this topic is discussed). The first example of a singular point of a Severi variety was given by Wahl [186]: it is a plane irreducible curve of degree 104 with 3636 nodes and 900 cusps. For other examples see [122], [80] and [183].

Example 4.7.10. If $\kappa > 3d$ then $\mathcal{V}_d^{\delta,\kappa}$ can be superabundant. The following classical example is due to B. Segre (see [159] and [189], p. 220). Consider plane curves of the following type:

$$C : [f_{2m}(x, y)]^3 + [f_{3m}(x, y)]^2 = 0$$

where $f_{2m}(x, y)$ and $f_{3m}(x, y)$ are general polynomials of the indicated degrees, and $m > 2$. Then $d = \deg(C) = 6m$, $\delta = 0$ and $\kappa = 6m^2$ because the only singularities of C are the points of intersection of the curves $f_{2m} = 0$ and $f_{3m} = 0$ and they are easily seen to be cusps. C is irreducible of geometric genus

$$g = \binom{6m-1}{2} - 6m^2$$

The dimension of the family of curves C is

$$R := \binom{2m+2}{2} + \binom{3m+2}{2} - 1 = \frac{1}{2}(13m+2)(m+1)$$

which is larger than

$$r = \frac{6m(6m-3)}{2} - 2\kappa = 6m^2 + 9m$$

In fact, $R - r = \binom{m-1}{2}$. Therefore $\mathcal{V}_{6m}^{0,6m^2}$ is superabundant at all points $[C]$.

Let's compute $h^1(C, N'_{C/\mathbb{P}^2})$. By the analysis of Example 4.7.1(iii) we know that $h^1(C, N'_{C/\mathbb{P}^2})$ equals the index of speciality ι of the linear system cut on the normalization $\tilde{C}$ of C by the curves of degree $6m$ passing through the cusps and tangent there to the cuspidal tangents. It is an easy computation (see [189] p. 220 for details) that

$$\iota = R - r$$

The conclusion is that each $[C]$ is a nonsingular point of a superbundant component of $\mathcal{V}_{6m}^{0,6m^2}$ of dimension R.

For a modern treatment of this example see [180].

4.7.3 Nonemptiness of Severi varieties

Even though we have proved precise results about the structure of the Severi varieties, it is not clear that nodal curves of given degree and number of nodes exist at all: the task of writing down explicitly the equation of such a curve is too concrete and precise to be within the reach of known techniques. Nevertheless Severi himself outlined a method to prove the existence of nodal curves. His approach is based on the notion of "analytic branch" and consists in analysing the local structure of $\overline{\mathcal{V}}_d^{\delta}$ along $\mathcal{V}_d^{\delta+1}$. While his proof makes perfectly good sense over the field of complex numbers, it is not straightforward to translate it into an algebraic proof. In this subsection we will show that the Severi varieties $\mathcal{V}_d^{\delta}$ of nodal curves are nonempty in the expected range of δ using a different method which is entirely algebraic and elementary. It is based on the theory of multiple point schemes, which we will now recall.

Consider a finite unramified morphism $f : X \to Y$ of algebraic schemes and let $N_k(f) \subset Y$ be the k-th multiple point scheme of f (for the definition see Example 4.2.9, page 200). Define

$$M_k(f) = f^{-1}(N_k(f)) \subset X$$

(scheme theoretic inverse image). Note that, in particular, $N_1(f)$ is supported on the image of f and $M_1(f) = X$. Let

$$X \times_Y X = \Delta \coprod X_2$$

(where Δ is the diagonal and the union is disjoint because f is unramified) and let $f_1 : X_2 \to X$ be the morphism induced by the first projection. Then f_1 is called *the first iteration morphism of* f.

Lemma 4.7.11. *Let* $f : X \to Y$ *be a finite unramified morphism of algebraic schemes. Then:*

(i) The first iteration morphism f_1 *is finite and unramified.*
(ii) $N_{k-1}(f_1) = M_k(f)$ *for all* $k \geq 2$.

Proof. (i) f_1 is finite and unramified because both properties are invariant under base change and composition with a closed embedding.

(ii) By the base change property of the Fitting ideals we have

$$M_k(f) = f^{-1}(N_k(f)) = N_k(\pi_1) = N_{k-1}(f_1)$$

where $\pi_1 : X \times_Y X \to X$ is the first projection. □

Lemma 4.7.12. *Let* $f : X \to Y$ *be a finite unramified morphism of purely dimensional algebraic schemes, such that* $\dim(X) = \dim(Y) - 1$ *and* Y *is nonsingular. Then every irreducible component* $\tilde{X}_2$ *of* X_2 *satisfies*

$$\dim(\tilde{X}_2) \geq \dim(Y) - 2$$

Proof. We have

$$X \times_Y X = (f \times f)^{-1}(\Delta_Y)$$

where $\Delta_Y \subset Y \times Y$ is the diagonal. Since Y is nonsingular, Δ_Y is regularly embedded of codimension $\dim(Y)$ in $Y \times Y$. It follows that every component of $X \times_Y X$ has codimension $\leq \dim(Y)$ in $X \times X$ (Lemma D.1.2). We deduce that every component $\tilde{X}_2$ of X_2 satisfies

$$\dim(\tilde{X}_2) \geq 2\dim(X) - \dim(Y) = \dim(Y) - 2$$ □

If the morphism f satisfies some further assumptions then one can describe its behaviour quite precisely.

Definition 4.7.13. *Let* $f : X \to Y$ *be a finite unramified morphism of algebraic schemes. Then* f *is called* self-transverse of codimension 1 *if* X *and* Y *are nonsingular and purely dimensional,* $\dim(X) = \dim(Y) - 1$, *and for any closed point* $y \in Y$ *and for any* r *distinct points* $x_1, \ldots, x_r \in f^{-1}(y)$ *the tangent spaces* $T_{x_1}X, \ldots, T_{x_r}X$, *viewed as subspaces of* T_yY, *are in general position (i.e. their intersection has codimension* r*).*

Lemma 4.7.14. *If* $f : X \to Y$ *is a self-transverse codimension 1 morphism then* $f_1 : X_2 \to X$ *is a self-transverse codimension 1 morphism.*

Proof. Let $(x_1, x_2) \in X_2$ and let $y = f_1(x_1, x_2) = f(x_1) = f(x_2)$. Then

$$T_{(x_1,x_2)}X_2 = T_{x_1}X \times_{T_yY} T_{x_2}X$$

Since f is self-transverse we have $\dim(T_{(x_1,x_2)}X_2) = \dim(Y)-2$. On the other hand, $\dim(X_2) \geq \dim(Y) - 2$, by Lemma 4.7.12. It follows that X_2 is nonsingular of pure dimension $\dim(Y) - 2$.

For any $x \in X$ and $(x, x') \in f_1^{-1}(x)$ the differential

$$df_{1(x,x')} : T_{(x,x')}X_2 \to T_xX$$

is an injection of codimension 1. For any $(x, x_2), \ldots, (x, x_s) \in f_1^{-1}(x)$ we have

$$df_{1(x,x_2)}(T_{(x,x_2)}X_2) \bigcap \cdots \bigcap df_{1(x,x_s)}(T_{(x,x_s)}X_2) = T_xX \bigcap T_{x_2}X \bigcap \cdots \bigcap T_{x_s}X$$

viewed as subspaces of T_yY. The self-transversality of f_1 now follows from the analogous property of f. □

With this terminology we can state the following useful result.

Proposition 4.7.15. *Assume that $f : X \to Y$ is a self-transverse codimension 1 morphism and that $N_r(f) \neq \emptyset$ for some $r \geq 2$. Then $N_r(f)$ has pure codimension r in Y and $N_s(f) \neq \emptyset$ for all $1 \leq s \leq r-1$. In particular, $N_s(f)$ has pure codimension s for all $1 \leq s \leq r - 1$.*

Proof. By induction on r. If $r = 2$ then $M_2(f) = N_1(f_1)$, by Lemma 4.7.11(ii), and has pure dimension equal to $\dim(X_2) = \dim(Y) - 2$ by Lemma 4.7.14: therefore $N_2(f)$ is of pure codimension 2. Moreover, $N_1(f) = f(X) \neq \emptyset$ and has pure dimension $\dim(X) = \dim(Y) - 1$.

Now assume $r \geq 3$. $N_r(f)$ has pure codimension r in Y if and only if $M_r(f)$ has pure codimension $r - 1$ in X. By Lemma 4.7.11(ii) we have

$$M_r(f) = N_{r-1}(f_1)$$

and by the inductive hypothesis $N_{r-1}(f_1)$ has pure codimension $r - 1$ because f_1 is self-transverse of codimension 1. Again by the inductive hypothesis

$$M_s(f) = N_{s-1}(f_1) \neq \emptyset$$

for all $1 \leq s - 1 \leq r - 2$. This implies that $N_s(f) \neq \emptyset$ for all $2 \leq s \leq r - 1$. The case $s = 1$ is a consequence of the first part of the proof. □

Next we will see how Proposition 4.7.15 can be applied to the study of the Severi varieties of nodal curves.

* * * * * *

Consider the universal family $\mathcal{H} \subset \mathbb{P}^2 \times \Sigma_d$ of plane curves of degree $d \geq 2$ and the cotangent sheaf $T^1_{\mathcal{H}/\Sigma_d}$. Since $\mathcal{H} \subset \mathbb{P}^2 \times \Sigma_d$ is a regular embedding of codimension 1, $T^1_{\mathcal{H}/\Sigma_d}$ is a quotient of $\mathcal{O}_{\mathbb{P}^2 \times \Sigma_d}$. Therefore we can identify

$T^1_{\mathcal{H}/\Sigma_d} = \mathcal{O}_Z$ where $Z \subset \mathcal{H}$ is a closed subscheme; evidently, the set of closed points of Z is

$$\{(p, s) : p \text{ is a singular point of } \mathcal{H}(s)\} \subset I\!P^2 \times \Sigma_d$$

Z can be described more precisely as follows. We let

$$F(X_0, X_1, X_2) = \sum_{n_0+n_1+n_2=d} A_{n_0,n_1,n_2} X_0^{n_0} X_1^{n_1} X_2^{n_2} = 0$$

be the equation of the universal curve $\mathcal{H}$ inside $I\!P^2 \times \Sigma_d$. Then Z is defined by the three equations:

$$\frac{\partial F}{\partial X_0} = 0, \quad \frac{\partial F}{\partial X_1} = 0, \quad \frac{\partial F}{\partial X_2} = 0$$

Denote by

$$I\!P^2 \xleftarrow{\pi_1} Z \xrightarrow{\pi_2} \Sigma_d$$

the projections.

Lemma 4.7.16. *(i) Z is irreducible, nonsingular, rational of codimension 3 in $I\!P^2 \times \Sigma_d$.*
(ii) π_2 maps Z birationally onto its image $W \subset \Sigma_d$, which is an irreducible rational divisor parametrizing all singular curves of degree d.

Proof. π_1 is surjective with fibres linear systems of dimension $\dim(\Sigma_d) - 3$ because for each $p \in I\!P^2$ the fibre $\pi_1^{-1}(p)$ is the linear system $\Sigma_d(-2p)$ of all curves of degree d which are singular at p. This proves (i). Moreover, by an easy application of Bertini's theorem one gets that a general element of $\Sigma_d(-2p)$ is a curve having a node at p as its only singular point. This takes care of (ii). □

Z contains an open subset Z_1 whose set of closed points is

$$\{(p, s) : \mathcal{H}(s) \text{ has only nodes as singularities and } p \text{ is a node of } \mathcal{H}(s)\}$$

From the proof of Lemma 4.7.16 it follows immediately that $Z_1 \subset Z$ is a nonempty (dense) open subset. Let

$$W_1 := \pi_2(Z_1)$$

W_1 is a dense open subset of W. It parametrizes all singular curves of degree d having only nodes as singularities; a general point of W_1 parametrizes a curve of degree d having one node and no other singularities.

Consider the closed subset $B_d := \pi_2(Z \backslash Z_1) \subset \Sigma_d$, which parametrizes all singular non-nodal curves of degree d, and the morphism

$$\pi : Z_1 \to \Sigma_d \backslash B_d$$

obtained by restricting π_2.

Proposition 4.7.17. *π is birational onto its image W_1, finite, unramified and self-transverse of codimension 1.*

Proof. If $s \in W_1 = \pi(Z_1)$ is a closed point then $\pi^{-1}(s)$ is the scheme of nodes of $\mathcal{H}(s)$ which is finite and reduced. Therefore π is unramified. Moreover, π is the restriction over an open subset of Σ_d of a projective morphism. Therefore π is finite. The birationality onto its image follows from that of π_2 (Lemma 4.7.16). In order to prove that π is self-transverse of codimension 1, let $X \subset I\!P^2$ be a curve of degree d having δ nodes $\{p_1, \ldots, p_\delta\}$ and no other singularity. Then $(p_i, [X]) \in Z_1, i = 1, \ldots, \delta$. The local analysis of Example 4.7.1(i) and the fact that π is unramified show that we have

$$\mathrm{Im}[d\pi_{(p_i,[X])}] = H^0(X, N_{X/I\!P^2}(-p_i))$$

where

$$N_{X/I\!P^2}(-p_i) = \ker[N_{X/I\!P^2} \to T^1_{X,p_i}]$$

Therefore

$$\bigcap_{i=1}^{\delta} \mathrm{Im}[d\pi_{(p_i,[X])}] = H^0(X, N'_{X/I\!P^2}) = T_{[X]}\mathcal{V}_d^\delta$$

Since $T_{[X]}\mathcal{V}_d^\delta$ has codimension δ by Corollary 4.7.8, this equality means that the subspaces $\mathrm{Im}[d\pi_{(p_i,[X])}]$ are hyperplanes of $T_{[X]}\Sigma_d = H^0(X, N_{X/I\!P^2})$ in general position. □

We can now prove the main result of this subsection.

Theorem 4.7.18 (Severi [173]). *Let $X \subset I\!P^2$ be a curve of degree d having δ nodes and no other singularity, i.e. such that $[X] \in \mathcal{V}_d^\delta$. Then*

$$\emptyset \neq \mathcal{V}_d^\delta \subset \overline{\mathcal{V}}_d^{\delta-1} \subset \cdots \subset \overline{\mathcal{V}}_d^1 \subset \overline{\mathcal{V}}_d^0 = \Sigma_d$$

and $\overline{\mathcal{V}}_d^s$ has pure codimension 1 in $\overline{\mathcal{V}}_d^{s-1}$, for every $s = 1, \ldots, \delta$.

Proof. From the previous analysis it follows that for each s

$$\mathcal{V}_d^s = N_s(\pi) \backslash N_{s+1}(\pi)$$

The existence of X implies that $\mathcal{V}_d^\delta \neq \emptyset$. The theorem is now an immediate consequence of Propositions 4.7.17 and 4.7.15. □

Corollary 4.7.19. *(i) For every $d \geq 2$ and*

$$0 \leq \delta \leq \binom{d}{2}$$

the Severi variety $\mathcal{V}_d^\delta$ is nonempty.

(ii) For every $d \geq 2$ and

$$0 \leq \delta \leq \binom{d-1}{2}$$

the Severi variety $\mathcal{V}_d^\delta$ contains irreducible curves.

Proof. (i) A curve X consisting of d distinct lines no three of which pass through the same point defines a point of $\mathcal{V}_d^{\delta_o}$ with $\delta_o = \binom{d}{2}$ and the non-emptiness of $\mathcal{V}_d^{\delta}$ for $\delta \leq \delta_o$ follows from the theorem.

(ii) A general projection in $\mathbb{P}^2$ of a rational normal curve $\Gamma_d \subset \mathbb{P}^d$ is an irreducible curve Y of degree d having $\binom{d-1}{2}$ nodes. The theorem implies that Y is in the closure of $\mathcal{V}_d^{\delta}$ for every $\delta \leq \binom{d-1}{2}$. The following lemma guarantees that for each such δ we can find irreducible curves in $\mathcal{V}_d^{\delta}$. □

Lemma 4.7.20. *Let $\mathcal{C} \subset \mathbb{P}^2 \times S$ be a flat family of plane curves of degree $d \geq 2$, with S an algebraic scheme, and let $o \in S$ be a* **k***-rational point. Assume that the fibre $\mathcal{C}(o)$ is reduced and irreducible. Then $\mathcal{C}(s)$ is reduced and irreducible for all $s \in S$ in an open neighbourhood of o.*

Proof. Denote by $f : \mathcal{C} \to S$ the projection. Since $H^i(\mathcal{C}(s), \mathcal{O}_{\mathcal{C}(s)}(m)) = 0$ for all $m \geq d-2, i \geq 1$, and for all $s \in S$, by Theorem 4.2.5

$$t_m^0(s) : f_*\mathcal{O}_{\mathcal{C}}(m)_s \otimes \mathbf{k}(s) \to H^0(\mathcal{C}(s), \mathcal{O}_{\mathcal{C}(s)}(m))$$

is bijective and $f_*\mathcal{O}_{\mathcal{C}}(m)_s$ is free for all $m \geq d-2$ and $s \in S$.

The multiplication map

$$\mu_s : H^0(\mathcal{O}_{\mathcal{C}(s)}(d-1)) \otimes H^0(\mathcal{O}_{\mathcal{C}(s)}(d-1)) \to H^0(\mathcal{O}_{\mathcal{C}(s)}(2d-2))$$

is surjective for all $s \in S$. Then the multiplication map

$$\mu : f_*\mathcal{O}_{\mathcal{C}}(d-1) \otimes f_*\mathcal{O}_{\mathcal{C}}(d-1) \to f_*\mathcal{O}_{\mathcal{C}}(2d-2)$$

is surjective and $\ker(\mu)$ is locally free.

Let $R \subset f_*\mathcal{O}_{\mathcal{C}}(d-1) \otimes f_*\mathcal{O}_{\mathcal{C}}(d-1)$ be the cone of reducible tensors. For each $s \in S$ it restricts to the cone $R(s)$ of reducible tensors in $H^0(\mathcal{O}_{\mathcal{C}(s)}(d-1)) \otimes H^0(\mathcal{O}_{\mathcal{C}(s)}(d-1))$, by the bijectivity of the maps $t_m^0(s)$ for $m = d-1, 2d-2$. The condition that $\mathcal{C}(o)$ is reduced and irreducible is equivalent to $R(o) \cap \ker(\mu_o) = \{0\}$. Therefore there is an open neighbourhood of U of $o \in S$ such that $R(s) \cap \ker(\mu_s) = \{0\}$, i.e. such that $\mathcal{C}(s)$ is reduced and irreducible, for all $s \in U$. □

Remarks 4.7.21. (i) It is easy to see that $\mathcal{V}_d^{\delta}$ is reducible in general. For example, $\mathcal{V}_4^3$ has two irreducible components and $\mathcal{V}_6^9$ has five irreducible components. An important classical problem, known as the "Severi problem", has been to decide about the irreducibility of the open set of $\mathcal{V}_d^{\delta}$ parametrizing *irreducible* nodal curves. This problem has been solved affirmatively in [82] and, independently, in [147] (see also [182]). A report of Harris' proof is given in [121]. It is known that the open set of $\mathcal{V}_d^{\delta,\kappa}$ parametrizing irreducible curves is reducible in general if $\kappa > 0$. For examples see [160] (such examples are also reported in [189]).

(ii) Theorem 4.7.18 and some generalizations of it have been reconsidered in [179], but the proof given there is based on infinitesimal considerations which seem to need a further insight. For an interesting discussion see [60]. The analogous problem for the varieties $\mathcal{V}_d^{\delta,\kappa}$ is still open, i.e. we don't know a characterization of the values of d, δ, κ for which $\mathcal{V}_d^{\delta,\kappa} \neq \emptyset$. For partial results see [95]; for a classical discussion see [159]. The results on multiple point schemes used here are special cases of a general theory for which we refer the reader to [99], [101] and to the references quoted there.

(iii) The proof of Theorem 4.7.3 can be easily modified to prove the existence of universal families of curves with nodes and cusps (generalized Severi varieties) on a projective nonsingular surface Y. In such a proof one replaces Σ_d by Hilb^Y and uses the existence and the universal property of Hilb^Y.

Such generalized Severi varieties behave in a way relatively similar to the $\mathcal{V}_d^{\delta,\kappa}$'s as long as Y has Kodaira dimension ≤ 0 (see [179], [112]). On surfaces of general type the situation changes radically. On such a surface Y the generalized Severi varieties can be superabundant even when $\kappa = 0$ and it is not known in which range of δ they are not empty. A systematic study of them has started relatively recently. We refer the reader to [34], [69], [32], [56], [57] for details.

A

Flatness

The algebraic notion of flatness, introduced for the first time in [164], is the basic technical tool for the study of families of algebraic varieties and schemes. In this appendix we will overview the main algebraic results needed. For the properties of flat morphisms between schemes we refer to [84]. See also § 4.2.

A module M over a ring A is *A-flat* (or *flat over A*, or simply *flat*) if the functor $N \mapsto M \otimes_A N$ from the category of A-modules into itself is exact. Since this functor is always right exact, the flatness means that it takes monomorphisms into monomorphisms. An *A-algebra B* is *flat over A* if B is flat as an A-module.

The A-module M is said to be *faithfully flat* if for every sequence of A-modules $N' \to N \to N''$ the sequence

$$M \otimes_A N' \to M \otimes_A N \to M \otimes_A N''$$

is exact if and only if the original sequence is exact. Obviously, if M is faithfully flat then it is flat. In a similar way we give the notion of faithfully flat A-algebra. It is straightforward to check that if $A \to B$ is a local homomorphism of local rings, then a B-module of finite type is faithfully A-flat if and only if it is A-flat and nonzero.

Recall that the flatness of an A-module M is equivalent to any of the following conditions:

(1) $\mathrm{Tor}_i^A(M, N) = (0)$ for all $i > 0$ and for every A-module N.
(2) $\mathrm{Tor}_1^A(M, N) = (0)$ for every A-module N.
(3) $\mathrm{Tor}_1^A(M, N) = (0)$ for every finitely generated A-module N.
(4) $\mathrm{Tor}_1^A(M, A/I) = (0)$ for every ideal $I \subset A$.
(5) $I \otimes_A M \to M$ is injective for every ideal $I \subset A$.
(6) $I \otimes_A M \to IM$ is an isomorphism for every ideal $I \subset A$.

Example A.1. Let k be a ring, u, v indeterminates and $f : k[u, uv] \to k[u, v]$ the inclusion. Then

$$\frac{k[u, uv]}{(uv)} = k[u] \xrightarrow{u} k[u] = \frac{k[u, uv]}{(uv)}$$

is injective. Tensoring by $\otimes_{k[u,uv]}k[u,v]$ we obtain:

$$\frac{k[u,v]}{(uv)} \xrightarrow{u} \frac{k[u,v]}{(uv)}$$

which is not injective. Therefore f is not flat.

We list without proof a few *basic properties of flat modules*:

Proposition A.2. *(I) M is A-flat if and only if M_p is A_p-flat for every prime ideal p.*
(II) Every projective module is flat.
(III) Assume M is finitely generated. Then M is flat if and only if it is projective; if A is local then M is flat if and only if it is free.
(IV) If $S \subset A$ is a multiplicative subset then A_S is A-flat.
(V) A direct sum $M = \oplus_{i \in I} M_i$ is flat if and only if all M_i's are flat.
(VI) Let

$$0 \to M' \to M \to M'' \to 0$$

be an exact sequence of A-modules with M'' flat. Then M is flat if and only if M' is flat.
(VII) Base change: if M is A-flat and $f : A \to B$ is a ring homomorphism, then $M \otimes_A B$ is B-flat.
(VIII) Transitivity: if B is a flat A-algebra and N is a flat B-module, then N is A-flat.
(IX) If A is a noetherian ring and I is an ideal, the I-adic completion $\hat{A}$ is a flat A-algebra. If I is contained in the Jacobson radical of A then $\hat{A}$ is a faithfully flat A-algebra.
(X) If B is an A-algebra and if there exists a B-module M which is faithfully flat, then the morphism $\mathrm{Spec}(B) \to \mathrm{Spec}(A)$ is surjective.
(XI) If $X_1, \dots, X_r$ are indeterminates, then $A[X_1, \dots, X_r]$ and $A[[X_1, \dots, X_r]]$ are A-flat.

The following result is frequently used:

Proposition A.3. *If A is an artinian local ring with residue field k the following are equivalent for an A-module M:*

(i) M is free
(ii) M is flat
(iii) $\mathrm{Tor}_1^A(M,k) = (0)$

Proof. $(i) \Rightarrow (ii) \Rightarrow (iii)$ are clear.
$(iii) \Rightarrow (ii)$. Let N be a finitely generated A-module and let

$$N = N_0 \supset \cdots \supset N_n = (0)$$

be a composition series for N such that

$$N_i/N_{i+1} \cong k$$

for $i = 0, \ldots, n-1$. Using the Tor exact sequences from hypothesis (iii) we deduce that $\mathrm{Tor}_1(M, N) = (0)$ and the flatness of M follows from (3).

Let's now prove $(ii) \Rightarrow (i)$. Let $\{e_j\}_{j \in J}$ be a system of elements of M which induces a basis of $M \otimes_A k$ over k. The system $\{e_j\}$ defines a homomorphism $f : A^J \to M$ which induces an isomorphism $k^J \to M \otimes_A k$. From the following lemma we find that f is an isomorphism, and therefore M is free. □

Lemma A.4. *Let R be a ring, I an ideal and $f : F \to G$ a homomorphism of R-modules with G flat. Assume that one of the following conditions is satisfied:*

(a) I is nilpotent.
(b) R is noetherian, I is contained in the Jacobson radical of R and F and G are finitely generated.

If the induced homomorphism $F/IF \to G/IG$ is an isomorphism, then f is an isomorphism.

Proof. Let $K = \mathrm{coker}(f)$. Tensoring the exact sequence

$$F \to G \to K \to 0$$

with R/I we get $K/IK = 0$: from Nakayama's lemma (which holds in either of hypotheses (a) and (b)) it follows that $K = 0$, and therefore F is surjective. Letting $H = \ker(f)$ we deduce an exact sequence

$$0 \to H/IH \to F/IF \to G/IG \to 0$$

using the flatness of G. By Nakayama again we deduce $H = 0$ and the conclusion follows. □

The following is a basic criterion of flatness.

Theorem A.5 (Local criterion of flatness). *Suppose that $\varphi : A \to B$ is a local homomorphism of local noetherian rings, and let $k = A/m_A$ be the residue field of A. If M is a finitely generated B-module, then the following conditions are equivalent:*

(i) M is A-flat.
(ii) $\mathrm{Tor}_1^A(M, k) = 0$.
(iii) $M \otimes_A (A/m_A^n)$ is flat over A/m_A^n for every integer $n \geq 1$.
(iv) $M \otimes_A (A/m_A^n)$ is free over A/m_A^n for every integer $n \geq 1$.

Proof. $(i) \Rightarrow (ii)$ is obvious.

$(ii) \Rightarrow (i)$ see [48], Th. 6.8, p. 167.

$(i) \Rightarrow (iii)$ is obvious.

$(iii) \Rightarrow (i)$ It suffices to show that for every inclusion $N' \to N$ of A-modules of finite type we have an inclusion $M \otimes_A N' \to M \otimes_A N$. For this purpose it suffices to show that the kernel of this last map is contained in

$$K_n := \ker[M \otimes_A N' \to M \otimes_A (N'/N' \cap m_A^n N)]$$

for all n, because $\bigcap_n K_n = (0)$. We have a commutative diagram with exact rows:

$$\begin{array}{ccccccc} 0 \to & K_n & \to & M \otimes_A N' & \to & M \otimes_A (N'/N' \cap m_A^n N) & \to 0 \\ & & & \downarrow & & \downarrow & \\ & & & M \otimes_A N & \to & M \otimes_A (N/m_A^n N) & \to 0 \end{array}$$

The last vertical arrow coincides with the map obtained from the injection

$$N'/N' \cap m_A^n N \to N/m_A^n N$$

after tensoring over A/m_A^n with the A/m_A^n-flat module $M \otimes_A (A/m_A^n)$, and therefore it is injective. The conclusion follows from the above diagram.

$(iii) \Leftrightarrow (iv)$ follows from Proposition A.3 because A/m_A^n is artinian. □

For a more general version of the local criterion we refer to [3], exp. IV, Théorème 5.6. Note that A.3 is a special case of A.5.

Corollary A.6. *Suppose that $\varphi : A \to B$ is a local homomorphism of local noetherian rings, let $k = A/m_A$ be the residue field of A, M, N two finitely generated B-modules, and suppose that N is A-flat. Let $u : M \to N$ be a B-homomorphism. Then the following are equivalent:*

(i) u is injective and coker(u) is A-flat.
(ii) $u \otimes 1 : M \otimes k \to N \otimes k$ is injective.

Proof. (i) $\Rightarrow$ (ii). Let $G = \text{coker}(u)$. Tensoring by k the exact sequence

$$0 \to M \xrightarrow{u} N \to G \to 0$$

by k we obtain the exact sequence:

$$\text{Tor}_1^A(G, k) \to M \otimes_A k \xrightarrow{u \otimes 1} N \otimes_A k \to G \otimes_A k \to 0$$

Since G is A-flat we have $\text{Tor}_1^A(G, k) = 0$, and it follows that $u \otimes 1$ is injective.

(ii) $\Rightarrow$ (i). Factor $u \otimes 1$ as

$$M \otimes_A k \xrightarrow{\alpha} \text{Im}(u) \otimes_A k \xrightarrow{\beta} N \otimes_A k$$

Then α is an isomorphism and β is injective. Tensoring by k the exact sequence

$$0 \to \text{Im}(u) \to N \to G \to 0 \tag{A.1}$$

we obtain the exact sequence:

$$\text{Tor}_1^A(N, k) \to \text{Tor}_1^A(G, k) \to \text{Im}(u) \otimes_A k \xrightarrow{\beta} N \otimes_A k \to G \otimes_A k \to 0$$

Since N is A-flat we have $\text{Tor}_1^A(N, k) = 0$; from the injectivity of β we deduce $\text{Tor}_1^A(G, k) = 0$ and from A.5 it follows that G is A-flat. Applying (VI) to the exact sequence (A.1) we deduce that $\text{Im}(u)$ is A-flat as well. Consider the exact sequence:

$$0 \to \ker(u) \to M \to \mathrm{Im}(u) \to 0$$

and tensor by k. We obtain the exact sequence:

$$0 \to \ker(u) \otimes_A k \to M \otimes_A k \xrightarrow{\alpha} \mathrm{Im}(u) \otimes_A k \to 0$$

Since α is an isomorphism we deduce that $\ker(u) \otimes_A k = 0$, and therefore $\ker(u) = 0$ by Nakayama's lemma. □

A related result is the following:

Lemma A.7. *Let B be a local ring with residue field K, and let $d : G \to F$ be a homomorphism of finitely generated B-modules, with F free. Then d is split injective if and only if $d \otimes_B K : G \otimes_B K \to F \otimes_B K$ is injective. In such a case G is also free.*

Proof. d is split injective if and only if coker(d) is free and d is injective. If this last condition is satisfied then clearly $d \otimes_B K$ is injective.

Conversely, assume that $d \otimes_B K$ is injective, and factor d as

$$G \to \mathrm{Im}(d) \to F$$

We see that

$$\begin{array}{ccll} G \otimes_B K & \to & \mathrm{Im}(d) \otimes_B K & \text{is bijective} \\ \mathrm{Im}(d) \otimes_B K & \to & F \otimes_B K & \text{is injective} \end{array}$$

From the exact sequence

$$0 \to \mathrm{Im}(d) \to F \to \mathrm{coker}(d) \to 0$$

we get

$$0 \to \mathrm{Tor}_1(\mathrm{coker}(d), K) \to \mathrm{Im}(d) \otimes_B K \to F \otimes_B K$$

so $\mathrm{Tor}_1(\mathrm{coker}(d), K) = (0)$ and this implies that coker(d) is free. From the above exact sequence we deduce that Im(d) is free as well, so that

$$0 \to \ker(d) \to G \to \mathrm{Im}(d) \to 0$$

is split exact. Recalling that $G \otimes_B K \cong \mathrm{Im}(d) \otimes_B K$ we deduce that $\ker(d) \otimes_B K = (0)$, hence $\ker(d) = (0)$ by Nakayama. □

For the reader's convenience we include the proof of the following well-known lemma:

Lemma A.8. *Let (B, m) be a noetherian local integral domain, with residue field K and quotient field L. If M is a finitely generated B-module and if*

$$\dim_K(M \otimes_B K) = \dim_L(M \otimes_B L) = r$$

then M is free of rank r.

Proof. Let $m_1, \ldots, m_r \in M$ be such that their images in $M \otimes_B K = M/mM$ form a basis. Then they define a homomorphism $\varphi : B^r \to M$ and we have an exact sequence:

$$0 \to N \to B^r \xrightarrow{\varphi} M \to Q \to 0$$

where N and Q are kernel and cokernel of φ. Since tensoring with K we get

$$K^r \xrightarrow{\bar{\varphi}} M/mM \to Q/mQ \to 0$$

and $\bar{\varphi}$ is surjective, we get $Q/mQ = (0)$ and from Nakayama's lemma it follows that $Q = (0)$: hence φ is surjective. Now we tensor the above exact sequence with L, which is flat over B (by (IV)), and we obtain the exact sequence:

$$0 \to N \otimes_B L \to L^r \xrightarrow{\tilde{\varphi}} M \otimes_B L \to 0$$

Since $M \otimes_B L \cong L^r$ and $\tilde{\varphi}$ is surjective, it follows that $N \otimes_B L = \ker(\tilde{\varphi}) = (0)$. Therefore N is a torsion module. But $N \subset B^r$ and therefore $N = (0)$. □

We have the following useful criterion:

Lemma A.9. *Let $A \to A'$ be a small extension in $\mathcal{A}$, and let $g : A \to R$ be a homomorphism of* **k***-algebras. Let $R_0 = R \otimes_A \mathbf{k}$. Then g is flat if and only if*

$$\ker(R \to R \otimes_A A') \cong R_0$$

and the homomorphism $g' : A' \to R \otimes_A A'$ induced by g is flat.

Proof. Assume that g is flat. Then since $R \otimes_A (\epsilon) \cong R \otimes_A \mathbf{k} = R_0$ and $\mathrm{Tor}_1^A(R, A') = 0$, from the exact sequence

$$0 \to \mathrm{Tor}_1^A(R, A') \to R \otimes_A (\epsilon) \to R \to R \otimes_A A' \to 0 \tag{A.2}$$

we deduce that the first condition is satisfied. The flatness of g' is obvious.

Assume conversely that the conditions of the statement are satisfied. Then the sequence (A.2) implies that $\mathrm{Tor}_1^A(R, A') = 0$. If $A' = \mathbf{k}$ the conclusion follows from A.3. If not, from the exact sequence

$$0 \to m_{A'} \to A' \to \mathbf{k} \to 0$$

one gets the exact sequence:

$$\begin{array}{ccccccccc} \mathrm{Tor}_1^A(R, A') & \to & \mathrm{Tor}_1^A(R, \mathbf{k}) & \xrightarrow{\partial} & R \otimes_A m_{A'} & \to & R' & \to & R \otimes_A \mathbf{k} \to 0 \\ \| & & & & \| & & & & \| \\ 0 & & & & R' \otimes_{A'} m_{A'} & & & & R' \otimes_{A'} \mathbf{k} \end{array}$$

From the flatness of R' over A' we deduce that $\partial = 0$, hence $\mathrm{Tor}_1^A(R, \mathbf{k}) = 0$, and we conclude by A.3. □

* * * * * *

Flatness in terms of generators and relations

Let P be a noetherian $\mathbf{k}$-algebra, $J \subset P$ an ideal. Let A be in $\mathrm{ob}(\mathcal{A})$, $P_A = P \otimes_{\mathbf{k}} A$, and $\mathbf{J} \subset P_A$ an ideal such that $(P_A/\mathbf{J}) \otimes_A \mathbf{k} \cong P/J$. We want to find the conditions $\mathbf{J}$ has to satisfy so that $P_A/\mathbf{J}$ is A-flat.

We have the following:

Theorem A.10. *Let*

$$\Pi_0 : \quad P^n \to P^N \to P \to P/J \to 0$$

be a presentation of P/J as a P-module. Then the following conditions are equivalent for an ideal $\mathbf{J} \subset P_A$:

(i) $P_A/\mathbf{J}$ is A-flat and $(P_A/\mathbf{J}) \otimes_A \mathbf{k} \cong P/J$.
(ii) There is an exact sequence

$$\Pi : \quad P_A^n \to P_A^N \to P_A \to P_A/\mathbf{J} \to 0$$

such that $\Pi_0 = \Pi \otimes_A \mathbf{k} \quad (= \Pi/m_A\Pi)$.
(iii) There is a complex

$$\Pi : \quad P_A^n \xrightarrow{\varphi} P_A^N \to P_A \to P_A/\mathbf{J} \to 0$$

which is exact except possibly at P_A^N, such that $\quad \Pi_0 = \Pi \otimes_A \mathbf{k}$.

Proof. $(ii) \Rightarrow (i)$. We have:

$$\mathrm{Tor}_1^A(P_A/\mathbf{J}, \mathbf{k}) = H_1(\Pi \otimes \mathbf{k}) = H_1(\Pi_0) = (0)$$

From A.3 it follows that $P_A/\mathbf{J}$ is A-flat. Moreover, (ii) implies that $(P_A/\mathbf{J}) \otimes_A \mathbf{k} \cong P/J$.

$(i) \Rightarrow (ii)$. Choose a P_A-homomorphism $p : P_A^N \to \mathbf{J}$ which makes the following diagram commute:

$$\begin{array}{cccc} p: & P_A^N & \to & \mathbf{J} \\ & \downarrow & & \downarrow \\ p_0: & P^N & \to & J \end{array}$$

where p_0 is the surjective homomorphism defined by the presentation Π_0. From the flatness of $P_A/\mathbf{J}$ it follows that $\mathrm{Tor}_1^A(P_A/\mathbf{J}, \mathbf{k}) = (0)$; hence the exact sequence

$$\begin{array}{ccccccccc} 0 \to & \mathrm{Tor}_1^A(P_A/\mathbf{J}, \mathbf{k}) & \to & \mathbf{J} \otimes \mathbf{k} & \to & P_A \otimes \mathbf{k} & \to & (P_A/\mathbf{J}) \otimes_A \mathbf{k} & \to 0 \\ & & & & & \| & & \| & \\ & & & & & P & & P/J & \end{array}$$

implies that $\mathbf{J} \otimes \mathbf{k} = J$. It follows that $p \otimes_A \mathbf{k} = p_0$ and therefore

$$\mathrm{coker}(p) \otimes_A \mathbf{k} = \mathrm{coker}(p_0) = (0)$$

so that $\mathrm{coker}(p) = (0)$ by Nakayama's lemma. Hence p is surjective.

Now consider the exact sequence

$$0 \to \ker(p) \to P_A^N \to \mathbf{J} \to 0$$

and the associated Tor sequence:

$$\mathrm{Tor}_1^A(\mathbf{J}, \mathbf{k}) \to \ker(p)/m_A\ker(p) \to P^N \to J \to 0 \tag{A.3}$$

From the flatness of $P_A/\mathbf{J}$ and from the exact sequence

$$0 \to \mathbf{J} \to P_A \to P_A/\mathbf{J} \to 0$$

we have $\mathrm{Tor}_1^A(\mathbf{J}, \mathbf{k}) = \mathrm{Tor}_2^A(P_A/\mathbf{J}, \mathbf{k}) = (0)$. Therefore from (A.3) we see that

$$\ker(p)/m_A\ker(p) \cong \ker(p_0)$$

Arguing as before we can find a surjective homomorphism $q : P_A^n \to \ker(p)$ which makes the following diagram commutative:

$$\begin{array}{ccc} P_A^n & \xrightarrow{q} & \ker(p) \\ \downarrow & & \downarrow \\ P^n & \to & \ker(p_0) \end{array}$$

$(ii) \Rightarrow (iii)$ is obvious.

$(iii) \Rightarrow (i)$ If Π is not exact at P_A^N then we can add finitely many generators of the kernel of $P_A^N \to P_A$ to obtain an exact sequence

$$\Pi' : \qquad P_A^{n'} \xrightarrow{\varphi'} P_A^N \to P_A \to P_A/\mathbf{J} \to 0$$

Then $\Pi' \otimes_A \mathbf{k}$ has the form:

$$P^{n'} \xrightarrow{\varphi'\otimes\mathbf{k}} P^N \to P \to P/J \to 0$$

Since

$$\mathrm{Im}(\varphi \otimes \mathbf{k}) \subset \mathrm{Im}(\varphi' \otimes \mathbf{k}) \subset \ker[P^N \to P]$$

we see that $\mathrm{Im}(\varphi' \otimes \mathbf{k}) = \ker[P^N \to P]$ and therefore $\Pi' \otimes_A \mathbf{k}$ is exact. Now (i) follows from A.3. □

Corollary A.11. *Assume that* $J = (f_1, \dots, f_N) \subset P$ *and that*

$$\mathbf{J} = (F_1, \dots, F_N) \subset P_A$$

with $f_j = F_j \pmod{m_A P_A}$, $j = 1, \dots, N$. *Then every relation among* $f_1, \dots, f_N$ *lifts to a relation among* $F_1, \dots, F_N$ *if and only if* $P_A/\mathbf{J}$ *is A-flat and* $(P_A/\mathbf{J}) \otimes_A \mathbf{k} \cong P/J$.

Proof. The condition that the F_j's reduce to the f_j's modulo $m_A P_A$ implies that the exact sequence

$$P_A^N \xrightarrow{\mathbf{F}} P_A \to P_A/\mathbf{J} \to 0$$

reduces to

$$P^N \xrightarrow{\mathbf{f}} P \to P/J \to 0 \tag{A.4}$$

when tensored by $\otimes_A \mathbf{k}$. Complete (A.4) to a presentation Π_0 of P/J. The condition that every relation among $f_1, \ldots, f_N$ lifts to a relation among $F_1, \ldots, F_N$ is a restatement of condition (iii) of A.10. Therefore the conclusion follows from Theorem A.10. □

Example A.12. Let A be in ob($\mathcal{A}$). Suppose that $f_1, \ldots, f_N \in P$ form a regular sequence, and let $F_1, \ldots, F_N \in P_A$ be any liftings of $f_1, \ldots, f_N$, i.e. such that $f_j = F_j \pmod{m P_A}$, $j = 1, \ldots, N$. Then $\mathbf{J} = (F_1, \ldots, F_N) \subset P_A$ defines a flat family of deformations of $X = \mathrm{Spec}(P/J)$, where $J = (f_1, \ldots, f_N)$.

In fact, every relation among $f_1, \ldots, f_N$ is a linear combination of the trivial ones

$$r_{ij} = (0, \ldots, f_j, \ldots, -f_i, \ldots, 0) \qquad 1 \leq i < j \leq N$$

and these can be lifted to the corresponding trivial relations

$$R_{ij} = (0, \ldots, F_j, \ldots, -F_i, \ldots, 0)$$

among $F_1, \ldots, F_N$. Applying Corollary A.11 it is easy to show that $F_1, \ldots, F_N$ form a regular sequence.

NOTES

1. In the proof of Theorem A.10 the condition that A is artinian has only been used in the proof of $(i) \Rightarrow (ii)$ in order to apply Nakayama's lemma. In particular, the implications $(ii) \Rightarrow (i)$, $(iii) \Rightarrow (i)$ and $(ii) \Rightarrow (iii)$ hold for any $A \in$ ob($\mathcal{A}^*$). Using the local criterion of flatness it is easy to verify that the implication $(i) \Rightarrow (ii)$ (and therefore the equivalence of the three conditions) holds as well if A is in ob($\hat{\mathcal{A}}$).

B

Differentials

Let $A \to B$ be a ring homomorphism. As usual, we will denote by $\Omega_{B/A}$ the *module of differentials of B over A*, and by $d_{B/A} : B \to \Omega_{B/A}$ the canonical A-derivation. Recall that

$$\Omega_{B/A} := I/I^2$$

where $I = \ker(B \otimes_A B \xrightarrow{\mu} B)$ is the natural map, and for each $b \in B$

$$d_{B/A}(b) = b \otimes 1 - 1 \otimes b$$

is called the *differential of b*. We have a natural isomorphism of B-modules

$$\mathrm{Der}_A(B, M) \cong \mathrm{Hom}_B(\Omega_{B/A}, M)$$

Note that the exact sequence

$$0 \to \Omega_{B/A} \to (B \otimes_A B)/I^2 \xrightarrow{\mu'} B \to 0 \tag{B.1}$$

where μ' is induced by μ, is an A-extension of B. The ring

$$P_{B/A} := (B \otimes_A B)/I^2$$

is called the *algebra of principal parts* of B over A. The A-extension (B.1) is trivial because we have splittings:

$$\lambda_1, \lambda_2 : B \to P_{B/A}$$

defined by $\lambda_1(b) = \overline{b \otimes 1}$, $\lambda_2(b) = \overline{1 \otimes b}$; note that $d_{B/A} = \lambda_1 - \lambda_2$. We will consider $P_{B/A}$ as a B-algebra via λ_1.

The following are some fundamental properties of the modules of differentials:

Proposition B.1. *(i) If*

$$\begin{array}{ccc} B & & \\ \uparrow & & \\ A & \longrightarrow & A' \end{array}$$

are ring homomorphisms, then:

$$\Omega_{B/A} \otimes_A A' \cong \Omega_{B\otimes_A A'/A'}$$

(ii) If $A \to B$ is a ring homomorphism and $\Delta \subset B$ is a multiplicative system, then:

$$\Omega_{\Delta^{-1}B/A} \cong \Delta^{-1}\Omega_{B/A}$$

(iii) Let $K \to L$ be a finitely generated extension of fields. Then

$$\dim_L(\Omega_{L/K}) \geq \operatorname{trdeg}(L/K)$$

and equality holds if and only if L is separably generated over K. In particular, $\Omega_{L/K} = (0)$ if and only if $K \subset L$ is a finite algebraic separable extension.

Proof. See [48]. □

We now introduce two standard exact sequences.

Theorem B.2 (Relative cotangent sequence). *Given ring homomorphisms*

$$A \xrightarrow{f} B \xrightarrow{g} C$$

there is an exact sequence of C-modules:

$$\Omega_{B/A} \otimes_B C \xrightarrow{\alpha} \Omega_{C/A} \xrightarrow{\beta} \Omega_{C/B} \to 0 \tag{B.2}$$

where the maps are given by:

$$\alpha(d_{B/A}(b) \otimes c) = c d_{C/A}(g(b)); \quad \beta(d_{C/A}(r)) = d_{C/B}(r) \quad b \in B, \ c \in C$$

Proof. See [48], prop. 16.2. □

When $B \to C$ is surjective we have $\Omega_{C/B} = (0)$ and the next theorem describes $\ker(\alpha)$.

Theorem B.3 (Conormal sequence). *Let*

$$A \xrightarrow{f} B \xrightarrow{g} C$$

be ring homomorphisms with g surjective, and let $J = \ker(g)$, so that $C = B/J$. Then:

(i) We have an exact sequence

$$J/J^2 \xrightarrow{\delta} \Omega_{B/A} \otimes_B C \xrightarrow{\alpha} \Omega_{C/A} \to 0 \tag{B.3}$$

where δ is the C-linear map defined by $\delta(\bar{x}) = d_{B/A}(x) \otimes 1$.

(ii) There is an isomorphism

$$\Omega_{(B/J^2)/A} \otimes_{(B/J^2)} C \cong \Omega_{B/A} \otimes_B C$$

In other words the conormal sequence (B.3) depends only on the first infinitesimal neighbourhood of Spec(C) *in* Spec(B).

(iii) The map δ is a split injection if and only if there is a map of A-algebras $C \to B/J^2$ splitting the projection $B/J^2 \to C$.

Proof. (i) see e.g. [48], prop. 16.3.

(ii) Comparing the exact sequence (B.3) with the analogous sequence associated to $A \to B/J^2 \to C$ we get a commutative diagram:

$$\begin{array}{ccccccc} J/J^2 & \to & \Omega_{B/A} \otimes_B C & \to & \Omega_{C/A} & \to 0 \\ \| & & \downarrow & & \| & \\ J/J^2 & \to & \Omega_{(B/J^2)/A} \otimes_{(B/J^2)} C & \to & \Omega_{C/A} & \to 0 \end{array}$$

and the vertical arrow, which is induced by $B \to B/J^2$, must be an isomorphism.

(iii) By (ii) we may assume that $J^2 = 0$, i.e. that $0 \to J \to B \to C \to 0$ is an A-extension. Assume that $\delta : J \to \Omega_{B/A} \otimes_B C$ is a split injection, and let $\sigma : \Omega_{B/A} \otimes_B C \to J$ be a splitting. Then the composition

$$B \xrightarrow{\bar{d}} \Omega_{B/A} \otimes_B C \xrightarrow{\sigma} J$$

is an A-derivation. It follows that $1 - \sigma\bar{d} : B \to B$ is an A-homomorphism such that $(1 - \sigma\bar{d})(J) = 0$ and therefore it induces an A-homomorphism $C \to B$ which splits g.

Conversely, assume that $g : B \to C$ has a section $\tau : C \to B$. Then we have a derivation

$$D : B \to J \oplus \Omega_{C/A}$$

given by $D(b) = (b - (\tau g)(b), d_{C/A}(g(b)))$. One easily checks that D induces an isomorphism $\Omega_{B/A} \otimes_B C \cong J \oplus \Omega_{C/A}$, thus proving the assertion. □

As an application we have the following:

Proposition B.4. *Let K be a field and (B, m) a local K-algebra with residue field $B/m = K'$. Then the map*

$$\delta : m/m^2 \to \Omega_{B/K} \otimes_B K'$$

in the exact sequence (B.3) relative to $K \to B \to K'$ is injective if and only if $K \subset K'$ is a separable field extension.
In particular, if $B/m = K$ then

$$\delta : m/m^2 \to \Omega_{B/K} \otimes_B K$$

is an isomorphism. Therefore

$$\dim(B) \leq \dim_K(\Omega_{B/K} \otimes_B K)$$

Proof. See [48], cor. 16.13. The last assertion follows from the conormal sequence relative to $K \to B \to K$. □

The following theorem describes the module of differentials for regular local rings.

Theorem B.5. *Assume that K is a field and B is a local noetherian K-algebra with residue field $B/m = K$. If $\Omega_{B/K}$ is a free B-module of rank equal to* $\dim(B)$ *then B is a regular local ring. If K is perfect (e.g. algebraically closed) and B is e.f.t. over K then the converse is also true.*

Proof. Assume first that $\Omega_{B/K}$ is free of rank equal to $\dim(B)$. Then

$$\dim_K(m/m^2) = \dim(B)$$

by B.4, so B is a regular local ring.

Assume conversely that K is perfect and that B is a regular local ring, e.f.t. over K. Then we have

$$\dim_K(\Omega_{B/K} \otimes_B K) = \dim_K(m/m^2) = \dim(B)$$

Let L be the quotient field of B. Then, by B.1(3), we have

$$\Omega_{B/K} \otimes_B L = \Omega_{L/K}$$

and

$$\dim_L(\Omega_{L/K}) = \text{trdeg}(L/K) = \dim(B)$$

because L is separably algebraic over K, since K is perfect. Therefore we have

$$\dim_K(\Omega_{B/K} \otimes_B K) = \dim(B) = \dim_L \Omega_{B/K} \otimes_B L$$

Since B is e.f.t. over K, $\Omega_{B/K}$ is a finitely generated B-module, and from Lemma A.8 it follows that it is free of rank equal to $\dim(B)$. □

In particular, we have the following:

Corollary B.6. *Let k be an algebraically closed field, and let B be an integral k-algebra of finite type. Then B is a regular ring if and only if $\Omega_{B/k}$ is a projective B-module of rank equal to* $\dim(B)$.

Proof. Both conditions are satisfied if and only if they are satisfied after localizing at the maximal ideals of B. For every maximal ideal $m \subset B$ the local ring B_m is a k-algebra e.f.t. with residue field k. By B.5, B_m is a regular local ring if and only if $\Omega_{B_m/k} = (\Omega_{B/k})_m$ is free of rank equal to $\dim(B)$. The conclusion follows. □

Proposition B.7. *If the ring homomorphism $A \to B$ is e.f.t. then $\Omega_{B/A}$ is a B-module of finite type.*
If, in particular, $B = S^{-1}A[X_1, \ldots, X_n]$ for some multiplicative system S, then $\Omega_{B/A}$ is a free B-module of rank n with basis $\{d_{B/A}(X_1), \ldots, d_{B/A}(X_n)\}$.

Proof. The last assertion is elementary (see [48]). To prove the first, let $B = (S^{-1}P)/J$, where $P = A[X_1, \dots, X_n]$ and $S \subset P$ is a multiplicative system. Then $\Omega_{B/A}$ is a quotient of $\Omega_{S^{-1}P/A} \otimes_{S^{-1}P} B$, by the conormal sequence. □

Remark B.8. If A and B are only assumed to be noetherian then $\Omega_{B/A}$ is not necessarily a B-module of finite type even if A is a field. An example is given by $\Omega_{\mathbf{Q}[[X]]/\mathbf{Q}}$ (see [1] ch. $\mathbf{0}_{IV}$, n. 20.7.16).

Examples B.9. (i) Assume that $B = S^{-1}A[X_1, \dots, X_n]$ for some multiplicative system S. Then $\mathrm{Der}_A(B, B) = \mathrm{Hom}_B(\Omega_{B/A}, B)$ is a free module of rank n with basis

$$\left\{\frac{\partial}{\partial X_1}, \dots, \frac{\partial}{\partial X_n}\right\}$$

which is the dual of the basis

$$\{d_{B/A}(X_1), \dots, d_{B/A}(X_n)\}$$

of $\Omega_{B/A}$, and where $\frac{\partial}{\partial X_j} : B \to B$ is the partial A-derivation with respect to X_j.

Let $Y_1, \dots, Y_n \in B$ be such that the jacobian determinant

$$\det\left(\frac{\partial Y_i}{\partial X_j}\right)$$

is a unit in B. Then

$$\{d_{B/A}(Y_1), \dots, d_{B/A}(Y_n)\}$$

is another basis of $\Omega_{B/A}$ and we have:

$$d_{B/A}(X_j) = \frac{\partial X_j}{\partial Y_1} d_{B/A}(Y_1) + \cdots + \frac{\partial X_j}{\partial Y_n} d_{B/A}(Y_n)$$

Dually:

$$\frac{\partial}{\partial X_j} = \frac{\partial Y_1}{\partial X_j}\frac{\partial}{\partial Y_1} + \cdots + \frac{\partial Y_n}{\partial X_j}\frac{\partial}{\partial Y_n} \tag{B.4}$$

The proof of these statements is straightforward.

(ii) Let k be a field and let $B = k[X, Y]/(XY)$, where X, Y are indeterminates. Then, since $\Omega_{k[X,Y]/K} \otimes B \cong BdX \oplus BdY$, using the conormal sequence we deduce that

$$\Omega_{B/k} \cong \frac{BdX \oplus BdY}{(YdX \oplus XdY)}$$

It follows that the element $YdX = -XdY$ is killed by the maximal ideal (X, Y) and therefore it generates a torsion submodule

$$T := (YdX) \cong k \subset \Omega_{B/k}$$

The quotient is

$$\frac{\Omega_{B/k}}{T} = \frac{BdX \oplus BdY}{(YdX, XdY)} \cong k[X]dX \oplus k[Y]dY \cong (X, Y) \subset B$$

where the last isomorphism is given by

$$f(X)dX \oplus g(Y)dY \mapsto f(X)X + g(Y)Y$$

Therefore we have an exact sequence:

$$0 \to T \to \Omega_{B/k} \to B \to k \to 0$$

(iii) Let k be a field and let $B = k[t, X, Y]/(f)$ where t, X, Y are indeterminates and $f = XY + t$. Then arguing as before we see that

$$\Omega_{B/k[t]} \cong \frac{BdX \oplus BdY}{(YdX \oplus XdY)}$$

The element $YdX = -XdY$ is not killed by any $b \in B$; therefore $\Omega_{B/k[t]}$ is torsion free of rank one. The homomorphism

$$\Omega_{B/k[t]} \to B$$

sending $f(t, X)dX \oplus g(t, Y)dY \mapsto f(t, X)X + g(t, Y)Y$ is bijective onto the maximal ideal (t, X, Y) so that we have an exact sequence:

$$0 \to \Omega_{B/k[t]} \to B \to k \to 0$$

(iv) Let k be a field and let $k[\epsilon] := k[t]/(t^2)$, where we have denoted by ϵ the class of t mod (t^2). Then the conormal sequence of $k \to k[t] \to k[\epsilon]$ is

$$(t^2)/(t^4) \xrightarrow{\delta} \Omega_{k[t]/k} \otimes_{k[t]} k[\epsilon] \to \Omega_{k[\epsilon]/k} \to 0$$

and the middle term is isomorphic to $k[\epsilon]$. The map δ acts as

$$\begin{array}{ccc} \bar{t}^2 & \mapsto & 2\epsilon \\ \bar{t}^3 & \mapsto & 0 \end{array}$$

In particular, we see that δ is not injective. Therefore

$$\Omega_{k[\epsilon]/k} = \begin{cases} kd\epsilon & \text{if } \mathrm{char}(k) \neq 2; \\ k[\epsilon]d\epsilon & \text{if } \mathrm{char}(k) = 2 \end{cases}$$

and $d : k[\epsilon] \to \Omega_{k[\epsilon]/k}$ acts as $d(\alpha + \epsilon\beta) = \beta d\epsilon$.

(v) An obvious generalization of the above computation shows that if $A = k[t]/(t^n)$, $n \geq 2$ and $\mathrm{char}(k) = 0$ or $\mathrm{char}(k) > n$ then

$$\Omega_{A/k} = A/(\bar{t}^{n-1})$$

(vi) If $B \in \mathrm{ob}(\mathcal{A}^*)$ then $t_B^\vee := m_B/m_B^2$ and $t_B := (m_B/m_B^2)^\vee$ are the (Zariski) *cotangent space*, respectively *tangent space* of B. We have $m_B/m_B^2 \cong \Omega_{B/\mathbf{k}} \otimes_B \mathbf{k}$ by Prop. B.4, and therefore

$$\mathrm{Der}_{\mathbf{k}}(B, \mathbf{k}) = \mathrm{Hom}_B(\Omega_{B/\mathbf{k}}, \mathbf{k}) = \mathrm{Hom}_{\mathbf{k}}(\Omega_{B/\mathbf{k}} \otimes_B \mathbf{k}, \mathbf{k}) = (m_B/m_B^2)^\vee$$

Moreover, there is a natural identification

$$\mathrm{Der}_{\mathbf{k}}(B, \mathbf{k}) = \mathrm{Hom}_{\mathbf{k}-alg}(B, \mathbf{k}[\epsilon])$$

which we leave to the reader to verify.

If $\mu : \Lambda \to B$ is a homomorphism in $\mathcal{A}^*$, the induced homomorphism

$$d\mu^\vee : m_\Lambda/m_\Lambda^2 \to m_B/m_B^2$$

is the *codifferential* of μ, while its transpose

$$d\mu : t_B \to t_\Lambda$$

is the *differential* of μ. We define the *relative cotangent space of B over Λ* to be

$$t_{B/\Lambda}^\vee := \mathrm{coker}(d\mu^\vee) = m_B/(m_B^2 + m_\Lambda B)$$

and the *relative tangent space of B over Λ* as its dual:

$$t_{B/\Lambda} = \ker(d\mu) = \left[m_B/(m_B^2 + m_\Lambda B)\right]^\vee$$

From the exact sequence

$$\Omega_{\Lambda/\mathbf{k}} \otimes_\Lambda B \to \Omega_{B/\mathbf{k}} \to \Omega_{B/\Lambda} \to 0$$

tensored by $\mathbf{k}$ we deduce an identification $t_{B/\Lambda}^\vee = \Omega_{B/\Lambda} \otimes_B \mathbf{k}$ and therefore

$$t_{B/\Lambda} = \mathrm{Hom}_B(\Omega_{B/\Lambda}, \mathbf{k}) = \mathrm{Der}_\Lambda(B, \mathbf{k}) = \mathrm{Hom}_{\Lambda-alg}(B, \mathbf{k}[\epsilon])$$

where the Λ-algebra structure on $\mathbf{k}[\epsilon]$ is defined by the composition $\Lambda \to \mathbf{k} \to \mathbf{k}[\epsilon]$ (the last equality is straightforward to verify).

The following lemma describes a situation where the conormal sequence is exact.

Lemma B.10. *Assume char*$(\mathbf{k}) = 0$. *Let*

$$e: \quad 0 \to (t) \to R' \to R \to 0$$

be a small extension in $\mathcal{A}$. Then the conormal sequence

$$\eta : 0 \to (t) \xrightarrow{\delta} \Omega_{R'/\mathbf{k}} \otimes_{R'} R \to \Omega_{R/\mathbf{k}} \to 0$$

is exact also on the left.

Proof. Assume first that e is trivial, so that $R' = R\tilde{\oplus}\mathbf{k}$. Then the codifferential

$$\begin{array}{ccc} \Omega_{R'/\mathbf{k}} \otimes_{R'} \mathbf{k} & \to & \Omega_{R/\mathbf{k}} \otimes_R \mathbf{k} \\ \| & & \| \\ m_{R'}/m_{R'}^2 & & m_R/m_R^2 \end{array}$$

has a nontrivial kernel (Example 1.1.2) so that a fortiori

$$\Omega_{R'/\mathbf{k}} \otimes_{R'} R \to \Omega_{R/\mathbf{k}}$$

has a nontrivial kernel.

Assume now that e is not trivial. Then, letting $m = \dim(t_{R'}) = \dim(t_R)$, we can write

$$R' = P/J', \quad R = P/J$$

where $P = \mathbf{k}[X_1, \ldots, X_m]$, a polynomial algebra, and $J', J \subset (\underline{X})^2 \subset P$ ideals such that $J' \subset J$ and $J/J' \cong (t)$. Let $T \in J$ be such that

$$t = T + J'$$

Since e is small we have $(\underline{X})J \subset J'$ and therefore $T \notin (\underline{X})J$.

Claim: We can choose T so that

$$\frac{\partial T}{\partial X_i} \notin J \quad \text{for some } i$$

If $\frac{\partial T}{\partial X_i} \in J$ then $X_i \frac{\partial T}{\partial X_i} \in J'$ so that we can replace T by

$$T_1 := T - X_i \frac{\partial T}{\partial X_i}$$

If

$$-X_i \frac{\partial^2 T}{\partial X_i^2} = \frac{\partial T_1}{\partial X_i} \notin J$$

we are done, otherwise we replace T_1 by

$$T_2 := T_1 - \frac{X_i}{2} \frac{\partial T_1}{\partial X_i} = T - X_i \frac{\partial T}{\partial X_i} + \frac{X_i^2}{2} \frac{\partial^2 T}{\partial X_i^2}$$

and we apply the same argument. After ν steps of this process we obtain

$$T_\nu := T_{\nu-1} - \frac{X_i}{\nu} \frac{\partial T_{\nu-1}}{\partial X_i} = T + \sum_{s \geq 1}^{\nu} (-1)^s \frac{X_i^s}{s!} \frac{\partial^s T}{\partial X_i^s}$$

Since $\frac{\partial T_\nu}{\partial X_i} = 0$ for $\nu \gg 0$ we see that either $\frac{\partial T_\nu}{\partial X_i} \notin J$ for some ν and we replace T by the first T_ν with this property, or we can replace T by a T_ν which is constant with respect to X_i. Repeating this process for every index i we will end up by replacing

T by a $\bar{T}$ having the required property or otherwise constant with respect to every variable, which is clearly a contradiction. The claim is proved.

From the claim we deduce that

$$dT = \sum_i \frac{\partial T}{\partial X_i} dX_i \notin J\Omega_{P/\mathbf{k}} \tag{B.5}$$

where $d = d_{P/\mathbf{k}} : P \to \Omega_{P/\mathbf{k}}$ is the universal derivation. But we have:

$$\Omega_{R'/\mathbf{k}} = \Omega_{P/\mathbf{k}}/\big(J'\Omega_{P/\mathbf{k}} + (dg')_{g' \in J'}\big)$$

so that

$$\Omega_{R'/\mathbf{k}} \otimes_{R'} R = \Omega_{P/\mathbf{k}}/\big(J\Omega_{P/\mathbf{k}} + (dg')_{g' \in J'}\big)$$

But since clearly $dt \neq dg'$ for all $g' \in J'$, (B.5) implies that

$$dt \neq 0 \in \Omega_{R'/\mathbf{k}} \otimes_{R'} R$$

□

* * * * * *

If $f : X \to Y$ is a morphism of schemes, we denote by $\Omega^1_{X/Y}$ the *sheaf of relative differentials*, or the *relative cotangent sheaf*, on X. It satisfies

$$\Omega^1_{X/Y,x} = \Omega_{\mathcal{O}_{X,x}/\mathcal{O}_{Y,f(x)}}$$

for all $x \in X$. If $f : \mathrm{Spec}(B) \to \mathrm{Spec}(A)$ is a morphism of affine schemes then

$$\Omega^1_{\mathrm{Spec}(B)/\mathrm{Spec}(A)} = (\Omega_{B/A})^{\sim}$$

We denote by

$$T_{X/Y} := Hom(\Omega^1_{X/Y}, \mathcal{O}_X)$$

the *sheaf of relative derivations*, or the *relative tangent sheaf* of f.

We will write Ω^1_X and T_X instead of $\Omega^1_{X/\mathrm{Spec}(\mathbf{k})}$ and $T_{X/\mathrm{Spec}(\mathbf{k})}$ respectively; they are the *cotangent sheaf* and the *tangent sheaf* of X, respectively (cotangent and tangent *bundles* if locally free).

If X is algebraic and $x \in X$ is closed then, by B.4:

$$\Omega^1_{X,x} \otimes \mathbf{k}(x) = \frac{m_{X,x}}{m^2_{X,x}}$$

is the cotangent space of X at x, and

$$T_x X := T_{X,x} \otimes \mathbf{k}(x) = \left(\frac{m_{X,x}}{m^2_{X,x}}\right)^{\vee} \cong \mathrm{Der}_{\mathbf{k}}(\mathcal{O}_{X,x}, \mathbf{k})$$

is the Zariski tangent space of X at x.

Let S be a scheme and

$$X \xrightarrow{g} Y$$

a morphism of S-schemes. The induced homomorphism of sheaves on X:

$$g^*\Omega^1_{Y/S} \to \Omega^1_{X/S}$$

is called the *relative codifferential* of g. The dual homomorphism:

$$T_{X/S} \to Hom(g^*\Omega^1_{Y/S}, \mathcal{O}_X)$$

is the *relative differential* of g. When $S = \mathrm{Spec}(\mathbf{k})$ we have $g^*\Omega^1_Y \to \Omega^1_X$, which is the *codifferential* of g, while its dual

$$dg : T_X \to Hom(g^*\Omega^1_Y, \mathcal{O}_X)$$

is the *differential* of g. Note that if $\Omega^1_{Y/S}$ is locally free then

$$Hom(g^*\Omega^1_{Y/S}, \mathcal{O}_X) = g^*Hom(\Omega^1_{Y/S}, \mathcal{O}_Y) = g^*T_{Y/S}$$

but in general the first and the second sheaves are different. The *relative cotangent sequence* is

$$g^*\Omega^1_{Y/S} \to \Omega^1_{X/S} \to \Omega^1_{X/Y} \to 0 \tag{B.6}$$

Conditions for the injectivity of the first map in this sequence are given in Theorem C.15, page 302 and Theorem D.2.8, page 310.

If $X \subset Y$ is an embedding of schemes and $\mathcal{I} = \mathcal{I}_{X/Y} \subset \mathcal{O}_Y$ is the ideal sheaf of X in Y, then $\mathcal{I}/\mathcal{I}^2$ is a sheaf of $\mathcal{O}_X$-modules in a natural way, called the *conormal sheaf* of X in Y. Its dual

$$N_{X/Y} := Hom_{\mathcal{O}_X}(\mathcal{I}/\mathcal{I}^2, \mathcal{O}_X) = Hom_{\mathcal{O}_Y}(\mathcal{I}, \mathcal{O}_X)$$

is called the *normal sheaf* of X in Y. $N_{X/Y}$ (resp. $\mathcal{I}/\mathcal{I}^2$) is called the *normal bundle* (resp. the *conormal bundle*) of X in Y if it is locally free. Given a closed embedding of S-schemes $i : X \subset Y$, we have an exact sequence of sheaves on X:

$$\mathcal{I}/\mathcal{I}^2 \to i^*\Omega^1_{Y/S} \to \Omega^1_{X/S} \to 0 \tag{B.7}$$

where $\mathcal{I} \subset \mathcal{O}_Y$ is the ideal sheaf of X in Y. (B.7) is called the *relative conormal sequence*. When $S = \mathrm{Spec}(\mathbf{k})$ we obtain the *conormal sequence*

$$\mathcal{I}/\mathcal{I}^2 \to i^*\Omega^1_Y \to \Omega^1_X \to 0$$

Conditions for the injectivity of the first map in these sequences are given in Proposition D.1.4, page 306, and Theorem D.2.7, page 310.

Examples B.11. In the following examples we will describe the global vector fields on the given schemes by exhibiting their restrictions to an affine open set. All will be done by explicit computation.

(i) $H^0(T_{I\!P^1})$ can be described explicitly as follows. Consider $I\!P^1 = U_0 \cup U_1$ where $U_0 = \mathrm{Spec}(\mathbf{k}[\xi])$ and $U_1 = \mathrm{Spec}(\mathbf{k}[\eta])$ with $\eta = \xi^{-1}$ on $U_0 \cap U_1$. We have

$$\frac{\partial}{\partial\eta} = \frac{\partial\xi}{\partial\eta}\frac{\partial}{\partial\xi} = -\frac{1}{\eta^2}\frac{\partial}{\partial\xi} = -\xi^2\frac{\partial}{\partial\xi}$$

on $U_0 \cap U_1$. Let $\theta \in H^0(T_{I\!P^1})$; then

$$\theta_{|U_0} = g(\xi)\frac{\partial}{\partial\xi} \qquad g(\xi) \in \mathbf{k}[\xi]$$

and

$$\theta_{|U_1} = h(\eta)\frac{\partial}{\partial\eta}, \quad h(\eta) \in \mathbf{k}[\eta]$$

On $U_0 \cap U_1$ we have

$$g(\xi)\frac{\partial}{\partial\xi} = h(\eta)\frac{\partial}{\partial\eta} = -h(\xi^{-1})\xi^2\frac{\partial}{\partial\xi}$$

and therefore $g(\xi) = -h(\xi^{-1})\xi^2$. It follows that $g(\xi) = a_0 + a_1\xi + a_2\xi^2$ and $h(\eta) = -(a_0\eta^2 + a_1\eta + a_2)$, with $a_0, a_1, a_2 \in \mathbf{k}$. In particular, $H^0(T_{I\!P^1}) \cong \mathbf{k}^3$.

Moreover, $H^i(T_{I\!P^1}) = 0$ if $i \geq 1$. For $i \geq 2$ it is obvious. Let $\theta \in H^1(T_{I\!P^1})$ be represented by a Čech 1-cocycle defined by $\theta_{01} \in \Gamma(U_0 \cap U_1, T_{I\!P^1})$. It can be written as

$$\theta_{01} = \sum_{i=-m}^{n} a_i\xi^i$$

Letting $\theta_1 = \sum_{i=-m}^{-1} a_i\eta^{-i}$ and $\theta_0 = -\sum_{i=0}^{n} a_i\xi^i$ we obtain:

$$\theta_{01} = \theta_1 - \theta_0$$

so (θ_{01}) is a coboundary.

(ii) We want to describe $H^0(T_{\mathbf{A}^1 \times I\!P^1})$. Let $\mathbf{A}^1 \times I\!P^1 = V_0 \cup V_1$ where

$$V_0 = \mathbf{A}^1 \times U_0 = \mathrm{Spec}[z, \xi])$$

$$V_1 = \mathbf{A}^1 \times U_1 = \mathrm{Spec}[z, \eta])$$

and $\eta = \xi^{-1}$ on $V_0 \cap V_1 = \mathrm{Spec}(\mathbf{k}[z, \xi, \xi^{-1}])$. We have

$$\frac{\partial}{\partial\eta} = \frac{\partial\xi}{\partial\eta}\frac{\partial}{\partial\xi} = -\frac{1}{\eta^2}\frac{\partial}{\partial\xi} = -\xi^2\frac{\partial}{\partial\xi}$$

on $V_0 \cap V_1$. Let $\theta \in H^0(T_{\mathcal{A}^1 \times I\!P^1})$; then

$$\theta_{|V_0} = g(z, \xi)\frac{\partial}{\partial z} + h(z, \xi)\frac{\partial}{\partial\xi} \qquad g(z, \xi), h(z, \xi) \in \mathbf{k}[z, \xi]$$

$$\theta_{|V_1} = \gamma(z, \eta)\frac{\partial}{\partial z} + \chi(z, \eta)\frac{\partial}{\partial\eta} \qquad \gamma(z, \eta), \chi(z, \eta) \in \mathbf{k}[z, \eta]$$

On $V_0 \cap V_1$ we have:

$$g(z,\xi) = \gamma(z,\xi^{-1})$$

and therefore $g(z,\xi) = g(z)$ is constant with respect to ξ. Moreover,

$$h(z,\xi)\frac{\partial}{\partial\xi} = \chi(z,\eta)\frac{\partial}{\partial\eta} = -\chi(z,\xi^{-1})\xi^2\frac{\partial}{\partial\xi}$$

and therefore

$$h(z,\xi) = -\chi(z,\xi^{-1})\xi^2$$

It follows that $h(z,\xi) = a(z) + b(z)\xi + c(z)\xi^2$, with $a(z), b(z), c(z) \in \mathbf{k}[z]$. In conclusion, every $\theta \in H^0(T_{\mathbf{A}^1\times I\!P^1})$ restricts to V_0 as a vector field of the form

$$\theta_{|V_0} = g(z)\frac{\partial}{\partial z} + (a(z) + b(z)\xi + c(z)\xi^2)\frac{\partial}{\partial\xi} \tag{B.8}$$

with $g(z), a(z), b(z), c(z) \in \mathbf{k}[z]$, and conversely, every such vector field is the restriction of a global section of $T_{\mathbf{A}^1\times I\!P^1}$. As in example (i) we also deduce that $H^i(T_{\mathbf{A}^1\times I\!P^1}) = 0$ if $i \geq 1$.

In a similar way, one describes $H^0(T_{(\mathbf{A}^1\setminus\{0\})\times I\!P^1})$ by showing that the image of the restriction

$$H^0(T_{(\mathbf{A}^1\setminus\{0\})\times I\!P^1}) \to H^0(T_{(\mathbf{A}^1\setminus\{0\})\times U_0})$$

consists of the vector fields of the form (B.8) with $g(z), a(z), b(z), c(z) \in \mathbf{k}[z, z^{-1}]$.

(iii) We now consider, for a given integer $m \geq 0$, the rational ruled surface

$$F_m = I\!P(\mathcal{O}_{I\!P^1}(m) \oplus \mathcal{O}_{I\!P^1})$$

Let $\pi : F_m \to I\!P^1$ be the projection. Then F_m can be represented as

$$F_m = \pi^{-1}(U) \cup \pi^{-1}(U') = (U \times I\!P^1) \cup (U' \times I\!P^1)$$

where $U = \mathrm{Spec}(\mathbf{k}[z])$, $U' = \mathrm{Spec}(\mathbf{k}[z'])$ and $z' = z^{-1}$ on $U \cap U'$. We consider the affine open sets

$$V_0 = \mathrm{Spec}(\mathbf{k}[z,\xi]) \subset U \times I\!P^1$$

$$V_0' = \mathrm{Spec}(\mathbf{k}[z',\xi']) \subset U' \times I\!P^1$$

where on $V_0 \cap V_0' = \mathrm{Spec}(\mathbf{k}[z, z^{-1}, \xi]) = \mathrm{Spec}(\mathbf{k}[z', z'^{-1}, \xi'])$ we have:

$$z' = z^{-1}, \quad \xi' = z^m\xi$$

Therefore we have:

$$\begin{aligned} \tfrac{\partial}{\partial z'} &= -z^2\tfrac{\partial}{\partial z} + mz\xi\tfrac{\partial}{\partial\xi} \\ \tfrac{\partial}{\partial\xi'} &= z^{-m}\tfrac{\partial}{\partial\xi} \end{aligned} \tag{B.9}$$

We will describe a typical element $\theta \in H^0(T_{F_m})$ by describing its restriction to the open sets V_0 and V_0'. We have, by example (ii) above:

$$\theta_{|V_0} = g(z)\frac{\partial}{\partial z} + (a(z) + b(z)\xi + c(z)\xi^2)\frac{\partial}{\partial \xi}$$

with $g(z), a(z), b(z), c(z) \in \mathbf{k}[z]$ and similarly

$$\theta_{|V_0'} = \rho(z')\frac{\partial}{\partial z'} + (\alpha(z') + \beta(z')\xi' + \gamma(z')\xi'^2)\frac{\partial}{\partial \xi'}$$

with $\rho(z'), \alpha(z'), \beta(z'), \gamma(z') \in \mathbf{k}[z']$. Imposing their equality on $V_0 \cap V_0'$ and using (B.9) we obtain the following conditions:

$$\begin{aligned} g(z) &= -\rho(z^{-1})z^2 \\ a(z) &= \alpha(z^{-1})z^{-m} \\ b(z) &= \beta(z^{-1}) + \rho(z^{-1})mz \\ c(z) &= \gamma(z^{-1})z^m \end{aligned} \tag{B.10}$$

We distinguish the cases $m = 0$ and $m > 0$. If $m = 0$ (B.10) give:

$$\begin{aligned} g(z) &= g_0 + g_1 z + g_2 z^2 \\ a(z) &= a \\ b(z) &= b \\ c(z) &= c \end{aligned} \qquad g_0, g_1, g_2, a, b, c \in \mathbf{k}$$

In the case $m > 0$ we have:

$$\begin{aligned} g(z) &= g_0 + g_1 z + g_2 z^2 \\ a(z) &= 0 \\ b(z) &= b - mz(g_1 + g_2 z) \\ c(z) &= c_0 + c_1 z + \cdots + c_m z^m \end{aligned} \qquad g_0, g_1, g_2, b, c_0, \ldots, c_m \in \mathbf{k}$$

Since the restriction $H^0(T_{F_m}) \to H^0(T_{V_0})$ is injective and we have described its image, we can conclude:

$$\begin{aligned} H^0(T_{F_0}) &\cong \mathbf{k}^6 \\ H^0(T_{F_m}) &\cong \mathbf{k}^{m+5} \end{aligned}$$

In particular, *F_m and F_n are not isomorphic if $m \neq n$* (note that $F_0 \cong \mathbb{P}^1 \times \mathbb{P}^1$ is not isomorphic to $F_1 \cong Bl_{(1,0,0)}\mathbb{P}^2$).

Since, by the calculations of the previous example (ii)

$$h^i(T_{U \times \mathbb{P}^1}) = h^i(T_{U' \times \mathbb{P}^1}) = h^i(T_{(U \cap U') \times \mathbb{P}^1}) = 0, \quad i \geq 1 \tag{B.11}$$

we deduce that:

$$H^1(T_{F_m}) = H^0(T_{(U \cap U') \times \mathbb{P}^1}) / H^0(T_{U \times \mathbb{P}^1}) + H^0(T_{U' \times \mathbb{P}^1})$$

An easy computation based on (B.10) shows that, for $m \geq 1$, $H^1(T_{F_m})$ consists of the classes, modulo $H^0(T_{U \times \mathbb{P}^1}) + H^0(T_{U' \times \mathbb{P}^1})$, of the vector fields

$$(b_1 z + \cdots + b_{m-1} z^{m-1})\frac{\partial}{\partial \xi}$$

In particular,

$$H^1(T_{F_m}) \cong \mathbf{k}^{m-1} \tag{B.12}$$

It also follows from (B.11) that

$$H^2(T_{F_m}) = (0) \tag{B.13}$$

NOTES

1. Let $X \to Y$ be a morphism of algebraic schemes. Prove that there is an exact sequence

$$0 \to \Omega^1_{X/Y} \to \mathcal{P}^1_{X/Y} \to \mathcal{O}_X \to 0 \tag{B.14}$$

which globalizes (B.1). $\mathcal{P}^1_{X/Y}$ is called the *sheaf of principal parts of X over Y*, denoted by $\mathcal{P}^1_X$ if $Y = \mathrm{Spec}(\mathbf{k})$.

Let $X = \mathbb{P}(V)$ for a finite-dimensional $\mathbf{k}$-vector space V. Then the exact sequence (B.14) is the dual of the Euler sequence; in particular,

$$\mathcal{P}^1_{\mathbb{P}(V)} \cong \mathcal{O}_{\mathbb{P}(V)}(-1) \otimes V^\vee$$

Therefore (B.14) is a generalization of the Euler sequence to any $X \to Y$.

2. Consider $\mathbb{P} = \mathbb{P}(V)$ for a finite-dimensional $\mathbf{k}$-vector space V and the *incidence relation*:

$$\mathbf{I} = \{(x, H) : x \in H\} \subset \mathbb{P} \times \mathbb{P}^\vee \tag{B.15}$$

Consider the twisted and dualized Euler sequence:

$$0 \to \Omega^1_{\mathbb{P}(V)}(1) \to \mathcal{O}_{\mathbb{P}(V)} \otimes V^\vee \to \mathcal{O}_{\mathbb{P}(V)}(1) \to 0$$

From its definition it follows that $\mathbf{I} = \mathbb{P}(\Omega^1_{\mathbb{P}(V)}(1))$ and $\mathbb{P} \times \mathbb{P}^\vee = \mathbb{P}(\mathcal{O}_{\mathbb{P}(V)} \otimes V^\vee)$ and the inclusion in (B.15) is induced by the first homomorphism in the above sequence.

C

Smoothness

The notion of "formal smoothness", introduced in [1], Ch. IV § 17, is of crucial importance in deformation theory, and therefore plays a special role in this book. In this appendix we introduce this concept from scratch, and we show how it is related to the notion of "smooth morphism" as introduced in [3] Exposé II, and [84]. We will not give a systematic treatment of the properties of smooth morphisms in algebraic geometry: the reader is referred to the above quoted references for them. For more details on the approach taken in this section the reader can also consult [13] and [94].

Definition C.1. *A ring homomorphism $f : R \to B$ is called* formally smooth, *and B is called a* formally smooth R-algebra, *if for every exact sequence:*

$$0 \to I \to A \xrightarrow{\eta} A' \to 0 \tag{C.1}$$

where A and A' are local artinian R-algebras, each R-algebra homomorphism $B \to A'$ has a lifting $B \to A$; equivalently, if the map:

$$\mathrm{Hom}_{R-alg}(B, A) \to \mathrm{Hom}_{R-alg}(B, A') \tag{C.2}$$

is surjective.
f is called smooth *if it is formally smooth and essentially of finite type (shortly e.f.t.). If the map (C.2) is bijective (instead of only being surjective) for all exact sequences (C.1), then f is* formally etale*; f is* etale *if it is formally etale and e.f.t..*

Recall that e.f.t. means that B is a localization of an R-algebra of finite type (see e.g. [127]). It is easy to prove by induction that it suffices to check the above conditions only for the exact sequences (C.1) such that $I^2 = (0)$, i.e. for *extensions* of local artinian R-algebras.

Proposition C.2. *(i) If B is a ring and $\Delta \subset B$ is a multiplicative system, then $B \to \Delta^{-1}B$ is formally etale. In particular, B is a formally etale B-algebra.*
(ii) The composition of formally smooth (resp. formally etale) homomorphisms is formally smooth (resp. formally etale).

(iii) If $f : R \to B$ is formally smooth (resp. formally etale) and C is an R-algebra, then $C \to C \otimes_R B$ is formally smooth (resp. formally etale).

(iv) A finitely generated field extension $K \subset L$ is smooth if and only if L is separable over K.

(v) Let $R \xrightarrow{f} B \xrightarrow{g} C$ be ring homomorphisms, and assume that f is formally etale. Then gf is formally smooth (resp. formally etale) if and only if g is formally smooth (resp. formally etale).

Proof. (i) Given an exact sequence (C.1) and a commutative diagram

$$\begin{array}{ccc} B & \to & \Delta^{-1}B \\ \downarrow \varphi' & & \downarrow \varphi \\ A & \underset{p}{\longrightarrow} & A' \end{array}$$

we must find $\tilde{\varphi} : \Delta^{-1}B \to A$ which makes it commutative. For every $s \in \Delta$ choose $a_s \in A$ such that $\varphi(s)^{-1} = p(a_s)$. Since

$$p(\varphi'(s)a_s) = \varphi(s)\varphi(s)^{-1} = 1_{A'}$$

we have

$$\varphi'(s)a_s = 1_A + i_s \qquad i_s \in I$$

for every $s \in \Delta$. Therefore

$$\varphi'(s)a_s(1_A - i_s) = 1_A$$

Hence $\varphi'(s) \in A$ is invertible. Now define $\tilde{\varphi}(r/s) = \varphi'(r)\varphi'(s)^{-1}$.

Noting that $\tilde{\varphi}$ is uniquely determined by φ' we get the assertion.

(ii) and (iii) are straightforward.

(iv) Assume first that $K \subset L$ is separable. By (ii) it suffices to consider the cases $L = K(X)$ and $L = K[X]/(f(X))$ where f is irreducible and $f'(x) \neq 0$. The first case is left to the reader (see remark C.3(i)).

In the second case consider an extension $\bar{A} = A/I$ of local artinian K-algebras, where $I \subset A$ is an ideal with $I^2 = (0)$. Let $\varphi : K[X]/(f(X)) \to \bar{A}$ be a homomorphism, sending $\bar{X} \mapsto \bar{\alpha}$. Choose arbitrarily $\alpha \in A$ such that $\bar{\alpha} = \alpha \bmod I$. It will suffice to find $e \in I$ such that

$$f(\alpha + e) = 0$$

We have $f(\alpha + e) = f(\alpha) + f'(\alpha)e$. Since $f'(\alpha)$ is a unit mod I it is also a unit in A, and therefore we can take $e = -f(\alpha)/f'(\alpha)$.

Assume conversely that $K \subset L$ is smooth. Then $L = F[X]/J$ where F is a purely transcendental extension of K and J is a principal ideal. We have an exact sequence of finite-dimensional L-vector spaces:

$$J/J^2 \to \Omega_{F[X]/K} \otimes L \to \Omega_{L/K} \to 0$$

where J/J^2 is one-dimensional. By the first part of the proof F is smooth over K and by B.3(ii) the left map is injective because, by the smoothness of L over K, the surjection $F[X]/J^2 \to L$ splits. It follows that

$$\dim(\Omega_{L/K}) = \dim(\Omega_{F[X]/K} \otimes L) - 1 = \text{trdeg}_K(F[X]) - 1$$

$$= \text{trdeg}_K(F) = \text{trdeg}_K(L)$$

From B.1(iii) it follows that $K \subset L$ is separable.

(v) "if" follows immediately from (ii); "only if" is left to the reader. □

Remarks C.3. (i) Any polynomial algebra $R[X_1, X_2 \ldots]$ is trivially a formally smooth R-algebra. From C.2(i) it follows that a localization of a polynomial R-algebra is also a formally smooth R-algebra.

More precisely, a localization $P = S^{-1}R[X_1, X_2, \ldots]$ of a polynomial algebra over a ring R satisfies the following condition, stronger than formal smoothness:

For every extension of R-algebras:

$$0 \to I \to A \to A' \to 0$$

where A and A' are R-algebras and $I^2 = 0$ the map

$$\text{Hom}_{R-alg}(P, A) \to \text{Hom}_{R-alg}(P, A')$$

is surjective.

Every R-algebra B is a quotient of a formally smooth R-algebra, because it is a quotient of a polynomial R-algebra. From C.2(i) it follows that every e.f.t. R-algebra is a quotient of a smooth R-algebra.

This is trivial for polynomial rings, and in the general case it can be proved adapting the proof of C.2(i) in an obvious way.

(ii) if R is in ob($\hat{\mathcal{A}}$) then every formal power series ring $R[[X_1, X_2, \ldots]]$ is a formally smooth R-algebra, because local artinian R-algebras are complete. More precisely, a formal power series ring $R[[X_1, X_2, \ldots]]$ satisfies the following condition, stronger than formal smoothness over R:

For every extension:

$$0 \to I \to A \to A' \to 0$$

of complete local R-algebras the map

$$\text{Hom}_{R-alg}(P, A) \to \text{Hom}_{R-alg}(P, A')$$

is surjective.

The proof is straightforward and is left to the reader.

The following result characterizes an important class of formally smooth algebras.

Theorem C.4. *Let k be a field and let (B, m) be a noetherian local k-algebra with residue field K. Suppose that K is finitely generated and separable over k. Then the following are equivalent:*

(i) B is regular.
(ii) $\hat{B} \cong K[[X_1, \ldots, X_d]]$, where $d = \dim(B)$.
(iii) B is a formally smooth k-algebra.

Proof. (i) $\Leftrightarrow$ (ii) is standard (see [48], prop. 10.16 and exercise 19.1).
(ii) $\Rightarrow$ (iii). It follows directly from the definition that B is formally smooth over k if and only if $\hat{B}$ is. Since $\hat{B}$ is formally smooth over K (remark C.3(ii)), and since K is smooth over k by C.2(iv), the conclusion follows by transitivity.
(iii) $\Rightarrow$ (i). Let $\{x_1, \ldots, x_d\}$ be a system of generators of m. Then, since B/m^2 is complete and K is separable over k, B/m^2 contains a coefficient field ([48], Theorem 7.8). Therefore there exists an isomorphism

$$v_1 : B/m^2 \cong K[X_1, \ldots, X_d]/M^2 \qquad M = (X_1, \ldots, X_d)$$

Let $v : B \to B/m^2 \xrightarrow{v_1} K[X_1, \ldots, X_d]/M^2$. By the formal smoothness of B and by induction we can find a lifting of v:

$$v_n : B \to K[X_1, \ldots, X_d]/M^{n+1}$$

for every $n \geq 2$. Consider the elements

$$v_n(x_1), \ldots, v_n(x_d) \in M/M^{n+1}$$

Their classes generate M/M^2, hence they generate M/M^{n+1}, by Nakayama. Then we have:

$$\begin{aligned} K[X_1, \ldots, X_d]/M^{n+1} &= v_n(B) + (M/M^{n+1}) \\ = v_n(B) + \textstyle\sum_i v_n(x_i)\big[v_n(B) + (M/M^{n+1})\big] &= v_n(B) + (M/M^{n+1})^2 = \cdots \\ = v_n(B) + (M/M^{n+1})^{n+1} &= v_n(B) \end{aligned}$$

hence v_n is surjective. Since $m^{n+1} \subset \ker(v_n)$ we have:

$$\ell(B/m^{n+1}) \geq \ell(K[X_1, \ldots, X_d]/M^{n+1}) = \binom{d+n}{d}$$

and this implies that $\dim(B) \geq d$. Since m is generated by d elements it follows that B is regular. □

For the reader's convenience we include the proof of the following well-known lemma:

Lemma C.5. *(i) A surjective endomorphism $f : A \to A$ of a noetherian ring is an automorphism.*

(ii) Let A be a complete noetherian local ring and $\psi : A \to A$ an endomorphism inducing an isomorphism $\psi_1 : A/m_A^2 \to A/m_A^2$. Then ψ is an automorphism.

Proof. (i) We have an ascending chain of ideals

$$\ker(f) \subseteq \ker(f^2) \subseteq \ker(f^3) \subseteq \cdots$$

Since A is noetherian we have $\ker(f^n) = \ker(f^{n+1}) = \ker(f^{n+2}) = \cdots$ for some n, and it suffices to prove that $\ker(f^n) = (0)$. After replacing f by f^n we may assume $\ker(f) = \ker(f^2)$. Let $a \in \ker(f)$; by assumption there exists $b \in A$ such that $a = f(b)$. Then $0 = f(a) = f^2(b)$ and therefore $b \in \ker(f^2) = \ker(f)$, i.e. $a = f(b) = 0$.

(ii) Let $gr(A) = A/m \oplus m/m^2 \oplus \cdots$ be the associated graded ring. Since $gr(A)$ is generated by m/m^2 over A/m the endomorphism $gr(\psi) : gr(A) \to gr(A)$ induced by ψ is surjective. It follows that ψ is also surjective. In fact, given $a \in A$ the surjectivity of $gr(\psi)$ implies that there are $a_1, a_2, a_3, \ldots, b_1, b_2, b_3, \ldots \in A$ such that $a_i \in m^{i-1}$, $b_i \in m^i$, and

$$a = f(a_1) + b_1, \; b_1 = f(a_2) + b_2, \; b_2 = f(a_3) + b_3, \ldots$$

We obtain a convergent power series $\bar{a} = a_1 + a_2 + a_3 + \cdots$ such that

$$a - \psi(a_1 + a_2 + \cdots + a_n) = b_n \in m^{n+1}$$

On the limit we therefore get $a = \psi(\bar{a})$. The conclusion is now a consequence of (i). □

Proposition C.6. *Let $f : R \to B$ be a local homomorphism of noetherian local rings containing a field k isomorphic to their residue fields. Then the following conditions are equivalent:*

(i) f is formally smooth.
(ii) $\hat{B}$ is isomorphic to a formal power series ring over $\hat{R}$.
(iii) The homomorphism $\hat{f} : \hat{R} \to \hat{B}$ induced by f is formally smooth.

Proof. (i) $\Rightarrow$ (ii). Let $m \subset B$ and $n \subset R$ be the maximal ideals. Choose elements $x_1, \ldots, x_d \in \hat{B}$ inducing a k-basis of $\hat{B}/(\hat{m}^2 + \hat{f}(\hat{n}))$, and let $F = \hat{R}[[X_1, \ldots, X_d]]$, where $X_1, \ldots, X_d$ are indeterminates. Denote by $M \subset F$ the maximal ideal.

The homomorphism

$$\begin{aligned} u : \quad F &\to \hat{B} \\ X_i &\mapsto x_i \end{aligned}$$

induces an isomorphism

$$u_1 : F/(M^2 + \hat{n}F) \to \hat{B}/(\hat{m}^2 + \hat{f}(\hat{n}))$$

By the formal smoothness of f the composition

$$v_1 : B \to \hat{B} \to \hat{B}/(\hat{m}^2 + \hat{f}(\hat{n})) \xrightarrow{u_1^{-1}} F/(M^2 + \hat{n}F)$$

can be lifted to an R-homomorphism

$$v_k : B \to F/M^k$$

for each $k \geq 2$. Therefore the sequence $\{v_k\}$ defines an $\hat{R}$-homomorphism

$$v : \hat{B} \to F$$

such that $vu : F \to F$ and $uv : \hat{B} \to \hat{B}$ induce isomorphisms $(vu)_1 : F/M^2 \to F/M^2$ and $(uv)_1 : \hat{B}/\hat{m}^2 \to \hat{B}/\hat{m}^2$ respectively. From Lemma C.5 it follows that u and v are isomorphisms.
(ii) $\Rightarrow$ (iii) is obvious.
(iii) $\Rightarrow$ (i) is left to the reader. □

Corollary C.7. *Let $f : R \to B$ be a local homomorphism of noetherian local rings containing a field k isomorphic to their residue fields. Then the following conditions are equivalent:*

(i) f is formally etale.
(ii) The homomorphism $\hat{f} : \hat{R} \to \hat{B}$ induced by f is an isomorphism.

Proof. Left to the reader.

Corollary C.8. *Let R be in* ob$(\mathcal{A}^*)$. *The inclusion $f : R \to \hat{R}$ is formally etale.*

The proof is obvious. We now restrict our attention to smooth homomorphisms, i.e. we add the condition that the homomorphism is e.f.t.. In this case the module of differentials comes into play; moreover, the defining condition of Definition C.1 can be replaced by the more general condition (i) in the following statement.

Theorem C.9. *Let $f : R \to B$ be an e.f.t. ring homomorphism. Then the following conditions are equivalent:*

(i) For every extension of R-algebras:

$$0 \to I \to A \to A' \to 0 \tag{C.3}$$

the map

$$\mathrm{Hom}_{R-alg}(B, A) \to \mathrm{Hom}_{R-alg}(B, A')$$

is surjective.
(ii) If $B = P/J$, where $P = S^{-1}R[X_1, \ldots, X_d]$, $S \subset R[X_1, \ldots, X_d]$ is a multiplicative system and $J \subset P$ is an ideal, the conormal sequence

$$0 \to J/J^2 \xrightarrow{\delta} \Omega_{P/R} \otimes_P B \to \Omega_{B/R} \to 0$$

is split exact. In particular, J/J^2 and $\Omega_{B/R}$ are finitely generated projective B-modules.

(iii) B is a smooth R-algebra.
(iv) (Jacobian criterion of smoothness) If P and J are as in (ii) the map

$$(J/J^2) \otimes_B K(p) \overset{\delta \otimes_B K(p)}{\longrightarrow} \Omega_{P/R} \otimes_P K(p) \qquad \textit{where } K(p) = B_p/m_{B_p}$$

is injective for every prime ideal $p \subset B$.

Proof. (i) $\Rightarrow$ (ii). The hypothesis implies that the extension:

$$0 \to J/J^2 \to P/J^2 \to B \to 0$$

splits. Therefore the conormal sequence is split exact by B.3(iii) and it follows that J/J^2 and $\Omega_{B/R}$ are finitely generated projective because the module $\Omega_{P/R} \otimes_P B$ is free of finite rank.

(ii) $\Rightarrow$ (i). Consider an exact sequence (C.3) and a homomorphism of R-algebras $f' : B \to A'$. By Remark C.3(i) there exists an R-homomorphism $g : P \to A$ making the following diagram commute:

$$\begin{array}{ccc} P & \to & B \\ \downarrow g & & \downarrow f' \\ A & \to & A' \end{array}$$

Since $g(J) \subset I$, we see that g factors through P/J^2, so that we have a commutative diagram:

$$\begin{array}{ccc} P/J^2 & \to & B \\ \downarrow \bar{g} & & \downarrow f' \\ A & \to & A' \end{array}$$

The hypothesis implies, via B.3(iii), that there exists $h : B \to P/J^2$ a splitting of $P/J^2 \to B$. The composition $f = \bar{g}h : B \to A$ gives a lifting of f'.

(i) $\Rightarrow$ (iii) is obvious.

(iii) $\Rightarrow$ (iv). We may assume B and P local with residue field K. To prove that $\delta \otimes_B K$ is injective, it suffices to show that for every K-vector space V the map induced by δ:

$$\begin{array}{ccc} \mathrm{Hom}_K(\Omega_{P/R} \otimes_P K, V) & \to & \mathrm{Hom}_K((J/J^2) \otimes_B K, V) \\ \| & & \| \\ \mathrm{Der}_R(P, V) & & \mathrm{Hom}_B(J/J^2, V) \end{array}$$

is surjective. Consider a homomorphism $g : J/J^2 \to V$, and the associated pushout diagram (see § 1.1 for the definition):

$$\begin{array}{lccccccccc} \Lambda : & 0 & \to & J/J^2 & \to & P/J^2 & \to & B & \to & 0 \\ & & & \downarrow g & & \downarrow & & \| & & \\ g_*(\Lambda) : & 0 & \to & V & \to & Q & \to & B & \to & 0 \end{array}$$

We can write $m_Q = V \oplus m'$, where $m' \subset Q$ is an ideal, because V is annihilated by m_Q. Therefore the previous diagram can be embedded in the following:

$$\begin{array}{ccccccccccc}
 & & & & & & P & & & & \\
 & & & & & & \downarrow & \searrow & & & \\
A: & & 0 & \to & J/J^2 & \to & P/J^2 & \to & B & \to & 0 \\
 & & & & \downarrow g & & \downarrow & & \| & & \\
g_*(A): & & 0 & \to & V & \to & Q & \to & B & \to & 0 \\
 & & & & \| & & \downarrow & & \downarrow \bar{v} & & \\
\eta: & & 0 & \to & V & \to & Q/m' & \to & K & \to & 0
\end{array}$$

where η is an extension of local artinian R-algebras. From the smoothness of B we deduce the existence of $v : B \to Q/m'$ lifting the projection $\bar{v} : B \to K$. Denoting by $r : P \to B$ the natural map, and by $w : P \to P/J^2 \to Q \to Q/m'$ the composition, consider the homomorphism:

$$d = w - vr : P \to V$$

It is easy to show that this is an R-derivation, which induces g.

(iv) $\Rightarrow$ (ii). From Nakayama's lemma it follows that $\ker(\delta) \otimes B_p = (0)$ for all prime ideals $p \subset B$ and therefore $\ker(\delta) = (0)$. Moreover, since $\Omega_{P/R} \otimes_B B_p$ is free and finitely generated it follows that $\mathrm{Tor}_1^{B_p}(\Omega_{B/R} \otimes_B B_p, K(p)) = 0$: it follows that $\Omega_{B/R} \otimes_B B_p$ is flat, and therefore, being finitely generated, it is free. Thus $\Omega_{B/R}$ is projective, δ has a splitting and J/J^2 is also projective and finitely generated. □

The following result follows easily from what we have seen so far.

Theorem C.10. *Let B be an integral* **k***-algebra of finite type and of dimension d. Then the following are equivalent:*

(i) B_p is smooth over **k** *for each prime ideal $p \in \mathrm{Spec}(B)$.*
(ii) B is a regular ring.
(iii) $\Omega_{B/\mathbf{k}}$ is projective of rank d.
(iv) B is smooth over **k**.

Proof. (ii) $\Leftrightarrow$ (iii) is Corollary B.6.

(i) $\Leftrightarrow$ (ii) follows from Theorem C.4.

(iv) $\Rightarrow$ (i). for each $p \in \mathrm{Spec}(B)$, B_p is smooth over B by Proposition C.2(i); from Proposition C.2(ii) it follows that B_p is smooth over **k**.

(i) $\Rightarrow$ (iv). (i) implies that condition (iv) of Theorem C.9 is satisfied for all $p \in \mathrm{Spec}(B)$, so that B is smooth by Theorem C.9. □

From now on we will freely replace the defining property for smooth homomorphisms given in Definition C.1 by condition (i) of Theorem C.9. Here is a first example.

Proposition C.11. *Let R be a ring, P an R-algebra and $B = P/J$ for an ideal $J \subset P$. If B is a smooth R-algebra the conormal sequence*

$$0 \to J/J^2 \to \Omega_{P/R} \otimes_P B \to \Omega_{B/R} \to 0$$

is split exact and $\Omega_{B/R}$ is projective and finitely generated. If, moreover, P is a smooth R-algebra then J/J^2 is finitely generated and projective as well.

Proof. Since B is smooth the R-algebra extension

$$0 \to J/J^2 \to P/J^2 \to B \to 0$$

splits. Therefore the conormal sequence splits by Theorem B.3(iii) and $\Omega_{B/R}$ is finitely generated and projective by Theorem C.9. If P is smooth then $\Omega_{P/R}$ is finitely generated and projective as well and so is J/J^2. □

Corollary C.12. *Let P be an e.f.t. $\mathbf{k}$-algebra and $B = P/J$ for an ideal $J \subset P$. Assume that B is reduced. Then in the conormal sequence*

$$J/J^2 \xrightarrow{\delta} \Omega_{P/\mathbf{k}} \otimes_P B \to \Omega_{B/\mathbf{k}} \to 0 \tag{C.4}$$

$\ker(\delta)$ *is a torsion B-module whose support is contained in the singular locus of* $\mathrm{Spec}(B)$.

If J/J^2 is torsion free then δ is injective.

Proof. Since B is reduced there is a dense open subset $U \subset \mathrm{Spec}(B)$ such that B_p is a regular local ring for all $p \in U$. From Theorem C.4 it follows that B_p is a smooth $\mathbf{k}$-algebra for all such p and, by Propositions C.11 and B.1(ii), the conormal sequence (C.4) localized at p is split exact. It follows that $\ker(\delta)_p = (0)$ for all $p \in U$ and the conclusion follows. The last assertion is an obvious consequence of the first part. □

The next result explains the relation between smoothness and the relative cotangent sequence.

Theorem C.13. *Let $K \xrightarrow{f} R \xrightarrow{g} B$ be ring homomorphisms, with g smooth. Then the relative cotangent sequence:*

$$0 \to \Omega_{R/K} \otimes_R B \xrightarrow{\alpha} \Omega_{B/K} \to \Omega_{B/R} \to 0$$

is split exact.

Proof. By Theorem B.2 it suffices to prove that α is a split injection; this is equivalent to showing that, for any B-module M, the induced map:

$$\begin{array}{ccc}
\mathrm{Hom}_B(\Omega_{B/K}, M) & \xrightarrow{\alpha^\vee} & \mathrm{Hom}_B(\Omega_{R/K} \otimes_R B, M) \\
\| & & \| \\
\mathrm{Der}_K(B, M) & & \mathrm{Der}_K(R, M) \\
& & \\
D' & \mapsto & D'g
\end{array}$$

is split surjective. Let $D : R \to M$ be a K-derivation and consider the commutative diagram:

$$\begin{array}{ccc} B & \xrightarrow{1_B} & B \\ \uparrow g & & \uparrow \\ R & \xrightarrow{\gamma} & B\tilde{\oplus}M \end{array}$$

where $\gamma(r) = (g(r), D(r))$, $r \in R$. By the smoothness of g we can find a homomorphism of R-algebras $\psi : B \to B\tilde{\oplus}M$ making the diagram

$$\begin{array}{ccc} B & \xrightarrow{1_B} & B \\ \uparrow g & \searrow \psi & \uparrow \\ R & \underset{\gamma}{\longrightarrow} & B\tilde{\oplus}M \end{array}$$

commutative. The homomorphism ψ is necessarily of the form:

$$\psi(b) = (b, D'(b))$$

and $D' : B \to M$ is a K-derivation such that $D = D'g$. This proves the surjectivity of $\alpha^\vee$. Now take $M = \Omega_{R/K} \otimes_R B$ and $D = d_{R/K} \otimes g : R \to \Omega_{R/K} \otimes_R B$ and let

$$\alpha' : \Omega_{B/K} \to \Omega_{R/K} \otimes_R B$$

be the B-linear map corresponding to $D' : B \to \Omega_{R/K} \otimes_R B$. Then $\alpha'\alpha = 1_M$ and this proves that α is split injective. □

Corollary C.14. *Let $K \xrightarrow{f} R \xrightarrow{g} B$ be ring homomorphisms, with g etale. Then*

$$\Omega_{R/K} \otimes_R B \cong \Omega_{B/K}$$

and

$$\Omega_{B/R} = (0)$$

Proof. By the relative cotangent sequence the two assertions are equivalent. We will prove the first. Keeping the notations of the proof of C.13, the hypothesis that g is etale implies that the derivation D' is unique and consequently α is an isomorphism. □

* * * * * *

A morphism $\varphi : X \to Y$ of algebraic schemes is *smooth at a point* $x \in X$ if $\mathcal{O}_{X,x}$ is a smooth $\mathcal{O}_{Y,\varphi(x)}$-algebra; φ is *smooth* if it is smooth at every point. The definition of *etale morphism* is given similarly. This definition is equivalent to the definition of smooth (resp. etale) morphism as given in [3] and in [84]. The equivalence can be seen by means of the jacobian criterion of smoothness, proved in Theorem C.9, and using [3], Exposé II, Corollaire 5.9.

By translating into geometrical language the algebraic results proved above we deduce in particular the following.

Theorem C.15. *Let S be an algebraic scheme, and $\varphi : X \to Y$ a morphism of algebraic S-schemes. Then:*

(i) If φ is smooth at $x \in X$ then the relative cotangent sequence

$$0 \to \varphi^* \Omega^1_{Y/S} \to \Omega^1_{X/S} \to \Omega^1_{X/Y} \to 0$$

is split exact at x and $\Omega^1_{X/Y}$ is locally free at x. The rank of the free module $\Omega^1_{X/Y,x}$ is called the relative dimension of φ at x.

(ii) φ is etale at $x \in X$ if and only if it is smooth of relative dimension zero at x. In particular, $\Omega^1_{X/Y,x} = 0$ (i.e. φ is unramified at x) and therefore we have an isomorphism

$$\varphi^* \Omega^1_{Y/S,x} \cong \Omega^1_{X/S,x}$$

(iii) If X is smooth over S at x and φ is a closed embedding with ideal sheaf $\mathcal{I} \subset \mathcal{O}_Y$ then the relative conormal sequence

$$O \to \mathcal{I}/\mathcal{I}^2 \to \varphi^* \Omega^1_{Y/S} \to \Omega^1_{X/S} \to 0$$

is exact at x and $\Omega^1_{X/S}$ is free at x; if, moreover, Y is also smooth over S at $\varphi(x)$ then $\mathcal{I}/\mathcal{I}^2$ is free at x as well.

The exactness of the relative conormal sequence in part (iii) holds under more general assumptions as well (see Theorem D.2.7).

D

Complete intersections

D.1 Regular embeddings

Definition D.1.1. *An embedding of schemes* $j : X \subset Y$ *is a* regular embedding of codimension n at the point $x \in X$ *if* $j(x)$ *has an affine open neighbourhood* $\mathrm{Spec}(R)$ *in* Y *such that the ideal of* $j(X) \cap \mathrm{Spec}(R)$ *in* R *can be generated by a regular sequence of length* n. *If this happens at every point of* X *we say that* j *is a* regular embedding of codimension n.

An open embedding is a regular embedding of codimension 0. If X and Y are both nonsingular then $X \subset Y$ is a regular embedding. The set of points of X where an embedding $j : X \subset Y$ is regular is open.

If $X \subset Y$ is a regular embedding of codimension n then $\mathcal{I}/\mathcal{I}^2$ and $N_{X/Y}$ are both locally free of rank n ([48], Exercise 17.12, p. 440). It follows from standard facts in commutative algebra (see [48], Exercise 17.16, p. 441) that $\mathcal{I}^k/\mathcal{I}^{k+1}$ is locally free as well for every $k \geq 2$.

A ring B is called a *complete intersection* if $\mathrm{Spec}(B)$ can be regularly embedded in $\mathrm{Spec}(R)$ where R is a regular ring.

A scheme X is a *local complete intersection* (l.c.i.) if every local ring $\mathcal{O}_{X,x}$ is a complete intersection ring.

A *nonsingular scheme* X, i.e. a scheme all of whose local rings are regular, is an example of an l.c.i. scheme. If $X \subset Y$ is a regular embedding and Y is an l.c.i. scheme, then X is an l.c.i. scheme.

Lemma D.1.2. *Let* $f : X \to Y$ *be a morphism of schemes and let* $Z \subset Y$ *be a regular embedding of codimension* n. *Then the induced embedding* $j : X \times_Y Z \subset X$ *has codimension* $\leq n$ *at every point and if equality holds at a point* $x \in X \times_Y Z$ *then* j *is regular at* x.

Proof. If $\mathcal{I}_Z \subset \mathcal{O}_Y$ is the ideal sheaf of Z in Y then the ideal sheaf $f^{-1}\mathcal{I}_Z$ of $X \times_Y Z \subset X$ is locally generated at a point x by the n images of the local generators of $\mathcal{I}_{Z,f(x)}$. The conclusion follows easily from this fact. □

If we have a flag of embeddings of schemes $X \subset Y \subset Z$ and $\mathcal{I}_Y \subset \mathcal{I}_X \subset \mathcal{O}_Z$ are the ideal sheaves of X and Y, we have the exact sequence

$$0 \to \mathcal{I}_Y \to \mathcal{I}_X \to \mathcal{I}_{X/Y} \to 0 \tag{D.1}$$

where $\mathcal{I}_{X/Y} \subset \mathcal{O}_Y$ is the ideal sheaf of X in Y. After tensoring by $\otimes_{\mathcal{O}_Z}\mathcal{O}_X$ we obtain an exact sequence of coherent $\mathcal{O}_X$-modules:

$$\frac{\mathcal{I}_Y}{\mathcal{I}_Y^2} \otimes \mathcal{O}_X \xrightarrow{\alpha} \frac{\mathcal{I}_X}{\mathcal{I}_X^2} \to \frac{\mathcal{I}_{X/Y}}{\mathcal{I}_{X/Y}^2} \to 0 \tag{D.2}$$

Its dual is the sequence:

$$0 \to N_{X/Y} \to N_{X/Z} \to N_{Y/Z} \otimes \mathcal{O}_X \tag{D.3}$$

Lemma D.1.3. *(i) If $f : X \subset Y$ and $g : Y \subset Z$ are regular embeddings of codimensions m and n respectively, then $gf : X \to Z$ is a regular embedding of codimension $m + n$.*
(ii) If the embeddings f and g are both regular then we have exact sequences of locally free sheaves on X:

$$0 \to \frac{\mathcal{I}_Y}{\mathcal{I}_Y^2} \otimes \mathcal{O}_X \xrightarrow{\alpha} \frac{\mathcal{I}_X}{\mathcal{I}_X^2} \to \frac{\mathcal{I}_{X/Y}}{\mathcal{I}_{X/Y}^2} \to 0 \tag{D.4}$$

$$0 \to N_{X/Y} \to N_{X/Z} \to N_{Y/Z} \otimes \mathcal{O}_X \to 0 \tag{D.5}$$

Proof. (i) Left to the reader.
(ii) All sheaves in (D.4) are locally free because they are conormal bundles of regular embeddings. Since $\mathrm{Im}(\alpha)$ is a torsion free sheaf of the same rank of $(\mathcal{I}_Y/\mathcal{I}_Y^2) \otimes \mathcal{O}_X$, it follows that α must be injective. The sequence (D.5) is exact because $Ext^1_{\mathcal{O}_X}(\mathcal{I}_{X/Y}/\mathcal{I}_{X/Y}^2, \mathcal{O}_X) = 0$. □

Proposition D.1.4. *Let $j : X \subset Y$ be an embedding of algebraic schemes, with X reduced and Y nonsingular. Consider the conormal sequence*

$$\mathcal{I}/\mathcal{I}^2 \xrightarrow{\delta} \Omega^1_{Y|X} \to \Omega^1_X \to 0 \tag{D.6}$$

(where $\mathcal{I} \subset \mathcal{O}_Y$ is the ideal sheaf of X). Then:

(i) The homomorphism δ is injective on the open set where j is a regular embedding.
(ii) If X and Y are nonsingular then the dual sequence

$$0 \to T_X \to T_{Y|X} \to N_{X/Y} \to 0 \tag{D.7}$$

is exact.

Proof. (i) It suffices to show that δ is injective under the assumption that j is a regular embedding. In this case the conormal sheaf $\mathcal{I}/\mathcal{I}^2$ is locally free of rank equal to the codimension of X. The sequence (D.6) is exact at every nonsingular point $x \in X$ by Theorem C.15(iii). Since X is reduced, this happens on a dense open subset so that $\ker(\delta)$ is supported on a nowhere dense subset. But X has no embedded points because it is regularly embedded in Y: it follows that $\ker(\delta) = 0$.

(ii) Under the stated hypothesis, j is a regular embedding and Ω^1_X is locally free, so we have $Ext^1(\Omega^1_X, \mathcal{O}_X) = 0$ and the exactness of (D.7) follows. □

Remark D.1.5. If we don't assume X reduced, part (i) of the proposition is false in general. An example is given by the closed regular embedding of codimension 1:

$$\mathrm{Spec}(\mathbf{k}[\epsilon]) \subset \mathrm{Spec}(\mathbf{k}[t]) = \mathbf{A}^1$$

(see Example B.9(iv)).

A morphism $f : X \to Y$ of schemes will be called a *cover* (or a *covering*) if it is finite and surjective.

Recall that a morphism of schemes $f : X \to Y$ is called *unramified at a point* $x \in X$ if $\Omega^1_{X/Y,x} = 0$; f is *unramified* if it is unramified at every $x \in X$. After identifying X with the diagonal $\Delta \subset X \times_Y X$, we see that $\Omega^1_{X/Y}$ gets identified with the conormal sheaf of this embedding. It follows that f is unramified at x if and only if $\Delta \subset X \times_Y X$ is an open embedding at x, and that the locus of $x \in X$ such that f is unramified at x is open. Moreover, f is unramified if and only if Δ is both open and closed in $X \times_Y X$.

D.2 Relative complete intersection morphisms

We now introduce a natural class of morphisms which generalize smooth morphisms and behave well with respect to differentials and base change.

Definition D.2.1. *A flat morphism of finite type $f : X \to S$ is called a* relative complete intersection (r.c.i.) morphism at the point $x \in X$ *if there is an open neighbourhood U of x such that the restriction of f to U can be obtained as a composition*

$$U \xrightarrow{j} V \xrightarrow{g} S$$

where j is a regular embedding and g is smooth. If f is an r.c.i. morphism at every point we call it an r.c.i. morphism, *and we call X a* complete intersection over S.

This definition is equivalent to Def. 19.3.6 of Ch. IV of [1]; the equivalence is proved in [19], Prop. 1.4. Note that in case $S = \mathrm{Spec}(\mathbf{k})$ the morphism f is an r.c.i. if and only if X is an l.c.i. of finite type.

If $X \to S$ is a flat morphism of finite type of nonsingular varieties then f is an r.c.i. because it factors as

$$X \to X \times S \to S$$

where the first morphism is the graph of f.

Before discussing the main properties of this notion we need two lemmas.

Lemma D.2.2. *Let $A \to B$ be a ring homomorphism, M a B-module and $f_1, \dots, f_n$ an M-regular sequence of elements of B. Assume that for each $i = 1, \dots, n$ the module $M/(\sum_{j=1}^{i-1} f_j M)$ is A-flat. Then, for every ring homomorphism $A \to A'$, letting $B' = B \otimes_A A'$, $M' = M \otimes_A A'$, and $f_i' = f_i \otimes 1$ ($1 \le i \le n$), the sequence $f_1', \dots, f_n'$ of elements of B' is M'-regular and the modules $M'/(\sum_{j=1}^{i-1} f_j' M')$ are A'-flat.*

Proof. Consider the exact sequence:

$$0 \to M \xrightarrow{f_1} M \to M/f_1 M \to 0$$

Since $M/f_1 M$ is A-flat, the sequence:

$$0 \to M \otimes_A A' \xrightarrow{f_1 \otimes 1} M \otimes_A A' \to (M/f_1 M) \otimes_A A' \to 0$$

is exact, and therefore f_1' is not a zero-divisor for M'. Let $M_i = M/(\sum_{j=1}^{i} f_j M)$, $M_i' = M'/(\sum_{j=1}^{i} f_j' M')$; then we have $M_i' = M_i \otimes_A A'$, $M_{i+1} = M_i/f_{i+1} M_i$, $M_{i+1}' = M_i'/f_{i+1}' M_i'$. Replacing M and f_1 by M_i and f_{i+1} in the above argument, one deduces that f_{i+1}' is not a zero-divisor for M_i', thereby proving the first assertion by induction. The last assertion follows from A.2(VII). □

Lemma D.2.3. *Let $A \to B$ be a local homomorphism of noetherian local rings, M a B-module of finite type, flat over A, and $f_1, \dots, f_n \in m_B$. For $1 \le i \le n$ let g_i be the image of f_i in $B \otimes_A k$, where $k = A/m_A$ is the residue field of A. Then the following conditions are equivalent:*

(i) $f_1, \dots, f_n$ is an M-regular sequence, and $M_i = M/(\sum_{j=1}^{i} f_j M)$ is A-flat for all $1 \le i \le n$.

(ii) $g_1, \dots, g_n$ is an $(M \otimes_A k)$-regular sequence.

Proof. (i) ⇒ (ii) follows from D.2.2 applied to $A' = k$.

(ii) ⇒ (i) Applying Corollary A.6, from the injectivity of $g_1 : M \otimes_A k \to M \otimes_A k$ we deduce that $f_1 : M \to M$ is injective and that $M_1 = M/f_1 M$ is A-flat. Proceeding by induction on i, assume M_i flat over A. Since $g_{i+1} : M_i \otimes_A k \to M_i \otimes_A k$ is injective from A.6, again we deduce that $f_{i+1} : M_i \to M_i$ is injective and that M_{i+1} is A-flat. □

In the next proposition some general properties of r.c.i. morphisms are proved.

Proposition D.2.4. *(i) An open embedding is an r.c.i. morphism. A smooth morphism of finite type is an r.c.i. morphism.*

(ii) If $f : X \to S$ is an r.c.i. morphism and $h : S' \to S$ is a morphism, then the morphism $f' : X \times_S S' \to S'$ induced by f after base change is an r.c.i. morphism.

Proof. (i) is an immediate consequence of the definition and (ii) follows easily from Lemma D.2.2. □

From D.2.4(ii) it follows in particular that if $f : X \to S$ is an r.c.i. morphism then X_s is an l.c.i. for every **k**-rational point $s \in S$. So for example, a non-l.c.i. algebraic scheme cannot be the fibre of a flat morphism of algebraic nonsingular varieties.

The next result gives a useful characterization of r.c.i. morphisms.

Proposition D.2.5. *Let*

$$\begin{array}{ccc} X & \stackrel{j}{\longrightarrow} & Y \\ & \searrow f \quad \swarrow g & \\ & S & \end{array} \qquad \text{(D.8)}$$

be a commutative diagram of morphisms of algebraic schemes, where f is flat, g is smooth and j is an embedding. Then the following conditions are equivalent for a **k***-rational point $x \in X$:*

(i) f is an r.c.i. morphism at x.
(ii) Letting $s = f(x)$, the fibre X_s is an l.c.i. at x.
(iii) j is a regular embedding at x.

Proof. (i) ⇒ (ii) follows from D.2.4(ii) and (iii) ⇒ (i) is obvious.

(ii) ⇒ (iii) From (ii) it follows that the embedding $j_s : X_s \subset Y_s$ is regular at x. Let $\mathcal{I} \subset \mathcal{O}_Y$ be the ideal sheaf of X. Tensoring the exact sequence

$$0 \to \mathcal{I} \to \mathcal{O}_Y \to \mathcal{O}_X \to 0$$

by $- \otimes_{\mathcal{O}_S} \mathbf{k}$ we obtain the sequence

$$0 \to \mathcal{I} \otimes_{\mathcal{O}_S} \mathbf{k} \to \mathcal{O}_{Y_s} \to \mathcal{O}_{X_s} \to 0$$

which is exact because f is flat. Therefore $\mathcal{I} \otimes_{\mathcal{O}_S} \mathbf{k}$ is the ideal sheaf of $j(X_s)$ in Y_s. Consider a sequence $f_1, \ldots, f_n$ of sections of $\mathcal{I}$ in an open neighbourhood of $j(x)$ which induce a basis of $\mathcal{I}_{j(x)}/(m_s\mathcal{I}_{j(x)} + \mathcal{I}^2_{j(x)})$ as a $\mathcal{O}_{Y,j(x)}/(m_s\mathcal{O}_{Y,j(x)} + \mathcal{I}_{j(x)})$-module. Then the images $f_1 \otimes 1 = g_1, \ldots, f_n \otimes 1 = g_n$ are generating sections of $\mathcal{I} \otimes_{\mathcal{O}_S} \mathbf{k}$ in an open neighbourhood of $j(x)$ in Y_s which form a regular sequence in $j(x)$. From Nakayama's lemma it follows that $f_1, \ldots, f_n$ generate $\mathcal{I}$ in an open neighbourhood of $j(x)$ in Y. From Lemma D.2.3 it follows that $f_1, \ldots, f_n$ form a regular sequence in $j(x)$ and therefore (iii) holds. □

Corollary D.2.6. *Under the hypothesis of Proposition D.2.5, the locus of points $x \in X$ such that f is an r.c.i. at x is open. If f is proper then the locus of points $s \in S$ such that X_s is an l.c.i. is open.*

Proof. The last assertion follows from the first because a proper map is closed. The first assertion can be proved using characterization D.2.5(iii) of r.c.i. morphism and the fact that the locus where an embedding is regular is open. □

We conclude this section with two results about the relative conormal sequence and cotangent sequence for r.c.i. morphisms.

Theorem D.2.7. *Let*

$$\begin{array}{ccc} X & \stackrel{j}{\longrightarrow} & Y \\ & \searrow f & \downarrow g \\ & & S \end{array}$$

be a commutative diagram of morphisms of algebraic schemes, with f an r.c.i., j an immersion and g smooth. Let $\mathcal{J} \subset \mathcal{O}_Y$ be the ideal sheaf of $j(X)$. If f is smooth on a dense open subset intersecting every fibre then the relative conormal sequence

$$0 \to \mathcal{J}/\mathcal{J}^2 \stackrel{\delta}{\longrightarrow} j^*\Omega^1_{Y/S} \to \Omega^1_{X/S} \to 0$$

is exact and $\mathcal{J}/\mathcal{J}^2$ is locally free.

Proof. From the equivalence $(i) \Leftrightarrow (iii)$ in Proposition D.2.5 it follows that j is a regular embedding and therefore $\mathcal{J}/\mathcal{J}^2$ is locally free. Moreover, the support of $\ker(\delta)$ does not contain any generic point of X nor any fibre of f because it is contained in the locus where f is not smooth. Since f is generically smooth and j is a regular embedding, X has no embedded components except possibly for some union of fibres. It follows that δ is injective. □

Theorem D.2.8. *Let $f : X \to S$ be an r.c.i morphism of algebraic schemes, and assume f smooth on a dense open subset intersecting every fibre. Then the relative cotangent sequence*

$$0 \to f^*\Omega^1_S \to \Omega^1_X \to \Omega^1_{X/S} \to 0 \tag{D.9}$$

is exact.

Proof. We only have to prove the injectivity of the left homomorphism and the question is local on X. Since all schemes are algebraic, locally on X we can construct the following commutative diagram:

$$\begin{array}{ccccc} X & \stackrel{j}{\longrightarrow} & V & \stackrel{i}{\longrightarrow} & U \\ & \searrow f & \downarrow \psi & & \downarrow \varphi \\ & & S & \stackrel{h}{\longrightarrow} & W \end{array}$$

where W, U, ψ, φ are smooth, i, h are closed embeddings and j is a regular closed embedding. From the smooth morphism φ we deduce the exact sequence of locally free sheaves on U:

$$0 \to \varphi^*\Omega^1_W \to \Omega^1_U \to \Omega^1_{U/W} \to 0$$

which restricts on X to the exact sequence:

$$0 \to (hf)^*\Omega^1_W \to (ij)^*\Omega^1_U \to j^*\Omega^1_{V/S} \to 0$$

Let $\mathcal{J} \subset \mathcal{O}_V$ and $\mathcal{I} \subset \mathcal{O}_U$ be the ideal sheaves of the embeddings j and ij respectively, and $\mathcal{H} \subset \mathcal{O}_W$ the ideal sheaf of the embedding h. Then we have an exact and commutative diagram of coherent sheaves on X:

$$\begin{array}{ccccccc}
 & & & & 0 & & \\
 & & & & \downarrow & & \\
 & f^*(\mathcal{H}/\mathcal{H}^2) & \to & \mathcal{I}/\mathcal{I}^2 & \to & \mathcal{J}/\mathcal{J}^2 & \to 0 \\
 & \downarrow & & \downarrow & & \downarrow \delta_j & \\
0 \to & (hf)^*\Omega^1_W & \to & (ij)^*\Omega^1_U & \to & j^*\Omega^1_{V/S} & \to 0 \\
 & \downarrow & & \downarrow & & \downarrow & \\
 & f^*\Omega^1_S & \xrightarrow{df^\vee} & \Omega^1_X & \to & \Omega^1_{X/S} & \to 0 \\
 & \downarrow & & \downarrow & & \downarrow & \\
 & 0 & & 0 & & 0 &
\end{array}$$

where the second and the third columns are the relative conormal sequences of ij and of j respectively; the first column is the pullback to X of the conormal sequence of h; the first row is exact because $\psi^*(\mathcal{H}/\mathcal{H}^2)$ is the conormal sheaf of i; the map δ_j is injective by Theorem D.2.7 and by the assumptions made on X and f. A diagram chasing shows that the codifferential $df^\vee$ is injective and proves the theorem. □

NOTES

1. An algebraic scheme can have different embeddings in $\mathbb{P}^r$, i.e. by means of nonisomorphic invertible sheaves, but with same normal sheaf. An example is given by a projective nonsingular curve C of genus 1, and by the embeddings in $\mathbb{P}^3$ given by two nonisomorphic invertible sheaves $L_1$and L_2 of degree 4 such that $L_1^2 = L_2^2$. Then C is embedded as a nonsingular complete intersection of two quadrics by both sheaves, and the normal bundles are $L_1^2 \oplus L_1^2 = L_2^2 \oplus L_2^2$.

2. Let S be a scheme, and X, Y smooth over S. Prove that every closed S-embedding $X \subset Y$ is regular. In particular, every section of a smooth morphism $f : Y \to S$ is a regular embedding of codimension equal to the relative dimension of f.

3. Let $f : \mathcal{X} \to S$ be a morphism of finite type and $s \in S$ a $\mathbf{k}$-rational point. Let $m_s \subset \mathcal{O}_{S,s}$ be the maximal ideal and $\mathcal{I} = \mathcal{I}_{\mathcal{X}(s)}$ the ideal sheaf of the fibre $\mathcal{X}(s)$ of f over s. Prove that we have a surjective homomorphism

$$\frac{m_s}{m_s^2} \otimes_{\mathbf{k}} \mathcal{O}_{\mathcal{X}(s)} \to \mathcal{I}/\mathcal{I}^2$$

and an injection:

$$N_{\mathcal{X}(s)/\mathcal{X}} \subset T_{S,s} \otimes_{\mathbf{k}} \mathcal{O}_{\mathcal{X}(s)}$$

If f is flat then they are isomorphisms; in particular, if f is flat then $N_{\mathcal{X}(s)/\mathcal{X}}$ is free.

E

Functorial language

Let $\mathcal{C}$ be a category. A covariant (resp. contravariant) functor F from $\mathcal{C}$ to (sets) is said to be *representable* if there is an object X in $\mathcal{C}$ such that F is isomorphic to the functor

$$Y \mapsto \mathrm{Hom}(X, Y) \tag{E.1}$$

(resp. $Y \mapsto \mathrm{Hom}(Y, X)$). We will denote by h_X a functor of the form (E.1). The representable functors are a full subcategory, isomorphic to $\mathcal{C}^\circ$ (resp. to $\mathcal{C}$ in the contravariant case), of the category $Funct(\mathcal{C}, \text{(sets)})$ of covariant functors (resp. $Funct(\mathcal{C}^\circ, \text{(sets)})$ of contravariant functors) from $\mathcal{C}$ to (sets).

To fix ideas let's consider covariant functors. In order to investigate conditions for the representability of a given functor F it is convenient to study functorial morphisms $h_X \to F$. Such morphisms turn out to be easy to describe, thanks to the following elementary lemma:

Lemma E.1 (Yoneda). *Let $F : \mathcal{C} \to$ (sets) be a covariant functor. For each object X in $\mathcal{C}$ there is a canonical bijection:*

$$\begin{array}{ccc} \mathrm{Hom}(h_X, F) & \leftrightarrow & F(X) \\ \Phi & \mapsto & \Phi(X)(1_X) \end{array}$$

Let's mention, in passing, that functorial morphisms $F \to h_X$ are much harder to control. They are related to the notion of "coarse moduli scheme".

We may consider *couples* of the form (X, ξ), where X is an object of $\mathcal{C}$ and $\xi \in F(X)$. Yoneda's lemma implies that to give such a couple is equivalent to giving a morphism of functors $h_X \to F$; if this morphism is an isomorphism then (X, ξ) is called a *universal couple*, and ξ a *universal element*, for F. The existence of a universal couple is equivalent to the representability of F.

The couples for F are the objects of a category in which a *morphism* $(X, \xi) \to (Y, \eta)$ between two couples is by definition a morphism $f : X \to Y$ in $\mathcal{C}$

such that $F(f)(\xi) = \eta$. We denote this category by I_F. A morphism $f : (X, \xi) \to (Y, \eta)$ in I_F corresponds to a commutative diagram of morphisms of functors:

$$\begin{array}{ccc} h_X & \stackrel{\xi}{\longrightarrow} & F \\ \uparrow f & \nearrow \eta & \\ h_Y & & \end{array}$$

We have an obvious "forgetful functor"

$$I_F \to \mathcal{C}$$

The fibres of this functor are precisely the sets $F(X)$, which are embedded as subcategories of I_F by $\xi \mapsto (X, \xi)$.
(Recall that, given a functor $G : \mathbf{C} \to \mathbf{D}$, the fibre $G^{-1}(D)$ of G over an object D of $\mathbf{D}$ is a subcategory of $\mathbf{C}$, consisting of all objects C such that $G(C) = D$ and of all morphisms f such that $G(f) = 1_D$. A set can be viewed as a category whose objects are its elements and the only morphisms are the identity morphisms.)

Lemma E.2. *The functor F is representable if and only if the category I_F has an initial object (X, ξ). If this is the case, (X, ξ) is a universal couple for F.*

The proof is immediate. Note that, since an initial object is unique up to isomorphism, it follows that a representable functor has a unique universal couple, up to isomorphism.

* * * * * *

Let I and $\mathcal{D}$ be two categories. Given an object A of $\mathcal{D}$, the constant functor $c_A : I \to \mathcal{D}$ is defined as $c_A(i) = A$ for each object i of I and $c_A(f) = 1_A$ for each morphism f in I. Note that c_A is both covariant and contravariant. Every morphism $\alpha : A \to B$ in $\mathcal{D}$ induces an obvious morphism of functors $c_\alpha : c_A \to c_B$. Consider a covariant functor $\Phi : I \to \mathcal{D}$. An *inductive limit* of Φ is an object A of $\mathcal{D}$ and a functorial morphism $\lambda : \Phi \to c_A$ such that for every other morphism $\mu : \Phi \to c_B$ there is a morphism $\alpha : A \to B$ such that $\mu = c_\alpha \lambda$.

$$\begin{array}{ccc} \Phi & \stackrel{\lambda}{\longrightarrow} & c_A \\ & \searrow \mu & \downarrow c_\alpha \\ & & c_B \end{array}$$

From the definition it follows that an inductive limit of Φ, if it exists, is unique up to unique isomorphism, and is denoted by

$$\lim_{\to} \Phi$$

In practice an inductive limit is an object A of $\mathcal{D}$ such that there is a morphism $\Phi(i) \to A$ for each $i \in Ob(I)$ with the condition that the diagram

$$\begin{array}{ccc} \Phi(i) & \to & A \\ \downarrow \Phi(f) & \nearrow & \\ \Phi(j) & & \end{array}$$

is commutative for each morphism $f : i \to j$ in I; moreover, these data must satisfy a universal property.

Dually, one has the notion of *projective limit* of a covariant functor $\Phi : I \to \mathcal{D}$: it is an object A of $\mathcal{D}$ and a morphism $\pi : c_A \to \Phi$ such that for every other morphism $\rho : c_B \to \Phi$ there is a morphism $\beta : B \to A$ such that $\rho = \pi c_\beta$. The projective limit of Φ, if it exists, is denoted by

$$\lim_{\leftarrow} \Phi$$

The above notions can be defined without changes replacing the covariant functor Φ by a contravariant one. We will write Φ_i for $\Phi(i)$, for each object i of I, and sometimes

$$\lim_{\rightarrow} \Phi_i \quad (\text{resp. } \lim_{\leftarrow} \Phi_i) \quad \text{instead of} \quad \lim_{\rightarrow} \Phi \quad (\text{resp. } \lim_{\leftarrow} \Phi)$$

Example E.3. Let J be a partially ordered set. We define a category $Ord(J)$ as follows. The objects of $Ord(J)$ are the elements of J; for any $i, j \in J$ the set $\mathrm{Hom}_{Ord(J)}(i, j)$ consists of one element if $i \leq j$ and is $\emptyset$ otherwise. A covariant (resp. contravariant) functor $\Phi : Ord(J) \to \mathcal{D}$ is called an *inductive system* (resp. a *projective system*) in $\mathcal{D}$ indexed by J; in the case $\mathcal{D}$ =(sets), we obtain the usual notions of inductive (projective) system and of inductive (projective) limit.
If I is a set and $\Phi : I \to \mathcal{D}$ is a functor, where $\mathcal{D}$ is a category with arbitrary coproducts, then

$$\lim_{\rightarrow} \Phi = \coprod_i \Phi_i$$

Similarly, if $\mathcal{D}$ has products then

$$\lim_{\leftarrow} \Phi = \prod_i \Phi_i$$

Proposition E.4. *The inductive limit and projective limit exist for every functor Φ : $I \to$ (sets) from any category I.*

Proof. We take

$$\lim_{\rightarrow} \Phi = \coprod_i \Phi_i / R$$

where R is the equivalence relation generated by pairs (x, y), $x \in \Phi_i$ and $y \in \Phi_j$, such that there exists $\varphi : i \to j$ with $\Phi(x) = y$. Similarly for the projective limit. □

Example E.5. Let $F : \mathcal{C} \to$ (sets) be a covariant functor, and let I_F be the category of couples for F. Then we have a contravariant functor

$$\Phi : I_F \to Funct(\mathcal{C}, (\text{sets}))$$

which sends a couple (X, ξ) to the functor $h_X : \mathcal{C} \to$ (sets), and a morphism $f : (X, \xi) \to (Y, \eta)$ to the functorial morphism $h_f : h_Y \to h_X$ induced by f.

By construction there is a morphism $\Phi \to c_F$. This morphism makes F the inductive limit of the functor Φ (the proof is an easy exercise). We will write:

$$F = \varinjlim_{(X,\xi)} h_X$$

Definition E.6. *A category I is* filtered *if:*

(a) for every pair of objects i, j in I there exists an object k in I and morphisms:

$$\begin{array}{ccc} & & i \\ & & \downarrow \\ j & \to & k \end{array}$$

(b) each pair of morphisms $i \rightrightarrows j$ has a coequalizer $i \rightrightarrows j \to k$.

The category I is cofiltered *if the dual category I° is filtered.*

Assume from now on that $\mathcal{C}$ is a category with products and fibred products.

Definition E.7. *A covariant functor $F : \mathcal{C} \to$ (sets) is called* left exact *if $F(B \times C) = F(B) \times F(C)$ and $F(B \times_A C) = F(B) \times_{F(A)} F(C)$ for each diagram*

$$\begin{array}{ccc} & & C \\ & & \downarrow \\ B & \to & A \end{array}$$

in $\mathcal{C}$ (i.e. F commutes with finite products and finite fibred products).

Every representable functor is left exact by definition of product and fibred product.

Lemma E.8. *Let I be a filtered category and $\Phi : I \to Funct(\mathcal{C},$ (sets)) a covariant functor. Then, for each diagram in $\mathcal{C}$:*

$$\begin{array}{ccc} & & C \\ & & \downarrow \\ B & \to & A \end{array}$$

there is a bijection:

$$\varinjlim \Phi_i(B) \times_{\lim \Phi_i(A)} \varinjlim \Phi_i(C) \cong \varinjlim [\Phi_i(B) \times_{\Phi_i(A)} \phi_i(C)]$$

The proof of this lemma is straightforward and we omit it. The following result is a useful characterization of left-exact functors.

Proposition E.9. *A covariant functor $F : \mathcal{C} \to$ (sets) is left exact if and only if the category I_F is cofiltered.*

Proof. Assume that I_F is cofiltered. Applying Lemma E.8 to the functor Φ of Example E.5, we see that the inductive limit $F = \lim_{(X,\xi)} h_X$ is left exact because each functor h_X is left exact.

Conversely, assume that F is left exact. Let $(X,\xi), (Y,\eta) \in Ob(I_F)$; we must find

$$\begin{array}{ccc} (Z,\zeta) & \rightarrow & (X,\xi) \\ \downarrow & & \\ (Y,\eta) & & \end{array}$$

Take $(Z,\zeta) = (X \times Y, (\xi,\eta))$. Now consider $(X,\xi) \overset{\rightarrow}{\underset{\rightarrow}{}} (Y,\eta)$ coming from $\phi, \psi : X \to Y$. We have

$$F(\phi)(\xi) = F(\psi)(\xi) = \eta$$

Consider the diagram:

$$\begin{array}{ccc} X & \xrightarrow{\Gamma_\phi} & X \times Y \\ \uparrow & & \uparrow \Gamma_\psi \\ K & \rightarrow & X \end{array}$$

where $\Gamma_\phi = (1_X, \phi)$ and $\Gamma_\psi = (1_X, \psi)$ and $K = X \times_{X \times Y} X$. Since F is left exact

$$F(K) = F(X) \times_{F(X \times Y)} F(X)$$

and there is $\chi \in F(K)$ corresponding to (ξ,ξ):

$$\begin{array}{ccc} \xi & \overset{F(\Gamma_\phi)}{\longmapsto} & (\xi,\eta) \\ \uparrow & & \uparrow F(\Gamma_\psi) \\ \chi & \longmapsto & \xi \end{array}$$

Then (K,χ) is the equalizer of ϕ and ψ. Therefore I_F is cofiltered. □

Let I be a category. A full subcategory J of I is *cofinal* if for each $i \in Ob(I)$ there is a morphism $f : i \to j$ for some $j \in Ob(J)$. It follows immediately from the definitions that if $\Phi : I \to \mathcal{D}$ is a covariant functor and $\Phi_J : J \to \mathcal{D}$ is its restriction, then

$$\varinjlim \Phi = \varinjlim \Phi_J$$

* * * * * *

Let Z be a scheme. In this subsection we will consider contravariant functors defined on (schemes/Z). All we will say holds, with obvious modifications, for functors defined on (algschemes/Z), the full subcategory of algebraic Z-schemes. A contravariant functor

$$F : (\text{schemes}/Z)^\circ \to (\text{sets})$$

defines on every Z-scheme S a presheaf of sets:

$$U \mapsto F(U)$$

for all open sets $U \subset S$. For this reason a functor as above is also called a *presheaf.* F is called a *sheaf* (more precisely a *sheaf in the Zariski topology*) if it defines a sheaf on every scheme; namely, if for all Z-schemes S and for all open coverings $\{U_i\}$ of S the following is an exact sequence of sets:

$$F(S) \to \prod_i F(U_i) \rightrightarrows \prod_{i,j} F(U_i \cap U_j)$$

The most important sheaves are the *representable functors*, i.e. functors isomorphic to one of the form:

$$S \mapsto \mathrm{Hom}_Z(S, X)$$

for some Z-scheme X. Such a functor is called the *functor of points* of X/Z.

It is very important to have conditions, easy to verify in practice, for a contravariant functor $F : (\text{schemes}/Z)^\circ \to (\text{sets})$ to be representable. Certainly a necessary condition is that F is a sheaf. Another necessary condition is the following.

Recall that a subfunctor G of F is said to be an *open* (resp. *closed*) *subfunctor* if for every scheme S and for every morphism of functors

$$\mathrm{Hom}(-, S) \to F$$

the fibred product $\mathrm{Hom}(-, S) \times_F G$, which is a subfunctor of $\mathrm{Hom}(-, S)$, is represented by an open (resp. closed) subscheme of S. A family of open subfunctors $\{G_i\}$ of F is a *covering* of F if for every Z-scheme S and for every morphism of functors $\mathrm{Hom}(-, S) \to F$ the family $\{\mathrm{Hom}(-, S) \times_F G_i\}$ of subschemes of S is an open covering of S.

An obvious example is obtained by considering an open (resp. closed) subscheme X' of a Z-scheme X: correspondingly, we obtain an open (resp. closed) subfunctor $\mathrm{Hom}(-, X')$ of $\mathrm{Hom}(-, X)$. An open cover $\{X_i\}$ of X defines a cover of $\mathrm{Hom}(-, X)$ by open subfunctors.

Therefore a second obvious necessary condition for a functor F to be representable is that it can be covered by representable open subfunctors. We will now show that these two necessary conditions are also sufficient.

Proposition E.10. *Let*

$$F : (\text{schemes}/Z)^\circ \to (\text{sets})$$

be a contravariant functor. Suppose that:

(a) F is a sheaf;
(b) F admits a covering by representable open subfunctors F_i.

Then F is representable.

Proof. Letting $F_{ij} = F_i \times_F F_j$, by (b) the projections $F_{ij} \to F_i$ correspond to open embedding of schemes $X_{ij} \to X_i$. Therefore the F_i's patch together to form a representable functor $\mathrm{Hom}(-, X)$, where X is the scheme obtained by patching the X_i's together along the X_{ij}'s. By (a), F and $\mathrm{Hom}(-, X)$ are isomorphic. □

The following is an easy but important remark.

Lemma E.11. *If F is a sheaf then F is determined by its restriction to the category of affine schemes.*

Proof. In fact, if S is any Z-scheme we can consider an affine open cover $\{U_i\}$. For any i, j we take an affine open cover $\{V_{i,j,\alpha}\}$ of $U_i \cap U_j$; composing the map

$$F(U_i) \rightrightarrows \prod_{i,j} F(U_i \cap U_j)$$

with the inclusions $F(U_i \cap U_j) \to \prod_\alpha F(V_{i,j,\alpha})$ we obtain the exact sequence:

$$F(S) \to \prod_i F(U_i) \rightrightarrows \prod_{i,j,\alpha} F(V_{i,j,\alpha})$$

which shows that $F(S)$ is determined by its values on affine schemes. □

This lemma implies that the functor of points of a scheme X

$$F = \mathrm{Hom}_Z(-, X) : (\text{schemes})^\circ \to (\text{sets})$$

is determined by its restriction to the category of affine schemes, or equivalently, by its covariant version:

$$F : (\mathbf{k}\text{-algebras}) \to (\text{sets})$$

Since the category of schemes is isomorphic to the category of functors of points, this means that we can *define* schemes as certain types of functors on (**k**-algebras). Thanks to Proposition E.10, we can say that these functors are precisely the sheaves admitting an open cover by affine schemes, i.e. by representable functors. This point of view is very fruitful because it gives the possibility of generalizing the notion of scheme by considering more general functors. The notion of *algebraic space* is such a generalization (see Artin [13]).

References

1. *Eléments de Géométrie Algébrique*, par A. Grothendieck, rédigés avec la collaboration de J. Dieudonné, Publ. Math. IHES, n. 4, 8, 11, 17, 20, 24, 32.
2. *Fondements de la Géométrie Algébrique*, by A. Grothendieck, exposés at Séminaire Bourbaki 1957–1962, Secretariat Math. Paris 1962 (some of them are listed separately below).
3. *Revetements Etales et Groupe Fondamentale*, Séminaire de Géométrie Algébrique du Bois Marie 1960/61, Springer Verlag Lecture Notes in Mathematics vol. 224.
4. André M.: *Homologie des algébres commutatives*, Springer Grundlehren b. 206 (1974).
5. Andreotti A.: On the complex structures on a class of simply-connected manifolds, *Algebraic Geometry and Topology – A symposium in honor of S. Lefschetz*, Princeton U.P. (1957).
6. Arakelov S.: Families of algebraic curves with fixed degeneracies, *Math. USSR Izv.* 5 (1971), 1277–1302.
7. Arbarello E., Cornalba M.: Su una proprietà notevole dei morfismi di una curva a moduli generali in uno spazio proiettivo. *Rend. Sem. Mat. Univers. Politecn. Torino* 38 (1980), 87–99.
8. Arbarello E., Cornalba M.: Su una congettura di Petri, *Comm. Math. Helvetici* 56 (1981), 1–38.
9. Arbarello E., Cornalba M., Griffiths Ph., Harris J.: *Geometry of Algebraic Curves, vol. I*, Springer Grundlehren b. 267 (1985).
10. Arbarello E., Sernesi E.: Petri's approach to the study of the ideal associated to a special divisor, *Inventiones Math.* 49 (1978), 99–119.
11. Artin M.: Algebraic approximation of structures over complete local rings, *Publ. Math. IHES* 36 (1969), 23–58.
12. Artin M.: Algebraization of formal moduli I, in *Global Analysis – Papers in honor of K. Kodaira*, Princeton Math. Series n. 29 (1969), 21–71.
13. Artin M.: *Theoremes de representabilite pour les espaces algebriques*, Les Presses de L'université de Montreal (1972).
14. Artin M.: *Deformations of Singularities*, TATA Lecture Notes vol. 54 (1976).
15. Badescu L.: *Algebraic Surfaces*, Springer Universitext (2001).
16. Badescu L.: *Projective Geometry and Formal Geometry*, Birkhauser (2004).
17. Bayer D., Eisenbud D.: Ribbons and their canonical embeddings, *Trans. Amer. Math. Soc.* 347 (1995), 719–756.

18. Behrend K.: Localization and Gromow-Witten invariants, in *Quantum Cohomology*, editors P. de Bartolomeis, B. Dubrovin, C. Reina, Springer Lecture Notes in Mathematics vol. 1776 (1998).
19. Berthelot P.: Le theoreme de Riemann-Roch, SGA 6, exp. VIII. Springer Lecture Notes in Mathematics vol. 225 (1971).
20. Birkenhake C., Lange H.: *Complex Abelian Varieties*, 2nd ed., Springer Grundlehren b. 302 (2004).
21. Bloch S.: Semiregularity and De Rham cohomology, *Inventiones Math.* 17 (1972), 51–66.
22. Bolondi G., Kleppe J.O., Mirò-Roig R.M.: Maximal rank curves and singular points of the Hilbert scheme, *Compositio Math.* 77 (1991), 269–291.
23. Buchweitz R.O., Flenner H.: A semiregularity map for modules and applications to deformations, *Compositio Math.* 137 (2003), 135–210.
24. Burns D., Wahl J.: Local contributions to global deformations of surfaces, *Inventiones Math.* 26 (1974), 67–88.
25. Caporaso L., Sernesi E.: Characterizing curves by their odd theta-characteristics, *J. fur die reine u. angew. Math.* 562 (2003), 101–135.
26. Castelnuovo G.: Sui multipli di una serie lineare di gruppi di punti appartenente ad una curva algebrica, *Rend. Circolo Mat. Palermo* 7 (1893), 89–110.
27. Catanese F.: Moduli of algebraic surfaces, in *Theory of Moduli – Montecatini Terme, 1985*, E. Sernesi Ed., Springer Lecture Notes in Mathematics vol. 1337 (1988), 1–83.
28. Catanese F.: Everywhere non-reduced moduli spaces, *Inventiones Math.* 98 (1989), 293–310.
29. Chang M.C.: Some obstructed manifolds with very ample canonical class, *J. Algebraic Geom.* 1 (1992), 1–4.
30. Cheah J.: Cellular decompositions for nested Hilbert schemes of points, *Pacific J. Math.* 183 (1998), 39–90.
31. Chiantini L., Ciliberto C.: A few remarks on the lifting problem, *Asterisque* 218 (1993), 95–109.
32. Chiantini L., Ciliberto C.: On the Severi varieties of surfaces in $\mathbb{P}^3$, *J. Algebraic Geom.* 8 (1999), 67–83.
33. Chiantini L., Lopez A.F.: Focal loci of families and the genus of curves on surfaces, *Proc. Amer. Math. Soc.* 127 (1999), 3451–3459.
34. Chiantini L., Sernesi E.: Nodal curves on surfaces of general type, *Math. Annalen* 307 (1997), 41–56.
35. Ciliberto C.: On the Hilbert scheme of curves of maximal genus in a projective space, *Math. Z.* 194 (1987), 351–363.
36. Ciliberto C., Sernesi E.: Families of varieties and the Hilbert scheme, in *Lectures on Riemann Surfaces (Trieste 1987)*, 428–499, World Sci. Publishing (1989).
37. Ciliberto C., Sernesi E.: Singularities of the theta divisor and congruences of planes, *J. Algebraic Geom.* 1 (1992), 231–250.
38. Ciliberto C., Sernesi E.: Singularities of the theta divisor and families of secant spaces to a canonical curve, *J. Algebra* 171 (1995), 867–893.
39. Ciliberto C., Sernesi E.: On the geometry of canonical curves of odd genus, *Comm. in Algebra* 28 (2000), 5993–6001.
40. Colmez P., Serre J.P.: *Correspondance Grothendieck-Serre*, Documents Mathématiques 2 (2001), Société Mathématique de France.
41. Curtin D.J.: Obstructions to deforming a space curve, *Trans. A.M.S.* 267 (1981), 83–94.
42. Debarre O.: *Higher-Dimensional Algebraic Geometry*, Springer Universitext (2001).

43. Deligne P., Mumford D.: The irreducibility of the space of curves of given genus, *Publ. Math. IHES* 36 (1969), 75–110.
44. Di Gennaro V.: Alcune osservazioni sulle singolarità dello schema di Hilbert che parametrizza le varietà lineari contenute in una varietà proiettiva, *Ricerche di Matematica* 39 (1990), 259–291.
45. Dolgachev I.: Weighted projective varieties, in *Group Actions and Vector Fields (Vancouver B.C. 1981)*, Springer Lecture Notes in Mathematics vol. 956 (1982), 34–71.
46. Edidin D.: Notes on the construction of the moduli space of curves, in *Recent Progress in Intersection Theory (Bologna, 1997)*, G. Ellingsrud, W. Fulton, A. Vistoli Eds., Birkhauser (2000), 85–113.
47. Ein L.: Hilbert schemes of smooth space curves, *Ann. Sci. E.N.S.* 19 (1986), 469–478.
48. Eisenbud D.: *Commutative Algebra with a View toward Algebraic Geometry*, Springer Graduate Texts in Mathematics vol. 150 (1995).
49. Eisenbud D., Van de Ven A.: On the normal bundles of smooth rational space curves, *Math. Annalen* 256 (1981) 453–463.
50. Ellia Ph., Fiorentini M.: Defaut de postulation et singularités du schema de Hilbert, *Annali Univ. Ferrara* 30 (1984), 185–198.
51. Ellia Ph., Hartshorne R.: Smooth specializations of space curves: questions and examples, in: *Commutative Algebra and Algebraic Geometry*, ed. by F. van Oystaeyen, Marcel Dekker Lecture Notes in Pure and Applied Math. n. 206 (1999), 53–79.
52. Enriques F.: *Le superficie algebriche*, Zanichelli, Bologna (1949).
53. Fantechi B.: Stacks for everybody, in *European Congress of Mathematics, Vol. I (Barcelona, 2000)*, Birkhauser (2001), 349–359.
54. Fantechi B., Manetti M.: Obstruction calculus for functors of Artin rings, I, *J. of Algebra* 202 (1998), 541–576.
55. Fantechi B., Pardini R.: On the Hilbert scheme of curves in higher-dimensional projective space, *Manuscripta Math.* 90 (1996), 1–15.
56. Flamini F.: Some results of regularity for Severi varieties of projective surfaces, *Comm. Algebra* 29 (2001), 2297–2311.
57. Flamini F.: Moduli of nodal curves on smooth surfaces of general type, *J. Algebraic Geom.* 11 (2002), 725–760.
58. Fogarty J.: Algebraic families on an algebraic surface, *Amer. J. Math.* 90 (1968), 511–521.
59. Friedman R.: *Algebraic Surfaces and Holomorphic Vector Bundles*, Springer Universitext (1998).
60. Fulton W.: On nodal curves, in *Algebraic Geometry – Open Problems (Ravello, 1982)*, 146–155, Springer Lecture Notes in Mathematics vol. 997 (1983).
61. Fulton W.: *Intersection Theory*, Springer Ergebnisse (3) b. 2 (1984).
62. Geertsen J.A., Hirschowitz A.: Saturation theory and very ample Hilbert functions, *Math. Z.* 245 (2003), 155–181.
63. Ghione F., Sacchiero G.: Normal bundles of rational curves in $\mathbb{P}^3$, *Manuscripta Math.* 33 (1980), 111–128.
64. Godement R.: *Topologie algébrique et théorie des fasceaux*, Hermann, Paris (1964).
65. Gómez T.L.: Algebraic stacks, *Proc. Indian Acad. Sci. Math. Sci.* 111 (2001), 1–31.
66. Gonzalez M.: *Alisamiento de cintas sobre curvas*, thesis, Univ. Complutense Madrid (2004).
67. Grauert H., Kerner H.: Deformationen von Singularitäten komplexer Räume, *Math. Annalen* 153 (1964), 236–260.

68. Greuel G.M.: A remark on the paper of A. Tannenbaum, *Compositio Math.* 51 (1984), 185–187.
69. Greuel G.M., Lossen C., Shustin E.: New asymptotics in the geometry of equisingular families of curves, *Intern. Math. Res. Notices* 13 (1997), 595–611.
70. Griffiths Ph.: Some remarks and examples on continuous systems and moduli, *J. Math. and Mech.* 16 (1967), 789–802.
71. Grothendieck A.: Géométrie formelle et géométrie algébrique, *Seminaire Bourbaki*, exp. 182 (1959).
72. Grothendieck A.: Le theoreme d'existence en theorie formelle des modules, *Seminaire Bourbaki*, exp. 195 (1960).
73. Grothendieck A.: Les schemas de Hilbert. *Seminaire Bourbaki*, exp. 221 (1960).
74. Grothendieck A.: Familles d'espaces complexes et fondements de la geometrie analytique, *Seminaire H. Cartan* 1960/61, exp. I-X, Secret. Math., Paris 1962.
75. Grothendieck A.: *Local Cohomology*, Springer Lecture Notes in Mathematics vol. 41 (1967).
76. Grothendieck A.: *Categories cofibrees additives et complexe cotangent relatif*, Springer Lecture Notes in Mathematics vol. 79 (1968).
77. Grothendieck A., Dieudonné J.A.: *Eléments de Géométrie Algébrique I*, Springer Grundlehren b. 166 (1971).
78. Gruson L., Peskine Ch.: Genre des courbes de l'espace projectif, in *Algebraic Geometry – Proceedings, Tromsø, Norway 1977*, L.D. Olson Ed., Springer Lecture Notes in Mathematics vol. 687 (1978), 31–59.
79. Gruson L., Peskine Ch.: Genre des courbes de l'espace projectif (II), *Ann. Sci. Ec. Norm. Sup.* 15 (1982), 401–418.
80. Guffroy S.: Sur l'incomplétude de la série linéare caractéristique d'une famille de courbes planes à noeuds et à cusps, *Nagoya Math. J.* 171 (2003), 51–83.
81. Gustavsen T.S., Laksov D., Skjelnes R.M.: An elementary, explicit, proof of the existence of the Hilbert schemes of points, AG/0506161.
82. Harris J.: On the Severi problem, *Inventiones Math.* 84 (1986), 445–461.
83. Hartshorne R.: Connectedness of the Hilbert scheme, *Publ. Math IHES* 29 (1966), 261–304.
84. Hartshorne R.: *Algebraic Geometry*, Springer Verlag Graduate Texts in Mathematics vol. 52 (1977).
85. Horikawa E.: On deformations of holomorphic maps I, *J. Math. Soc. Japan* 25 (1973), 372–396.
86. Horikawa E.: On deformations of holomorphic maps II, *J. Math. Soc. Japan* 26 (1974), 647–667.
87. Horikawa E.: On deformations of quintic surfaces, *Inventiones Math.* 31 (1975), 43–85.
88. Horikawa E.: On deformations of holomorphic maps III, *Math. Annalen* 222 (1976), 275–282.
89. Horikawa E.: On deformations of rational maps, *J. Fac. Sci. Univ. Tokyo* 23 (1976), 581–600.
90. Huybrechts D., Lehn M.: *The Geometry of Moduli Spaces of Sheaves*, Aspects of Mathematics vol. E31 (1997), Vieweg.
91. Iarrobino A.: Hilbert scheme of points: overview of last ten years, in *Algebraic Geometry – Bowdoin 1985*, Proc. of Symposia in Pure Math. 46, part 2, 297–320 (1987).
92. Iarrobino A., Kanev V.: *Power sums, Gorenstein Algebras, and Determinantal Loci*, Springer Lecture Notes in Mathematics vol. 1721 (1999).
93. Illusie L.: *Complexe cotangent et deformations I, II*. Springer Lecture Notes in Mathematics vol. 239, 283 (1971).

94. Iversen B.: *Generic Local Structure in Commutative Algebra*, Springer Lecture Notes in Mathematics vol. 310 (1973).
95. Kang P.: A note on the variety of plane curves with nodes and cusps, *Proceedings AMS* 106 (1989), 309–312.
96. Kas A.: On obstructions to deformations of complex analytic surfaces, *Proc. Nat. Acad. Sci. USA* 58 (1967), 402–404.
97. Kleiman S.: Geometry on grassmannians and applications to splitting bundles and smoothing cycles, *Publ. Math. IHES* 36 (1969), 281–298.
98. Kleiman S.: Les théorèmes de finitude pour le foncteur de Picard. SGA 6, exp. XIII. Springer Lecture Notes in Mathematics vol. 225 (1971).
99. Kleiman S.: Multiple point formulas I: iteration, *Acta Mathematica* 147 (1981), 13–49.
100. Kleiman S.: The Picard scheme, AG/0504020.
101. Kleiman S., Lipman J., Ulrich B.: The multiple point schemes of a finite curvilinear map of codimension one, *Ark. Mat.* 34 (1996), 285–326.
102. Kleppe J.O.: The Hilbert-flag scheme, its properties and its connection with the Hilbert scheme. Applications to curves in 3-space. Preprint, Inst. of Math. Univ. Oslo (1981).
103. Kleppe J.O.: Non-reduced components of the Hilbert scheme of smooth space curves. Springer Lecture Notes in Mathematics vol. 1266 (1987), 181–207.
104. Kodaira K.: On stability of compact submanifolds of complex manifolds, *Amer. J. Math.* 85 (1963), 79–94.
105. Kodaira K.: *Complex Manifolds and Deformations of Complex Structures*, Springer Grundlehren b. 283 (1986).
106. Kodaira K., Nirenberg L., Spencer D.C.: On the existence of deformations of complex analytic structures, *Annals of Math.* 68 (1958), 450–459.
107. Kodaira K., Spencer D.C.: On deformations of complex analytic structures I, II. *Ann. of Math.* 67 (1958), 328–466.
108. Kodaira K., Spencer D.C.: A theorem of completeness of characteristic systems of complete continuous systems, *Amer. J. Math.* 81 (1959), 477–500.
109. Kollar J.: *Rational Curves on Algebraic Varieties*, Springer Ergebnisse b. 32 (1996).
110. Kollar J.: Non quasi-projective moduli spaces, AG/0501294.
111. Kuranishi M.: On the locally complete families of complex analytic structures, *Ann. of Math.* 75 (1962), 536–577.
112. Lange H., Sernesi E.: Severi varieties and branch curves of abelian surfaces of type $(1,3)$, *International J. of Math.* 13 (2002), 227–244.
113. Lange H., Sernesi E.: On the Hilbert scheme of a Prym variety, *Annali di Matematica* 183 (2004), 375–386.
114. Laudal O.A.: *Formal Moduli of Algebraic Structures*, Springer Lecture Notes in Mathematics vol. 754 (1979).
115. Laumon G., Moret-Bailly L.: *Champs Algébriques*, Springer Ergebnisse b. 39 (2000).
116. Lazarsfeld R.: *Positivity in Algebraic Geometry*, vol. I, II, Springer Ergebnisse b. 48, 49 (2004).
117. Lehn M.: On the cotangent sheaf of Quot schemes, *Intern. J. of Math.* 9 (1998), 513–522.
118. Le Potier J.: *Lectures on Vector Bundles*, Cambridge U.P. (1997).
119. Levelt A.H.M.: Foncteurs exacts à gauche, *Inventiones Math.* 8 (1969), 114–140.
120. Lichtenbaum S., Schlessinger M.: The cotangent complex of a morphism, *Transactions AMS* 128 (1967), 41–70.
121. Loeser F.: Déformations de courbes planes (d'aprés Severi et Harris), Séminaire Bourbaki 1986/87, *Astérisque* 152–153 (1987), 187–205.
122. Luengo I.: On the existence of complete families of projective plane curves which are obstructed, *J. London Math. Soc.* 36 (1987), 33–43.

123. MacLane S.: *Homology*, Springer Grundlehren b. 114 (1967).
124. Manetti M.: *Lectures on Deformations of Complex Manifolds*, Rendiconti di Matematica s. VII n. 24 (2004), 1–183.
125. Martin-Deschamps M., Perrin D.: Le schéma de Hilbert des courbes gauches localement Cohen-Macaulay n'est (presque) jamais réduit, *Ann. Scient. Ec. Norm. Sup.* 29 (1996), 757–785.
126. Maruyama M.: On a generalization of elementary transformations of algebraic vector bundles, *Rend. Sem. Mat. Univ. Politec. Torino*, special issue (1986), 1–13.
127. Matsumura H.: *Commutative Ring Theory*, Cambridge University Press (1986).
128. Morrow J., Kodaira K.: *Complex Manifolds*, Holt, Rinehart & Winston, New York (1971).
129. Mumford D.: Further pathologies in algebraic geometry, *Amer. J. of Math.* 84 (1962), 642–648.
130. Mumford D.: *Lectures on Curves on an Algebraic Surface*, Annals of Mathematics Studies vol. 59, Princeton U.P. (1966).
131. Mumford D.: A remark on the paper of M. Schlessinger, *Rice Univ. Studies* 59 (1973), 113–117.
132. Mumford D., Fogarty J.: *Geometric Invariant Theory*, second enlarged edition, Springer Ergebnisse b. 34 (1982).
133. Nakajima H.: *Lectures on Hilbert Schemes of Points on Surfaces*, AMS Univ. Lecture Series, vol. 18 (1999).
134. Namba M.: *Families of Meromorphic Functions on Compact Riemann Surfaces*, Springer Lecture Notes in Mathematics vol. 767 (1979).
135. Norman P., Oort F.: Moduli of abelian varieties, *Ann. of Math.* (2) 112 (1980), 413-439.
136. Okonek C., Schneider M., Spindler H.: *Vector Bundles on Complex Projective Space*, Progress in Math. vol. 3, Birkhauser (1980).
137. Oort F.: Finite group schemes, local moduli for abelian varieties, and lifting problems, *Compositio Math.* 23 (1971), 265-296.
138. Palamodov V.P.: Deformations of complex spaces, in *Several Complex Variables IV: Algebraic Aspects of Complex Analysis*, Encyclopaedia of Mathematical Sciences, vol. 10, 105–194. Springer (1990).
139. Parshin A.N.: Algebraic curves over function fields I, *Math. SSSR Izv.*, 2 (1968), 1145–1170.
140. Peeva I., Stillman M.: Connectedness of the Hilbert schemes, *J. Algebraic Geom.* 14 (2005), 193–211.
141. Petri K.: Über Spezialkurven I, *Math. Ann.* 93 (1925), 182–209.
142. Piene R., Schlessinger M.: On the Hilbert scheme compactification of the space of twisted cubics, *Amer. J. Math.* 107 (1985), 761–774.
143. Pinkham H.: Deformations of cones with negative grading, *J. of Algebra* 30 (1974), 92–102.
144. Pinkham H.: Deformations of algebraic varieties with G_m-action, *Asterisque* 20 (1974).
145. Popp H.: *Moduli Theory and Classification Theory of Algebraic Varieties*, Springer Lecture Notes in Mathematics vol. 620 (1977).
146. Quillen D.: On the (co)-homology of commutative rings. *Proc. Symp. in Pure Math.* 17 (1970), 65–87.
147. Ran Z.: On nodal plane curves, *Inventiones Math.* 86 (1986), 529–534.
148. Ran Z.: Deformations of maps, in *Algebraic Curves and Projective Geometry – Proceedings, Trento 1988*, E. Ballico and C. Ciliberto Eds., Springer Lecture Notes in Mathematics vol. 1389 (1989), 246–253.

149. Ran Z.: Hodge theory and the Hilbert scheme, *J. Diff. Geometry* 37 (1993), 191–198.
150. Ran Z.: Hodge theory and deformations of maps, *Compositio Math.* 97 (1995), 309–328.
151. Ran Z.: A note on Hilbert schemes of nodal curves, AG/0410037.
152. Reid M.: Chapters on algebraic surfaces, in *Complex Algebraic Geometry (Park City, UT, 1993)*, 3–159, Amer. Math. Soc., Providence, RI (1997).
153. Sacchiero G.: Fibrati normali di curve razionali dello spazio proiettivo, *Annali Univ. Ferrara* (VII) 26 (1980) 33–40.
154. Schlessinger M.: *Infinitesimal deformations of singularities*, PhD thesis, Harvard University (1964).
155. Schlessinger M.: Functors of Artin rings, *Transactions AMS* 130 (1968), 208–222.
156. Schlessinger M.: Rigidity of quotient singularities, *Inventiones Math.* 14 (1971), 17–26.
157. Schlessinger M.: On rigid singularities, *Rice University Studies* vol. 59 (1973), No. 1, 147–162.
158. Segre C.: Sui fuochi di secondo ordine dei sistemi infiniti di piani, e sulle curve iperspaziali con una doppia infinità di piani plurisecanti, *Atti R. Accad. Lincei* (5) 30 (1921), 67–71.
159. Segre B.: Esistenza e dimensione dei sistemi continui di curve piane algebriche con dati caratteri, *Rendiconti Accad. Naz. Lincei* (6) 10 (1929), 31–38.
160. Segre B.: Esistenza e dimensione dei sistemi continui di curve piane algebriche con dati numeri plueckeriani, *Rendiconti Accad. Naz. Lincei* (6) 10 (1929), 557–560.
161. Sernesi E.: Small deformations of global complete intersections, *Bollettino UMI* (4) 12 (1975), 138–146.
162. Sernesi E.: Un esempio di curva ostruita in $\mathbb{P}^3$, *Seminario di variabili complesse*, Bologna (1981).
163. Sernesi E.: *Topics on Families of Projective Schemes*, Queen's Papers in Pure and Applied Mathematics, vol. 73, Kingston, Ontario, Canada (1986).
164. Serre J.P.: Géométrie algébrique et géométrie analytique, *Ann. Inst. Fourier*, Grenoble 6 (1955-56), 1–42.
165. Serre J.P.: Espaces fibrés algébriques, *Seminaire Chevalley* exp. 1 (1958).
166. Serre J.P.: *Groupes algébriques et corps de classes*, Hermann, Paris (1959).
167. Serre J.P.: *Local Algebra*, Springer Monographs in Mathematics (2000).
168. Seshadri C.S.: Theory of moduli. In *Algebraic Geometry*, Proc. Symposia Pure Math., vol. 29 (1975), pp. 263–304. Amer. Math. Soc., Providence, R. I.
169. Severi F.: Sulla classificazione delle curve algebriche e sul teorema di esistenza di Riemann, *Rend. Acc. Naz. Lincei* (5) 241 (1915), 877–888 and 1011–1020.
170. Severi F.: Nuovi contributi alla teoria dei sistemi continui di curve appartenenti ad una superficie algebrica, *Rend. Acc. Naz. Lincei* (5) 25 (1916), 459–471.
171. Severi F.: *Vorlesungen über Algebraische Geometrie*, Teubner, Leipzig (1921).
172. Severi F.: Sul teorema fondamentale dei sistemi continui di curve sopra una superficie algebrica, *Annali di matematica* (4) 23 (1944), 149–181.
173. Severi F.: *Geometria dei sistemi algebrici sopra una superficie e sopra una varietà algebrica II*, Cremonese, Roma, 1958.
174. Sheets R.: A criterion for the rigidity of fibered products of rigid algebras, *J. of Algebra* 46 (1977), 37–53.
175. Stevens J.: *Deformations of Singularities*, Springer Lecture Notes in Mathematics vol. 1811 (2003).
176. Sundararaman D.: *Moduli, Deformations and Classification of Compact Complex Manifolds*, Pitman Research Notes in Math. vol. 45 (1980).
177. Svanes T.: Some criteria for rigidity of noetherian rings, *Math. Z.* 144(1975), 135–145.

178. Szpiro L.: Sur le théorème de rigidité de Parshin et Arakelov, *Asterisque* 64 (1978), 169–202.
179. Tannenbaum A.: Families of algebraic curves with nodes, *Compositio Math.*, 41 (1980), 107–126.
180. Tannenbaum A.: On the classical characteristic linear series of plane curves with nodes and cuspidal points: two examples of Beniamino Segre, *Compositio Math.* 51 (1984), 169–183.
181. Tate J.: Homology of noetherian rings and local rings, *Illinois J. Math.* 1 (1957), 14–25.
182. Treger R.: The local Severi problem, *Bull. AMS* 19 (1988), 325–327.
183. Vakil R.: Murphy's law in algebraic geometry: badly-behaved deformation spaces. AG/0411469.
184. Viehweg E.: *Quasi-Projective Moduli for Polarized Manifolds*, Springer Ergebnisse (3) b. 30 (1995).
185. Vistoli A.: Intersection theory on algebraic stacks and on their moduli spaces, *Inventiones Math.* 97 (1989), 613–670.
186. Wahl J.: Deformations of plane curves with nodes and cusps, *Amer. J. of Math.* 96 (1974), 529–577.
187. Walter C.: Some examples of obstructed curves in $\mathbb{P}^3$, *London Math. Soc. LNS* 179 (1992), 325–340.
188. Zappa G.: Sull'esistenza, sopra le superficie algebriche, di sistemi continui completi infiniti, la cui curva generica è a serie caratteristica incompleta, *Pont. Acad. Sci. Acta* 9 (1945), 91–93.
189. Zariski O.: *Algebraic Surfaces*, second supplemented edition, Springer Ergebnisse b. 61 (1971).
190. Zariski O., Samuel P.: *Commutative Algebra*, Van Nostrand, Princeton (1960).

List of Symbols

Index

Grundlehren der mathematischen Wissenschaften

A Series of Comprehensive Studies in Mathematics

A Selection

246. Naimark/Stern: Theory of Group Representations
247. Suzuki: Group Theory I
248. Suzuki: Group Theory II
249. Chung: Lectures from Markov Processes to Brownian Motion
250. Arnold: Geometrical Methods in the Theory of Ordinary Differential Equations
251. Chow/Hale: Methods of Bifurcation Theory
252. Aubin: Nonlinear Analysis on Manifolds. Monge-Ampère Equations
253. Dwork: Lectures on ρ-adic Differential Equations
254. Freitag: Siegelsche Modulfunktionen
255. Lang: Complex Multiplication
256. Hörmander: The Analysis of Linear Partial Differential Operators I
257. Hörmander: The Analysis of Linear Partial Differential Operators II
258. Smoller: Shock Waves and Reaction-Diffusion Equations
259. Duren: Univalent Functions
260. Freidlin/Wentzell: Random Perturbations of Dynamical Systems
261. Bosch/Güntzer/Remmert: Non Archimedian Analysis – A System Approach to Rigid Analytic Geometry
262. Doob: Classical Potential Theory and Its Probabilistic Counterpart
263. Krasnosel'skiĭ/Zabreĭko: Geometrical Methods of Nonlinear Analysis
264. Aubin/Cellina: Differential Inclusions
265. Grauert/Remmert: Coherent Analytic Sheaves
266. de Rham: Differentiable Manifolds
267. Arbarello/Cornalba/Griffiths/Harris: Geometry of Algebraic Curves, Vol. I
268. Arbarello/Cornalba/Griffiths/Harris: Geometry of Algebraic Curves, Vol. II
269. Schapira: Microdifferential Systems in the Complex Domain
270. Scharlau: Quadratic and Hermitian Forms
271. Ellis: Entropy, Large Deviations, and Statistical Mechanics
272. Elliott: Arithmetic Functions and Integer Products
273. Nikol'skiĭ: Treatise on the shift Operator
274. Hörmander: The Analysis of Linear Partial Differential Operators III
275. Hörmander: The Analysis of Linear Partial Differential Operators IV
276. Liggett: Interacting Particle Systems
277. Fulton/Lang: Riemann-Roch Algebra
278. Barr/Wells: Toposes, Triples and Theories
279. Bishop/Bridges: Constructive Analysis
280. Neukirch: Class Field Theory
281. Chandrasekharan: Elliptic Functions
282. Lelong/Gruman: Entire Functions of Several Complex Variables
283. Kodaira: Complex Manifolds and Deformation of Complex Structures
284. Finn: Equilibrium Capillary Surfaces
285. Burago/Zalgaller: Geometric Inequalities
286. Andrianaov: Quadratic Forms and Hecke Operators
287. Maskit: Kleinian Groups
288. Jacod/Shiryaev: Limit Theorems for Stochastic Processes

289. Manin: Gauge Field Theory and Complex Geometry
290. Conway/Sloane: Sphere Packings, Lattices and Groups
291. Hahn/O'Meara: The Classical Groups and K-Theory
292. Kashiwara/Schapira: Sheaves on Manifolds
293. Revuz/Yor: Continuous Martingales and Brownian Motion
294. Knus: Quadratic and Hermitian Forms over Rings
295. Dierkes/Hildebrandt/Küster/Wohlrab: Minimal Surfaces I
296. Dierkes/Hildebrandt/Küster/Wohlrab: Minimal Surfaces II
297. Pastur/Figotin: Spectra of Random and Almost-Periodic Operators
298. Berline/Getzler/Vergne: Heat Kernels and Dirac Operators
299. Pommerenke: Boundary Behaviour of Conformal Maps
300. Orlik/Terao: Arrangements of Hyperplanes
301. Loday: Cyclic Homology
302. Lange/Birkenhake: Complex Abelian Varieties
303. DeVore/Lorentz: Constructive Approximation
304. Lorentz/v. Golitschek/Makovoz: Construcitve Approximation. Advanced Problems
305. Hiriart-Urruty/Lemaréchal: Convex Analysis and Minimization Algorithms I. Fundamentals
306. Hiriart-Urruty/Lemaréchal: Convex Analysis and Minimization Algorithms II. Advanced Theory and Bundle Methods
307. Schwarz: Quantum Field Theory and Topology
308. Schwarz: Topology for Physicists
309. Adem/Milgram: Cohomology of Finite Groups
310. Giaquinta/Hildebrandt: Calculus of Variations I: The Lagrangian Formalism
311. Giaquinta/Hildebrandt: Calculus of Variations II: The Hamiltonian Formalism
312. Chung/Zhao: From Brownian Motion to Schrödinger's Equation
313. Malliavin: Stochastic Analysis
314. Adams/Hedberg: Function spaces and Potential Theory
315. Bürgisser/Clausen/Shokrollahi: Algebraic Complexity Theory
316. Saff/Totik: Logarithmic Potentials with External Fields
317. Rockafellar/Wets: Variational Analysis
318. Kobayashi: Hyperbolic Complex Spaces
319. Bridson/Haefliger: Metric Spaces of Non-Positive Curvature
320. Kipnis/Landim: Scaling Limits of Interacting Particle Systems
321. Grimmett: Percolation
322. Neukirch: Algebraic Number Theory
323. Neukirch/Schmidt/Wingberg: Cohomology of Number Fields
324. Liggett: Stochastic Interacting Systems: Contact, Voter and Exclusion Processes
325. Dafermos: Hyperbolic Conservation Laws in Continuum Physics
326. Waldschmidt: Diophantine Approximation on Linear Algebraic Groups
327. Martinet: Perfect Lattices in Euclidean Spaces
328. Van der Put/Singer: Galois Theory of Linear Differential Equations
329. Korevaar: Tauberian Theory. A Century of Developments
330. Mordukhovich: Variational Analysis and Generalized Differentiation I: Basic Theory
331. Mordukhovich: Variational Analysis and Generalized Differentiation II: Applications
332. Kashiwara/Schapira: Categories and Sheaves. An Introduction to Ind-Objects and Derived Categories
333. Grimmett: The Random-Cluster Model
334. Sernesi: Deformations of Algebraic Schemes
335. Bushnell/Henniart: The Local Langlands Conjecture for GL(2)

Printing: Krips bv, Meppel
Binding: Stürtz, Würzburg